LEARNING RESOURCE CENTRE
GRANT MacEWAN
COMMUNITY COLLEGE

S0-EAI-525

INTRODUCTION
TO WORKING CONDITIONS
AND ENVIRONMENT

INTRODUCTION TO WORKING CONDITIONS AND ENVIRONMENT

Edited by J.-M. Clerc

INTERNATIONAL LABOUR OFFICE GENEVA

ISBN 92-2-105125-0 (limp cover)
ISBN 92-2-105124-2 (hard cover)

First published 1985

Printed in Switzerland

PREFACE

This book has been designed to meet a need.

The subject of working conditions and the working environment has, for the most part, been dealt with piecemeal in the past. This was mainly because the field was so vast, because many of the themes were *sui generis* and because the highly technical aspects had to be fully grasped if they were to be correctly mastered, with nothing left to chance. However, man's working conditions and their consequences form a whole which, like man himself, cannot be broken down into their component parts without being deformed. Careful and accurate observation requires both a wide-angle and a close-up lens. Scientific and technological progress compel us to make even greater efforts to understand precisely what the effects of working conditions on the worker are, and to develop the necessary safeguards. However, we must also maintain an overall view, with due perspective, to give prominence to the many inter-relationships that make the individual components of man's working environment inseparable, interdependent and interactive. Therefore, despite the problems inherent in such an undertaking, there was a clear need for a general work covering, to the extent possible, the whole gamut of safety, health and general working conditions and the ties that link them.

A second, no less urgent, reason for this publication was the alarming state of working conditions and environment throughout the world. Significant progress in some fields is counterbalanced by a deterioration in others, and redoubled efforts are called for. The ILO could not remain indifferent to a situation which affects the lives, health and welfare of millions of workers. It was this concern in the face of growing danger that stimulated the search for a major breakthrough, and led eventually to the launching, in 1976, of the International Programme for the Improvement of Working Conditions and Environment, with the unanimous support of ILO member governments, employers and workers. It was clear from the start that one key component of this Programme would have to be an ambitious education and training effort, since teachers and trainers in this area are sorely lacking, especially in the developing countries. However, there is a vast wealth of men and women in undertakings, trade unions, administrations and various other bodies throughout the world, waiting to disseminate the necessary basic knowledge; all they need is a reference book containing all the basic concepts which they can add to the knowledge they have gained through personal experience.

The authors of this book, working in collaboration and with the assistance of numerous consultants, have therefore not been content merely to assemble isolated items of information. What would be the use of a book which was too technical or too abstract? Their aim has been above all to produce a book that is easily understood – a book that will be not only read but also used as a tool.

The publication of a readily understood treatise on working conditions and the working environment was therefore a task which the ILO could not but assume. In attempting to ensure the widest possible dissemination of knowledge about working conditions and the working environment, the ILO is firmly expressing its clear conviction that improvements cannot be achieved by specialists alone – what is needed is a joint endeavour by all. The task is an urgent one if we hope thereby to increase the number of those involved in the improvement of working conditions and environment. This is the message contained in this publication, and I earnestly hope that it will be widely heard.

Francis Blanchard
Director-General
International Labour Office

CONTENTS

Preface, by Francis Blanchard . . . V

Introduction . . . XV
Objectives . . . XV
Scope . . . XV
Target audience . . . XVI
Coverage . . . XVII
How the book was written . . . XIX
Structure . . . XX

1. Importance and unity of working conditions and environment . . . 1
An unacceptable situation . . . 1
A heavy price to pay . . . 1
A worrying trend . . . 3
Promoting work adapted to the worker . . . 3
The need for a global approach . . . 7
Working conditions and environment form a whole . . . 8
Inter-relations and interactions . . . 8
The comprehension of work situations . . . 8
Working conditions and environment in their context . . . 9
Working conditions and environment, and conditions of life 9
Wage earners in legislation and in practice . . . 14
Economic constraints . . . 15
Aims and scope of the global approach . . . 20
Meaning and objectives of the global approach . . . 22
Practical applications . . . 22
Implementation and impact . . . 23

2. Occupational safety and health . . . 27
Extent of the problem . . . 27
Introduction . . . 27
Still cause for concern . . . 27
Considerable efforts have been made . . . 28

The rise of new concepts 28
Material factors 30
Machinery and plant 30
Harmful agents in the work environment 30
Fatigue 32
Multiple facets of fatigue 32
The contribution of science and technology 33
Hours of work and work organisation 34
Social problems 34
Nutrition 34
Rest, housing and transport 35
Health conditions 36
Occupational safety and health in agriculture 38
Extent and characteristics of farming communities . . . 38
Risks and patterns of accidents and diseases 38
Accident and disease control 39
Occupational risk control: Safety and health measures in the plant 42
Risk analysis 44
Installation, layout and maintenance of premises and workplaces 46
Plant and equipment hazards 47
Dangerous substances 49
Hazards related to working procedures and work organisation 50
Risks related to the individual 51
Other risk factors 53
Safety and health measures 53
Obligations of employers and workers 55
Buildings, workplaces and means of access 55
Machinery guarding 60
Dangerous substances and agents 65
Operating methods and work organisation 69
Measures involving the worker 72
Overall occupational safety and health organisation 77
Selecting the means of action 78
Workers' participation 81
Evaluation of results 84
Occupational risk control: Safety and health measures at the national level 87
Need for a coherent policy 87
Legislation, standards and directives 90
Enforcement of regulations and advisory activities 91
Education and training 95
Research 97
Information 98
Conclusions 99

3. Working time 101
The importance of working time 101
Collective bargaining 102
Legislation 102
International action 103
International labour standards 103
Hours of work 104
Normal hours of work 105
Legal limits 105
Length of the working day 108
Flexible hours schemes 108
Overtime 109
Definitions 109
Reasons for overtime work 110
Advantages and disadvantages 112
Measures to reduce or limit overtime 113
Excessive hours in poorly regulated sectors 115
Rest periods and breaks 118
Shift and night work 120
Shift systems 120
Reasons for shift work 122
Effects of shift work on workers 124
Improving the conditions of shiftworkers 126
Improving shift-work arrangements 126
Improving conditions of work and life 136
Should shift work be encouraged or discouraged? 137
Night work 138
Holidays and leave 139
Annual holidays with pay 139
International labour standards on holidays and leave . 139
Length of annual holidays 140
Holiday pay 142
Other problems of definition and regulation 143
Public holidays 146
Other forms of leave 146
Paid educational leave 147
Reduction of working time 149
Should working time be reduced? 149
What is meant by reduction of working time? 150
Effects of reducing working time 151
New patterns in working time 153
Compressed working week 154
Flexible hours 155
Staggered hours and staggered holidays 158
Part-time work 158
Towards a more integrated approach 159

4. Wages . . . 163
Wage-fixing machinery . . . 164
Individual contracts . . . 164
Collective agreements . . . 164
Voluntary or compulsory arbitration . . . 165
Minimum wage . . . 166
Wage fixing in the public sector . . . 167
Wage-fixing parameters or criteria . . . 168
A living wage . . . 169
Cost of living . . . 170
Wage comparisons with employment market rates . . . 171
An acceptable wage (for the undertaking and the economy) 171
An equitable wage . . . 172
Wage systems . . . 174
Time rate system . . . 175
Payment by results . . . 175
Straight piece-rate systems . . . 176
Differential piece-rate systems . . . 177
Premium and task bonus systems . . . 177
Merit rating . . . 177
Profit-sharing and co-partnership schemes . . . 178
Protection of wages . . . 178
Protection from the employer . . . 178
Wage protection in the event of bankruptcy of the undertaking . . . 181
Protecting the worker against himself . . . 182
Wages and other conditions of work and life . . . 182
Wage level and subsistence . . . 182
Wages, skills and employment . . . 182
Wages and the improvement of working conditions . . . 183

5. Organisation of work and job content . . . 185
Efficiency principles of work organisation . . . 186
Division of labour . . . 186
Hierarchies . . . 188
Scientific management . . . 189
Social principles of work organisation . . . 190
Methods of improving work organisation and job content . . . 191
Decoupling people from machines . . . 191
Optimisation of cycle time . . . 192
Job rotation . . . 195
Job enrichment . . . 195
Group work . . . 197
Semi-autonomous groups . . . 197
Matrix groups . . . 198
Organisation of industrial work: Flow grouping versus functional grouping . . . 201

Organisation of office work: Forms of decentralisation . . . 204
Other techniques affecting work organisation and job content . . . 204
Automation and its effects on work organisation . . . 206
Introducing improvements in work organisation and job content . . . 208
Worker and trade union participation . . . 208
Sources of pressure for change . . . 209
Conclusion: Old and new design principles . . . 211

6. Workers' welfare facilities . . . 213
Facilities for workers' welfare during working hours . . . 213
Catering services . . . 214
Commuting facilities . . . 218
Welfare facilities to improve the living conditions of workers and their families . . . 220
Employers' action with respect to housing . . . 222
At the level of the undertaking . . . 223
Beyond the level of the undertaking . . . 227
Employers' action in the field of provisions . . . 228
Works stores . . . 229
Consumer co-operative stores . . . 231
Other forms of assistance . . . 231
Employers' action in the field of health . . . 233
At the level of the undertaking . . . 233
Beyond the level of the undertaking . . . 236
Conclusion . . . 236

7. Workers in the rural and urban informal sectors in developing countries . . . 237
Working conditions in rural areas of developing countries . . . 237
A complex situation . . . 237
Deep and growing impoverishment . . . 238
Need for a wider analytical perspective . . . 238
Nature of agricultural work . . . 238
Physical and living environment . . . 239
Economic environment . . . 241
Agricultural modernisation and working conditions . . . 242
Technological choice and working conditions . . . 242
Some policy implications and issues . . . 244
Need for rural development strategies directed against poverty . . . 244
Need to promote participation . . . 245
Rural development, technological choice and working conditions . . . 246
The urban informal sector . . . 246
Characteristics and contribution . . . 247
Earnings and conditions of work . . . 247

Occupational and community health hazards 248
Problems and policies 249
Increasing income-earning opportunities 249
Improved access to social services and physical infrastructure 250
Basic measures relating to safety and health 251

8. How to improve working conditions and environment 255
The men and the means 255
Public institutions 258
Labour administration 258
Labour legislation 260
Labour inspection 261
Role of employers and workers 262
Employers' and workers' organisations 262
Collective bargaining 263
Specialists and educational, research and advisory bodies . 264
Action at the level of the work unit 264
Need for a global approach to the problem 265
A multidisciplinary approach required 265
Working conditions and productivity 265
The global approach and specific problems 267
Two basic principles 268
Enforcement of labour legislation 268
A vigilant approach to the problems of man at work . 269
Key figures in the undertaking 270
Action at the national level 272
Assessing the situation 272
Knowledge of the problems 272
Obstacles and means of overcoming them 273
Selecting priorities 274
Guiding principles 276
A coherent policy 276
A co-ordinated policy 277
A concerted policy 278
Diversified action 278
Strengthening the institutions 278
General and specific measures 279
Direct and indirect measures 279
Coercive and incentive measures 279
Measures at the design and planning stage and corrective action 280
Technological choice 281
Specific categories of workers and their problems 282
Workers underprivileged or vulnerable owing to personal characteristics 282

Workers underprivileged by the *de jure* or *de facto* absence of social protection 286
Education and training 287
Role and objectives 287
Importance 288
Function 288
In-plant training 291
Workers' representatives 291
Workers in general 291
Management, engineers, technicians and foremen . . . 292
Designers and planners 292
Heads of undertakings 292
Training of specialists 293
Labour inspectorate staff 293
Occupational health and hygiene staff 294
Specialists in occupational safety and ergonomics . . . 294
Dissemination of scientific training and the multidisciplinary approach 295
Training in teaching establishments 295
Vocational and technical training 295
Training of other professionals 296
Training in schools 297
Training policies and training the trainers 297
Decision-makers (senior civil servants and planners) . . 298
Employers' and workers' organisations 298
Public information 299
Training the trainers 299

Conclusion 305

Appendices
A. The International Labour Organisation 309
B. The International Programme for the Improvement of Working Conditions and Environment (PIACT) 311
C. International labour Conventions and Recommendations concerning working conditions and environment 313
D. Guide to further reading 316

INTRODUCTION

Training is an essential component of any strategy for improving working conditions and environment.

It is a major factor in the implementation of legislation and regulations for the protection of workers; it makes official bodies effective, nourishes concerted efforts and augments the results of day-to-day actions at the workplace that are essential if regular improvements in the worker's lot are to be made.

Objectives

There will be no improvement in working conditions and environment unless, at every step in the design, planning, organisation, distribution and execution of work, consideration is given to the conditions under which man actually carries out his tasks. What improvement would be possible if the majority were not sufficiently aware of the problems, if their resources were not mobilised and if they did not know the basic principles needed to achieve this improvement? These principles need to be widely understood and disseminated.

The aim of this book is to bring all these concepts together. It is an ambitious aim, since the field is vast and since the readers for whom it is intended are drawn from widely differing backgrounds and circumstances.

Scope

The term "working conditions and environment" covers occupational safety and health and general working conditions. Although safety and health are concepts that are quite clear to all, general working conditions are vague and may mean different things depending on the countries or persons in question. They may be defined as the factors determining the situation in which the worker lives, and are commonly seen to include hours of work, work organisation, job content and welfare services. However, when this book was planned it was considered that wages, in view of their importance and relations with other working conditions, should also be covered.

Safety and health and general working conditions are interlinked in many ways, and it was to emphasise these relationships and the need for a global approach that, in adopting the International Programme for the Improvement of Working Conditions and Environment (known for short as PIACT, from its French acronym), the International Labour Conference decided to bring the two concepts together under the single heading of "working conditions and environment". Chapter 1 deals with this new concept.

Social security, contracts of employment and, with the exception of a few brief references, labour administration and labour relations are not dealt with in this book.

Target audience

This book is intended for "trainers" in the widest sense, i.e. those who, in one way or another, are responsible for educational programmes about working conditions and environment and related subjects.

The majority of those who teach the subject of working conditions and environment deal with one specialised field, such as occupational safety, health or medicine. Industrial physicians, safety engineers in industry, accident prevention specialists, inspectors from social security funds and others may find this book a useful supplement to the documents they normally consult. Their own speciality will not be considered in detail, but it is hoped that the book will draw their attention to the relationships between their own discipline and other fields.

This book is, perhaps above all, aimed at those who have little access to the basic reference documents and who teach, either full time, part time or even as a sideline to their main job, about one or more aspects of working conditions and environment as a part of:

(a) the preliminary or advanced training of those engaged in some way on improving working conditions and environment: experienced labour inspectors or staff members of a ministry of labour or social affairs running preliminary or advanced training courses for their younger colleagues; trade union educators responsible for training trade unionists; administrators in public or private undertakings responsible for training industrial social service workers. This list is endless, since trainers may be as varied as the needs they meet: those training extension workers in the rural sector; members of safety and health committees or works committees; social workers;

occupational health nurses; those running workers' agricultural or service co-operatives, etc.;

(b) the preliminary or advanced training of those such as industrial, agricultural, commercial, service or plantation managers who, although not actively engaged in promoting the improvement of working conditions and environment, may still be able to exert an indirect influence on the conditions of others in the course of their jobs. The target audience will therefore also include teachers in general or technical schools, public or private vocational training centres, engineering schools, management training colleges, and so on, and those in charge of in-service training for foremen, technical or personnel managers, time and methods specialists, and so on;

(c) the training of "decision-makers": the target audience here will be teachers in national administration schools and higher education establishments and institutes.

This book is also aimed at future trainers, i.e. those who, as the result of decisions in their own countries or perhaps as the result of reading this book, will be given responsibility for education and training. Certainly, the need for training exists and is widely perceived; what is often lacking is the stimulus of adequate documentation.

More generally, the book is also intended for those who have been looking for a general publication on working conditions and environment to help them in their day-to-day programmes, be they employers, heads of personnel, trade unionists, field workers in production or services, or officials in "social", economic or technical ministries whose decisions may affect workers directly or indirectly.

Coverage

This book could not hope to cover the vast field of working conditions and environment in its extreme diversity and its multiple relationships with the economic, physical, social and cultural environment. It does not *describe* working conditions and environment; it is not a philosophical treatise on the role of work or its meaning in society. Nor is it a manual bringing together everything that should be known on a given subject, and this for two reasons: first, the subject is too large; and second, a manual should be designed for a well-defined teaching approach and provide educational guidance relevant to the audience's needs.

This book is intended as a reference work and a guidebook. It therefore has a twofold approach:

(a) it presents an overall picture of a field which until now has been dealt with only in separate studies and publications; the unity of working conditions and environment – dealt with in Chapter 1 –

justifies this approach. Individual components of working conditions and environment are examined, but not in the detail required by an in-depth study; and an attempt is made to highlight the relationships between the various facets of man's working conditions; and

(b) it is designed to provide general information, and to give background support to attempts to improve working conditions and environment. It tries to present the concepts as simply as possible, and emphasises basic guiding principles by separating them from excessively technical or legal matters which may tend to obscure them and should, in any case, follow and not precede an understanding of them.

The book is therefore an "introduction" from both points of view: it does not claim to be comprehensive; and it attempts to give an overall picture and to delineate the main features.

These two approaches may appear contradictory. However, the first emphasises reality in its true complexity (the unity of the worker's life is made up of numerous interactions); the second encourages action to change current reality, and action is generated by simple ideas.

It is these two concerns that together have been the inspiration behind this book. Wide-ranging action is urgently needed to change working conditions and environment and to make them more humane, but the authors felt it was not possible to give priority to simplicity: this would have meant concealing the complexities of the situation and the difficulties entailed in certain actions, and making choices on alternative actions in the reader's stead. It is the reader's task to determine the priority needs and action for his own particular case. It was the authors' task not to betray the reader's confidence and to avoid hasty assessments and false concepts. In this regard, the book attempts to make people think. Occupational safety and health and welfare will not improve of their own accord; purpose and effort will be needed, and will stem from a realisation of the unacceptability of a situation and the freedom to condemn it. This has been the approach of all those who have striven to improve the worker's lot. In this respect, the aim of the book is to stimulate people to act.

To conclude, the reader should be given three brief warnings that stem from what has been said above.

First, this book may seem to give importance to problems, situations and, consequently, prospects of change that some readers may not look on as equally acute or serious. It is for readers to establish their own order of priorities in any given situation, but under no circumstances should they view any of the problems of safety, health and working conditions dealt with here as a "luxury"; each one deserves careful assessment before priority ratings are established.

Second, action-oriented as it is, the book reviews some basic concepts that may appear elementary and self-evident to certain readers. However, situations abound (and not just in developing countries) where these concepts are not as self-evident as they may appear at first sight; although they may have formed part of accepted principles for many years, they have never been fully implemented.

Third, an attempt has been made to avoid abstract statements. However, generalisations – which are to some extent inevitable when the scope is so wide – result in a schematic presentation, and, for fear of over-simplification, may imperfectly describe specific situations and lead to an understatement of reality. So, if the book is to achieve its objective and be of practical value, readers must adapt it to the realities in which they and, in particular, their audience live. This will enable readers, better than any book, to convince people that working conditions and environment must and can be changed. The authors sincerely hope that this is the way the book will be used.

How the book was written

The manuscript was prepared over a number of years. In the first phase, a provisional draft was produced by the ILO's Working Conditions and Environment Department with the collaboration of Mr. J. Carpentier, a consultant in ergonomics and working conditions at the European Human Ecology Centre of the University of Geneva. This draft was based on a preliminary outline drawn up by the Centre d'études supérieures industrielles de l'Est (CESI-Est), Obernai (France), assisted by the ILO's International Advanced Technical and Vocational Training Centre in Turin (Italy).

In a second phase, this provisional draft was examined at four meetings of experts and field workers from various occupational backgrounds, organised in Yaoundé (December 1979), Lima (May 1980), Turin (October 1980) and Bangkok (November 1980), and comments were obtained from some 30 principal technical consultants and experts involved in various projects in the developing countries (labour administration, workers' education, management training, vocational training, occupational safety and health). With this harvest of opinions and suggestions it was possible to resume the work.

In the third and last phase, final drafting was entrusted to a team of ILO staff and external collaborators under the direction of Mr. G. Spyropoulos, Head of the Working Conditions and Environment Department. Dr. A. Annoni, a former member of the Occupational Safety and Health Branch, wrote the first part of Chapter 2, and Mr. M. Robert, a former Chief of that Branch, wrote the second part, whilst collaborating with Dr. G. Coppée of the same Branch to ensure the chapter's overall integration; the section on safety and health in agriculture was written by Dr. M. Stilon de Piro, also of the Occupational Safety and Health Branch. Mr. A. Taqi, Chief of the Conditions of Work and Welfare Facilities Branch, was the author of Chapter 3, while Chapter 4 was written by Mr. J. Carpentier, in collaboration with Mr. H. Suzuki, of the Labour Law and Labour Relations Branch. Mr. J. Carpentier compiled the material for Chapter 5, the final drafting of which was entrusted to Mr. J. Thurman, of the Conditions of Work and Welfare Facilities Branch. Chapter 6 was written by Mr. C. Dumont, of the Conditions of Work and Welfare Facilities Branch, and Chapter 7 was contributed by Mr. A. Bequele, also of the Conditions of Work and Welfare Facilities Branch.

Mr. J.-M. Clerc, of the Conditions of Work and Welfare Facilities Branch, wrote Chapters 1 and 8; he undertook the co-ordination of and assumed editorial responsibility for the book as a whole.

Since this is the product of a collective endeavour, the style of writing may vary between individual sections: wealth and variety of experience and competence in the treatment of content have been given preference over homogeneity of form. Moreover, there was a need for internal cohesion between the subject-matter and the way it is presented. An attempt has been made to ensure that the individual chapters complement each other and form a cohesive whole through the adoption of a common approach and, in particular, through the content of the opening and closing chapters.

The various contributors would like to express their sincere gratitude to everyone who helped in the preparation of the volume by reviewing the drafts and communicating their comments and impressions.

Structure

Chapter 1 is more than an introduction, since it also shows how the various facets of the worker's life are inter-related, how they form part of the overall pattern and, consequently, why a global approach must be adopted.

Chapters 2 to 6 examine in turn the main features of: working conditions and environment; occupational safety and health; working time; wages and income; work organisation and job content; and welfare services.

Chapter 7 is devoted to workers in the rural and the informal sectors in developing countries.

Chapter 8 attempts to set the stage for decision-making and action for the improvement of working conditions and environment.

Readers are advised first to read Chapter 1 as a whole and then to follow the sequence that best suits their own needs or interests.

IMPORTANCE AND UNITY OF WORKING CONDITIONS AND ENVIRONMENT 1

This chapter has two distinct yet inter-related objectives.

The first and main objective is to review working conditions and environment, to emphasise their importance, to draw attention to the inadmissible situations only too frequently encountered – in spite of the progress achieved in recent years – and to stress the urgent need to intensify efforts to improve them. This is the underlying purpose behind the International Programme for the Improvement of Working Conditions and Environment (PIACT) of the International Labour Organisation, reviewed briefly in this opening chapter and referred to throughout the book in general and in the last chapter in particular.

The second objective is to arouse an awareness of the oneness of the worker's situation. Each component in the improvement of working conditions and environment – occupational safety and health, working time, wages and work organisation in particular – has its own special features, characteristics, methods and even techniques, and is usually studied separately (as in the following chapters); nevertheless, it is still necessary to point to the numerous inter-relationships between the different facets of working conditions and environment and to review their major environmental effects (economic, climatic, social, cultural, political, and so on).

An unacceptable situation

A heavy price to pay

Some 50 million accidents occur every year in industry alone, i.e. an average of 160,000 each day. Of these, around 100,000 are fatal – an estimate that is probably on the low side. Each year, industrial accidents also disable 1.5 million workers for the rest of their lives; in addition, millions suffer from disabling occupational diseases.

These figures would be swelled, if the statistics were available, by the many serious accidents that occur in agricultural and forestry work

PANEL 1

The International Labour Organisation (ILO), pledged to promote better and safer working conditions, is using every means at its disposal to help nations fight death and disablement at the workplace. These dangers are a surcharge on the price of progress that the worker and his family ought not to have to pay. For example —

A coalminer is crushed to death by a fallen boulder because the gallery roof is not properly supported.

A young girl is painfully mutilated by nylon underwear that fuses in the heat of a factory explosion.

A man's body is shorn in two by an unprotected lift car.

A farmer dies under the weight of his overturned tractor because it has no safety bar.

A stonecutter dies slowly from silicosis because he has had prolonged occupational exposure to stone dust.

None of these things need have happened if enough foresight had been exercised, and if enough priority had been given to ensuring that it was exercised.

To encourage such foresight and a proper sense of priorities is among the most important tasks of the ILO.

ILO Fact Sheet No. 5: *Protecting the worker.*

and on plantations. On average, one out of every ten industrial workers will be injured at work each year in the industrialised countries, and it is estimated that the world-wide figure for certain sectors is one in three: in other words, each worker in these sectors will be injured every third year on average. These are the completely pointless human tragedies that occur daily throughout the world. The examples in Panel 1 are just some of the more harrowing episodes in a true story of bad working conditions that adversely affect the worker's health, equilibrium and, consequently, productivity: long working hours; non-existent or inadequate rest; repetitive tasks; exhaustion caused by heavy physical work, a hostile environment or strenuous postures; fatigue and premature ageing caused by a fast work pace and the need for intense vigilance. For many years the ILO has been warning that working conditions are only too often not adapted to the body's physical and mental capacities. Not only do bad working conditions cause occupational accidents and diseases; they are also the source of tension, fatigue and dissatisfaction leading to poor health, high absenteeism, rapid labour turnover and mediocre productivity.

Are there at least some favourable signs to offer hope that technological progress and past endeavours will gradually improve the situation? Unfortunately, there are serious doubts that this is the case – which is the second reason for our concern.

A worrying trend

Two types of hazard currently confront us: first, conventional hazards, which continue to claim many victims; and second, new hazards, resulting from technological change, new products or new forms of work organisation which modify even job content.

The first is just as burning a question as it was 20 or 50 years ago. Although, especially in industrialised countries, the severity and frequency rates of occupational accidents are levelling out, few countries can actually point to even a modest downturn. In the developing countries, despite the incompleteness of statistics, there are indications that the number of occupational accidents is probably rising with increasing production. Occupational diseases due to silica or asbestos dust exposure or to the intake of lead, mercury, solvents, etc., were first documented long ago but are still widespread. Long working hours and the disease, accidents and premature ageing that they cause are still only too common in many countries.

Technological progress has brought with it new, even more insidious, risks – such as those found in chemical exposure – with medium- and long-term effects on health that are still poorly understood. Nervous tension is more common, and the job content of some types of industrial or service work is being lessened; although some progress has been made, growing numbers of workers are having to do simple, repetitive tasks that in no way match their training and capabilities. The trend towards shift work may have effects on, for instance, health and family and social equilibrium, especially if suitable medical and welfare measures are not taken. The importation of new technologies and foreign life-styles into rural communities may increase working hours and initiate chain reactions in working and living conditions and the organisation of society. Finally, concern is being expressed, in both the developing and the industrialised countries, at the growth in the number of workers in marginal employment who have no social protection.

Promoting work adapted to the worker

Even though the above findings are sketchy and narrow in their coverage, they cannot be ignored. A more human approach to work is essential: [1] we cannot passively accept that man's most treasured possessions – health, physical integrity (and even his life), aptitudes, professional skills, dignity – should be endangered by his employment. As well as providing for the necessities of life, work should offer a means

PANEL 2

Constitution of the International Labour Organisation

PREAMBLE

Whereas universal and lasting peace can be established only if it is based upon social justice;

And whereas conditions of labour exist involving such injustice, hardship and privation to large numbers of people as to produce unrest so great that the peace and harmony of the world are imperilled; and an improvement of those conditions is urgently required; as, for example, by the regulation of the hours of work, including the establishment of a maximum working day and week, the regulation of the labour supply, the prevention of unemployment, the provision of an adequate living wage, the protection of the worker against sickness, disease and injury arising out of his employment, the protection of children, young persons and women, provision for old age and injury, protection of the interests of workers when employed in countries other than their own, recognition of the principle of equal remuneration for work of equal value, recognition of the principle of freedom of association, the organisation of vocational and technical education and other measures;

Whereas also the failure of any nation to adopt humane conditions of labour is an obstacle in the way of other nations which desire to improve the conditions in their own countries;

The High Contracting Parties, moved by sentiments of justice and humanity as well as by the desire to secure the permanent peace of the world, and with a view to attaining the objectives set forth in this Preamble, agree to the following Constitution of the International Labour Organisation:

CHAPTER I — ORGANISATION

Article 1

1. A permanent organisation is hereby established for the promotion of the objects set forth in the Preamble to this Constitution and in the Declaration concerning the aims and purposes of the International Labour Organisation adopted at Philadelphia on 10 May 1944 the text of which is annexed to this Constitution.[1]

[1] See Panel 6, p. 18.

of personal achievement; under no circumstances should it be degrading. To diminish man's stature is to weaken the structures and equilibrium of society.

The founding fathers of the International Labour Organisation had a clear understanding of this, as is shown by the reference, in the Preamble to the Constitution of the ILO, to the existence of conditions of labour "involving such injustice, hardship and privation to large numbers of people as to produce unrest so great that the peace and harmony of the world are imperilled" (Panel 2). Since its creation in 1919, the International Labour Organisation [2] has been actively involved in improving conditions of work and life and safety and health, as explicitly required by its Constitution. The very first international standard adopted by the International Labour Conference dealt with hours of work,[3] and this was followed by a long series of international labour Conventions and Recommendations, many of which deal specifically with these important subjects.[4] The improvement of working conditions and environment is scarcely a new mission for the ILO.

Why, then, was a new international programme considered necessary? The Chief of the ILO's Working Conditions and Environment Department at the time of the launching of PIACT wrote: "despite considerable efforts made to improve occupational safety and health, the situation in the world as a whole remains disturbing. . . . In a period of mass unemployment and stagflation there is a temptation to put off until a better tomorrow the solution of apparently less urgent issues. The improvement of the quality of working life has so far fallen into this category in most places. This must change." [5] In June 1975 the International Labour Conference adopted a resolution (see Panel 3) charging the Director-General of the International Labour Office to set up a new international programme "to strengthen and revitalise the Organisation's action on working conditions and environment". Thus, PIACT, which – it must be stressed – was adopted unanimously by government, employers' and workers' delegates of the ILO's member States, was designed to give a new and vigorous stimulus to the ILO's future work in this field.[6]

Action is needed if work is to be adapted to the worker instead of causing death, disease, exhaustion, physical or mental mutilation – an assault on human dignity. And action is possible. Progress can be achieved by even the simplest measures, especially in accident prevention. Undertakings with a methodical safety and health programme have shown that most accidents *can* be prevented.

PANEL 3

Resolution concerning future action of the International Labour Organisation in the field of working conditions and environment [1]

The General Conference of the International Labour Organisation,

Considering that the improvement of working conditions and the protection of the physical and mental health of workers constitute an essential and permanent mission of the International Labour Organisation,

Considering the slow and uneven progress realised concerning hours of work and associated problems concerning work safety and health primarily because of the absence of a general strategy relating to an improvement of working conditions and environment,

Noting that the utilisation of scientific research and technology, without taking into account considerations of a social nature, could not only create dangers at the workplace but could also have an adverse effect on the human environment generally,

Considering that changes in techniques, production methods and the importance of transfers of technology and the evolution of human society and of social aspirations place the improvement of working conditions and environment in a new and sometimes different context according to the countries, branches, occupations and categories of workers,

Referring to the resolutions adopted by the International Labour Conference in 1972 and 1974 relating to the working environment,

Having noted the activities anticipated in the ILO's Programme and Budget for 1976-77,

Having received with satisfaction the Report submitted by the Director-General of the International Labour Office to the 60th Session of the International Labour Conference and the determination indicated therein to reinforce and renew ILO action in the field of working conditions and environment,

Considering that ILO action concerning working conditions and environment should, in taking into account aspirations for a better quality of life, be closely joined with other activities relating to the protection of the human environment;

1. Solemnly reaffirms that the improvement of working conditions and environment and the well-being of workers remains the first and permanent mission of the ILO.

2. Earnestly invites member States — ▷

(1) to promote the objectives of an improvement of working conditions and environment with all aspects of their economic, educational and social policy;

(2) to set periodically for themselves a number of definite objectives designed to reduce as far as possible certain industrial accidents and occupational diseases or the most unpleasant and tedious of jobs;

(3) to normalise the application of scientific research so that it is carried out for man, and not against him and against his environment.

3. Supports the world-wide action suggested by the Director-General of the International Labour Office in his Report with a view to reconsidering the current activities of the ILO and to launching an international programme for the improvement of working conditions and environment which is designed to promote or support activities of member States in this field.

4. Invites the Governing Body of the International Labour Office to instruct the Director-General, as soon as resources permit —

(1) to prepare and submit to it such a programme based on the general discussion of his Report to the 60th Session of the International Labour Conference and after consultation with the competent international organisations as well as with the national, regional and international bodies specialised in working conditions and environment. . . .

[1] Adopted on 24 June 1975.

The need for a global approach

Working conditions and environment are not isolated phenomena unconnected with each other and with other aspects of the worker's life. This commonsense approach is sometimes overlooked and, consequently, a feature of PIACT is its emphasis on a global approach to working conditions and environment. In 1974, the year before PIACT was launched, the International Labour Conference emphasised that "the improvement of the working environment should be considered a global problem in which the various factors affecting the physical and mental well-being of the worker are inter-related"; [7] and one of the needs that PIACT is intended to meet is "the fact that problems of working conditions and environment should be approached globally within the framework of all aspects of economic, educational and social policy".[8]

Working conditions and environment form a whole

Inter-relations and interactions

Each component is related to the others in a multitude of ways. For example, accidents, diseases and fatigue may be caused just as much by bad working conditions, long or poorly scheduled working hours, too rapid a work pace and piece-work rates, as by dangerous machines or a hazardous physical environment. In the same way, the length and arrangement of working time may condition not only free time and the quality of life in general but also wages and safety and health. The level, calculation and regularity of wages are also directly related to other working conditions: "danger money" or "hard-work" money are still common, as are higher rates for overtime, piece-work or other types of payment system that may encourage workers to ignore the hazards to which they may be exposed. Many factors may contribute to fatigue: physical effort, inadequate work breaks or rest, microclimatic temperature, noise, lighting, work organisation and so on.

The comprehension of work situations

Two basic conclusions summarise the findings outlined above:

(a) working conditions and environment form an entity for each worker. Work is perceived and experienced as the convergence and accumulation of a series of factors which, although individually distinguishable, are assessed as a single whole; and

(b) these factors form a complex system:

(i) they interact, and consequently to assess each in isolation would give only an imprecise evaluation of a given situation: a proper appreciation of the factors and their significance often entails a search for the basic causes behind them, which may not be immediately apparent. Only by a global approach can diagnostic errors be avoided and the true causes identified (e.g. the reason why a piece-worker does not use his machine guard);

(ii) the total effect of working conditions and environment may be greater than the sum of the individual components. Panel 4 gives an example of bad working conditions in the restaurant and hotel industry, where the various factors interact to cause health effects that may explain the reduced life expectancy of these workers.

Clearly the realisation that working conditions and environment form a whole must be reflected in the approach to improvements, and this will be dealt with in the section on "Aims and scope of the global approach", below.

PANEL 4

An example of stress-factor interaction

Some time ago the Norwegian Central Bureau of Statistics compiled figures on the relationship between mortality and occupational background. Surprisingly to some, hotel and restaurant workers emerged as one of the groups with the shortest average life span.

Why should this be so? At the request of the Hotel and Restaurant Workers' Union the Work Research Institute undertook a survey of a representative sample of the union's members. The overall picture to emerge from the survey is shown in the figure overleaf.

The main point to note is that a number of factors in the work environment *combine* to produce complex and many-sided effects on health. These include cardiovascular diseases, cancer due to stress control through smoking, and a slow but steady deterioration of functional ability due to physiological problems of various kinds, e.g. pains in the neck, back and shoulders. No single factor in the environment can be pointed to as *the* cause. Looked at in isolation, each of the contributing factors can appear harmless enough. Nor is the impact on health limited to a single, easily identifiable type of effect.

A further point of major importance is that feedback effects also occur. When negative effects on health and welfare start to make themselves felt, workloads will often seem heavier, speed requirements harder to meet, the shift system more inconvenient, and so on, with consequent reinforcement of the adverse effects on health. In practice we are dealing with a highly complex network of circular processes where "causes" and "effects" are intermingled.

▷

Working conditions and environment in their context

Working conditions and environment are not a closed system isolated from the general environment in which they are located. It may therefore be useful to review and re-emphasise some apparently self-evident features.

Working conditions and environment, and conditions of life

Each worker recognises intuitively the existence of factors linking various aspects of working conditions and environment with conditions of life. There is a constant multiple interplay – be it conscious or unconscious, simple or complex – between man and the physical, social

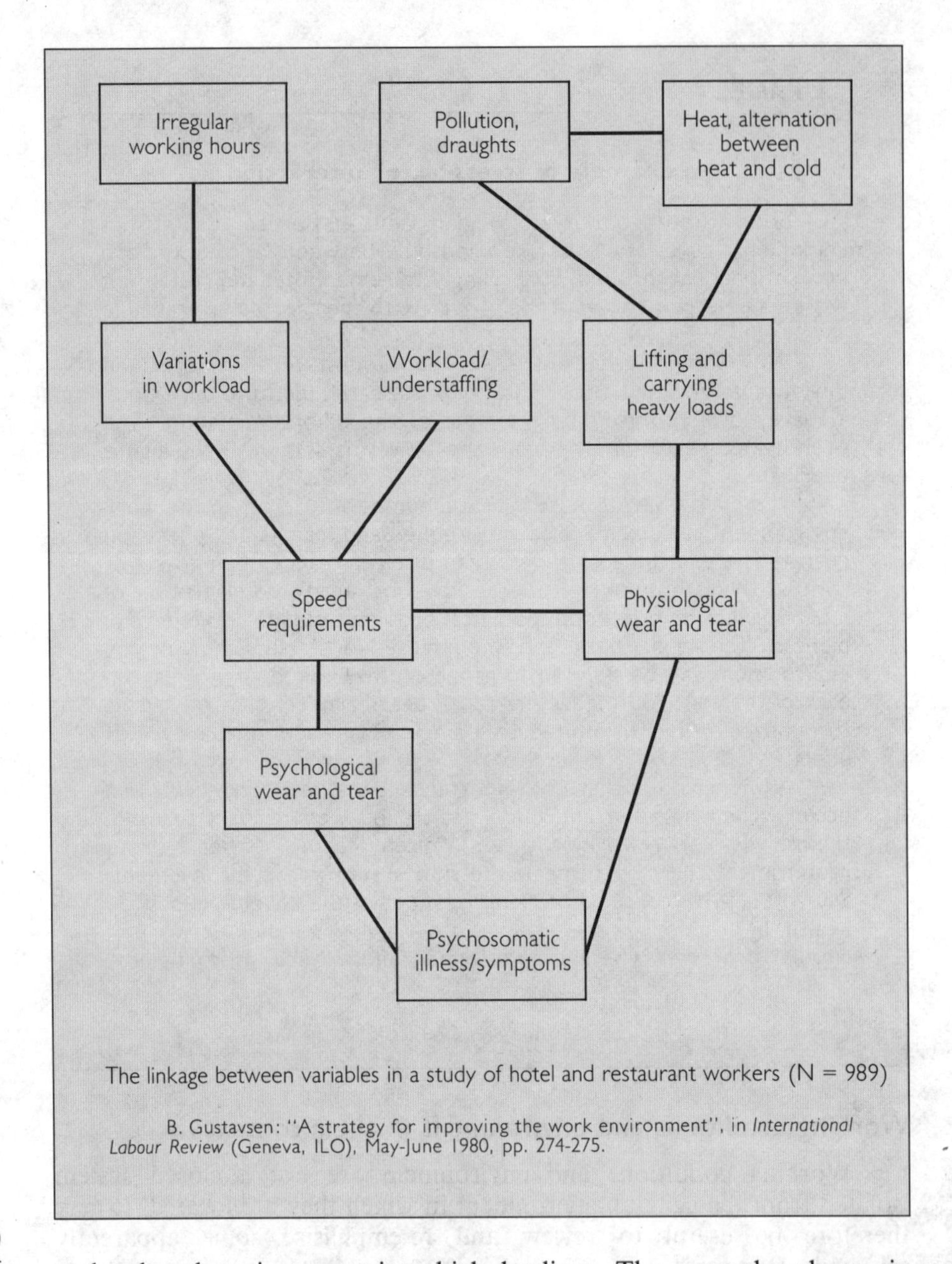

The linkage between variables in a study of hotel and restaurant workers (N = 989)

B. Gustavsen: "A strategy for improving the work environment", in *International Labour Review* (Geneva, ILO), May-June 1980, pp. 274-275.

and cultural environment in which he lives. The examples shown in figure 1 illustrate this. The relationships vary considerably from case to case, and merely listing them would not faithfully describe all those that directly or indirectly affect the various aspects of a given worker's personal, family and social life; the examples provided are merely illustrations. Food, housing and climate are the only major conditions of life that will be dealt with here, and then only in a summary fashion.

They are the subject of more specialised publications,[9] and will also be referred to in subsequent chapters, in particular Chapter 6 on social services.

1. Nutrition is particularly important both generally and in relation to work: "good nutrition not only contributes to the greater welfare and happiness of mankind but also enables people to work and produce more. . . . The evidence suggests . . . that poor nutrition does restrain productivity, not only in the conventional, narrow sense – output per unit of input (units produced per man-hour, for example) – but also as a result of increased absenteeism, lowered resistance to disease, lethargy, lack of drive." [10] If calorie intake does not match work energy expenditure, the number of actual working hours will be reduced accordingly; for example, a food intake of 2,000 kilocalories per day will allow 4.5 hours of moderate physical work in light industry or 2 hours of heavier work in agriculture, forestry or manual handling. Acute or chronic undernutrition is an insidious factor in the causes of occupational accidents, and has been classified among the factors which may cause fatigue and lessen concentration. What length, intensity and type of work will a given level of nutrition permit? For example, if hours of work exceed those permitted by a given level of food intake, should they be considered excessive? Food composition is another factor to be considered. The problem is one of significant proportions, but an appreciable improvement could probably be achieved if each undertaking made an adequate effort to ensure additional calorie intake through the provision of free meals in a works canteen. This may also enhance productivity: "Undertakings or industries in developing countries which have devoted to this nutrition problem the importance it merits, have seen a spectacular increase in their workers' working capacity and an improvement in their general health status" [11] (see also Chapter 6).

2. Housing is also closely related to working conditions, and social legislation throughout the world contains clauses on the rest needed to compensate for work-induced fatigue. The quality of rest will depend largely on housing conditions. Sleep is an essential component, but it will not restore the sleeper unless it is sound and long enough. There is a clear link between the qualitative and quantitative aspects of housing, since overcrowding (for instance) may make otherwise suitable accommodation unhealthy and inadequate (see also Chapter 6).

3. The role of climate in conditions of work and life is often underestimated. This is not the place to review climatic conditions or the ways in which temperature, humidity, solar radiation, and so on, may affect man,[12] but this aspect is of particular importance in relation to working time and schedules, nutrition and occupational safety and health. Work in hot conditions is common in many countries, and all physical work raises body temperature.

Figure 1. Typical inter-relationships between various facets of working conditions and environment. Only a few examples have been given and these cannot, of course, portray all possible situations or all the possible inter-relations and interactions.
→ designates an effect or consequence.

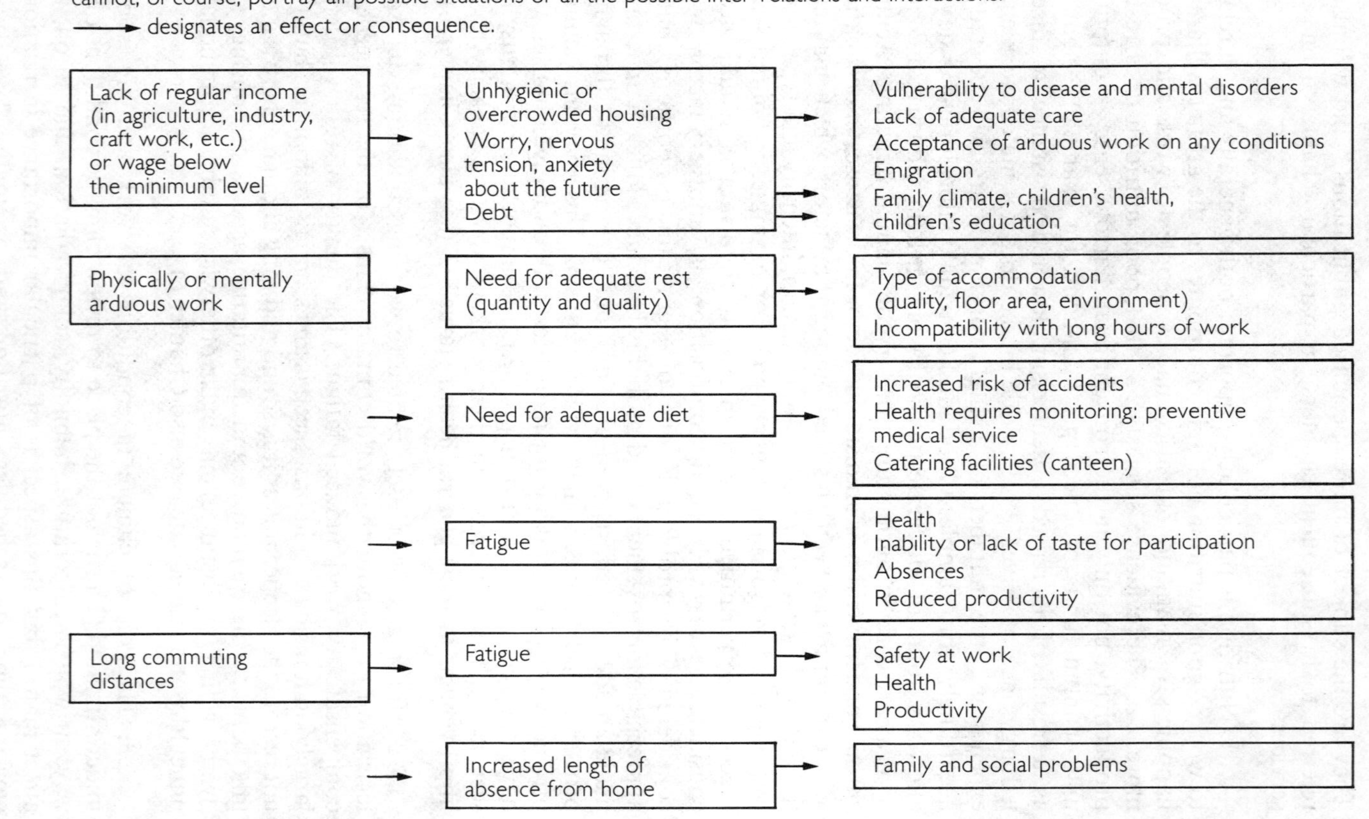

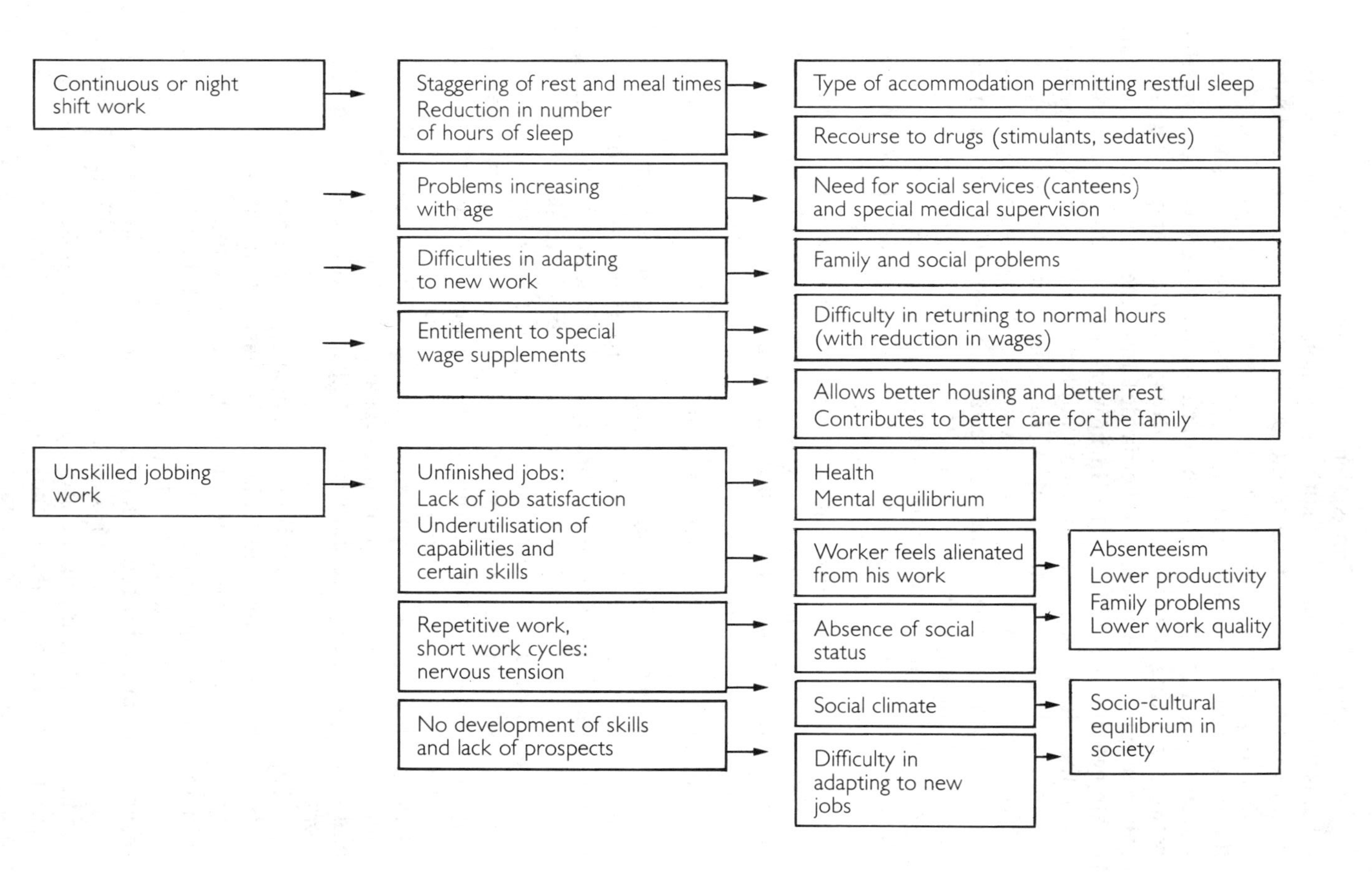

Continuous or night shift work
Staggering of rest and meal times
Reduction in number of hours of sleep
Problems increasing with age
Difficulties in adapting to new work
Entitlement to special wage supplements
Type of accommodation permitting restful sleep
Recourse to drugs (stimulants, sedatives)
Need for social services (canteens) and special medical supervision
Family and social problems
Difficulty in returning to normal hours (with reduction in wages)
Allows better housing and better rest
Contributes to better care for the family
Unskilled jobbing work
Unfinished jobs:
Lack of job satisfaction
Underutilisation of capabilities and certain skills
Repetitive work, short work cycles: nervous tension
No development of skills and lack of prospects
Health
Mental equilibrium
Worker feels alienated from his work
Absence of social status
Social climate
Difficulty in adapting to new jobs
Absenteeism
Lower productivity
Family problems
Lower work quality
Socio-cultural equilibrium in society

4. In general, the unprecedented growth of industrialisation and urbanisation is profoundly changing the conditions of work and life of urban workers in developing countries. In addition to their exposure to occupational hazards and stress, these workers also have to cope with the adverse effects of environmental pollution, noise, inadequate housing, transport, sanitation, and so forth. In rural areas, there is a large overlap between working and living environments; and, for example, the use (and common misuse) of pesticides, herbicides and other farm chemicals may create new hazards not only for farm workers but also for the crops on which their livelihood depends (see Chapter 2, section on "Occupational safety and health in agriculture"). Likewise, the over-exploitation of scarce natural resources, such as water and vegetation in semi-arid pasture lands, or large-scale deforestation of mountain and tropical areas, may cause irreversible damage to life-supporting ecosystems. There is growing recognition of the complementarity of and interaction between the working and living environments, and the ILO Working Environment (Air Pollution, Noise and Vibration) Recommendation, 1977 (No. 156), provides that "the competent authority should take account of the relationship between protection and the working environment and the protection of the general environment".

Wage earners in legislation and in practice

The protection of wage earners is a central theme in social legislation throughout the world. Apart from regulations designed to protect both employers and workers or to control labour relations (redundancy or resignation notices, labour tribunals, freedom of association and collective bargaining, etc.), the main provisions derive from the wage earner's position of subordination to the employer, and they consequently establish rules and obligations that give the wage earner certain minimum guarantees (minimum wage, maximum hours of work, presentation of a pay sheet, safety and health protection, etc.). Other equally important provisions guarantee freedom of association aimed at permitting equitable negotiation between employers and workers.

The Preambles to the ILO instruments on freedom of association and labour inspection (see Panel 5) emphasise that the wage earner's working conditions and environment depend directly on basic freedom of association being achieved and on labour legislation being properly enforced. Wherever social legislation and regulations are not enforced or only partially enforced, where trade union rights are little or poorly respected, where labour inspection is weak or ineffective, working conditions and environment are threatened and probably bad. These hazards are more acute if the economic situation and the risk of unemployment makes workers fear anything that may imperil their jobs.

PANEL 5

Considering that the Preamble to the Constitution of the International Labour Organisation declares "recognition of the principle of freedom of association" to be a means of improving conditions of labour and of establishing peace . . . (Preamble to the Freedom of Association and Protection of the Right to Organise Convention, 1948 (No. 87)).

Whereas the Constitution of the International Labour Organisation includes among the methods and principles of special and urgent importance for the physical, moral and intellectual welfare of the workers the principle that each State should make provision for a system of inspection in which women should take part, in order to ensure the enforcement of the laws and regulations for the protection of the workers . . . (Preamble to the Labour Inspection Recommendation, 1923 (No. 20)).

Economic constraints

These are a major and permanent concern for workers, undertakings and the public authorities.

Economic constraints on wage earners and other workers

As has already been indicated, workers striving to increase their wages may endanger their safety and health – especially when there is a danger of the workers' losing their jobs. The spectre of unemployment and of being unable to provide for the family can only paralyse any endeavours to improve safety and working conditions. Contesting or merely pointing out dangerous or abnormal working conditions may be a risk that the worker cannot afford to take too often, even if basic rights such as the minimum wage and paid overtime and leave are guaranteed – all the more so where there is no unemployment pay or when job prospects are poor. Often, it is only after dismissals have taken place that abuses of this type are reported to the labour inspectorate.

In developing countries, this fear may be so acute that the workers themselves have little interest in safety and working conditions: accidents and diseases may seem less of an imminent threat than unemployment and poverty.

Economic constraints may be even more striking in the case of self-employed workers, i.e. the great mass of rural workers or craftsmen

Figure 2. The "economic cycle of disease"

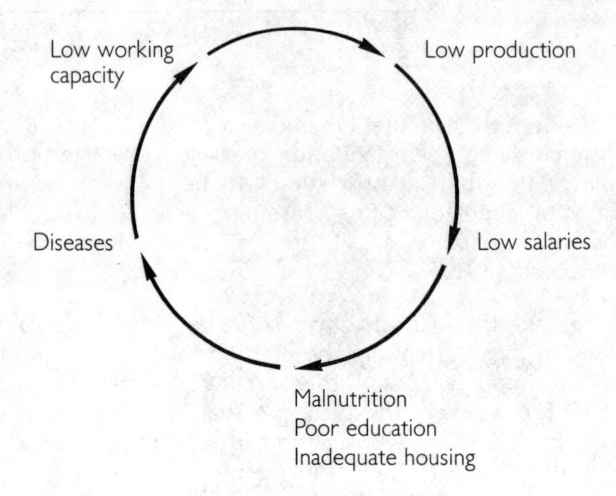

Source. K. Elgstrand: "Teaching ergonomics in tropical countries", in J. H. van Loon et al. (eds.): *Ergonomics in tropical agriculture and forestry*, Proceedings of the Fifth Joint Ergonomic Symposium organised by the Ergonomic Commissions of IAAMRH, CIGR and IUFRO, Wageningen, Netherlands, 14-18 May 1979 (Wageningen, Centre for Agricultural Publishing and Documentation, 1979).

throughout the world, who may be on or near the poverty line (see also Chapter 7). Debt may become a true psychosis for smallholders in developing countries, and many economists believe that these workers' situation (and, consequently, agricultural production) will not be improved until the debt problem is solved. Want, even if it has not struck, is a constant threat, and attempts at escape may produce dangerous decisions. This precarious situation may initiate a chain reaction of cause and effect and a vicious circle (as shown in figure 2) from which it is difficult to escape.

Also on or near the poverty line are the numerous independent craftsmen or proprietors of small family businesses for whom economic pressure may be compounded by ignorance of occupational risks or simple preventive measures. A similarly marginal situation is characteristic of various casual workers, those working for temporary employment agencies or labour subcontractors, clandestine workers and homeworkers.

To summarise, where there is economic constraint, the worker needs job and wage security; this is what lies behind the governmental wage protection measures examined in Chapter 4.

Economic constraints on undertakings

The economy may have major direct or indirect effects on the undertaking and on working conditions and environment: costs, competition, viability, profit, pressure to meet deadlines or win contracts, and so forth. The centre of decision-making may be in another country, and this makes the consideration of factors affecting working conditions and environment all the more difficult and uncertain. States themselves are subject to similar constraints; in crisis situations or difficult economic conditions, these constraints are clearly much stronger. The growing internationalisation of trade has started a wild race for contracts from which no country can escape and which gives economic factors and material and commercial preoccupations priority over social and environmental factors; it also increases dependence on investors who can influence production conditions and employment.

Is there conflict between economic and social factors?

The term "economic concern" may be used to describe very different attitudes, ranging from the self-interest that may be at the root of the most flagrant abuses to the most legitimate objectives; these should not be confused. However, no matter how reasonable legitimate economic objectives may be, they usually seem to be in conflict with social preoccupations. An employer's need to make a profit, or a government's need to balance a budget and increase the general wealth of the population, may all seem urgent objectives with priority over social objectives. It would seem that the confrontation between economic and social objectives is an unequal conflict in which the laws of economics must necessarily dominate. Is this in fact the case?

The problem is far from new. As Albert Thomas, the first Director of the ILO, stated:

> The discussions at our Conferences . . . have shown with increasing clearness that the immediate future of our work and even the very future of our institution depend on the solution of this question and on the spirit which is to inspire national and international economic policy. One possible solution is that, having established a minimum of protection for humanity – what has been called "the sanitary cordon" [13] – we may subordinate the ever-evolving conceptions of, and the ever-recurring aspirations after social justice to "economic necessity" and "economic laws", the "free play" of which must not be further disturbed. Or we may decide that, just because of these ideals of justice and because of the growing clearness of the voice of the human conscience, and sometimes even in the face of economic laws, which (to put it mildly) have perhaps not always the full force of natural laws, human intelligence must make every possible effort to organise the economic system and has in fact the power to do so. The social factor must take precedence over the economic factor; it must regulate and guide it in the highest cause of justice.[14]

This categoric statement of principle is faithful to the ideal on which the ILO was established, and should not be forgotten. The Declaration of Philadelphia – an integral part of the Constitution of the ILO – just as clearly accords "material progress" and "spiritual development" priority over programmes "in finance and economics" (see Panel 6).

However, the contrast and conflict between economic and social factors are not so clear as might appear at first sight. Moreover, the

PANEL 6

Declaration of Philadelphia

. . . .

II

Believing that experience has fully demonstrated the truth of the statement in the Constitution of the International Labour Organisation that lasting peace can be established only if it is based on social justice, the Conference affirms that:

(a) all human beings, irrespective of race, creed or sex, have the right to pursue both their material well-being and their spiritual development in conditions of freedom and dignity, of economic security and equal opportunity;

(b) the attainment of the conditions in which this shall be possible must constitute the central aim of national and international policy;

(c) all national and international policies and measures, in particular those of an economic and financial character, should be judged in this light and accepted only in so far as they may be held to promote and not to hinder the achievement of this fundamental objective;

(d) it is a responsibility of the International Labour Organisation to examine and consider all international economic and financial policies and measures in the light of this fundamental objective;

. . . .

economic advantages of social measures should be carefully considered, even though the measurement techniques currently available do not make their quantification possible. Francis Blanchard, the present Director-General of the ILO, in his introduction to the report *Making work more human*, gave two reasons for this: the relationships between different concepts and different problems; and the fact that the effects of these inter-relationships are not necessarily felt in the very short term (see Panel 7).

There is no lack of examples of social measures which, although first delayed for economic reasons and then finally forced through by — for example — social pressure, have (at least as far as the economist is concerned) had unexpectedly favourable economic effects, as is the case in particular of increases in the minimum wage. Job creation that distributes earning power also helps to expand sales markets and economic activity. Similarly, certain improvements in conditions of work

PANEL 7

Another reason is that, in choosing the theme of working conditions and environment, I am sure that I am not losing sight of the problems of employment and of income. First of all, because the relationship between questions such as satisfactory conditions and hours of work and the optimum level of employment is obvious. But, above all, because those who are responsible for social policies today are increasingly aware of the interdependence of all the elements which make up these policies. It can indeed be maintained that the success or failure of modern societies will depend on how they solve this key problem of the inter-relationship between employment, remuneration, working conditions and environment, education, health and leisure.

Apart from these considerations, there is a further reason for dealing now with the vast problem of working conditions and environment despite the fact that most countries are having to face a slackening of economic activity.

Mesmerised as we often are by short-term considerations, unaware at times of the connections between the various elements I have just mentioned, we are tempted in times of stagnation or crisis to put off to a better tomorrow the solving of issues that appear less urgent: the improvement of conditions of work has until now been one of them.

We have already seen, in other fields, the consequences of "reflexes" of this kind: industrialisation policies planned or applied without thought for the environment; development policies that overlook a fairer distribution of the fruits of growth; policies designed to encourage migration for economic reasons without taking sufficient account of the social consequences.

One of the central arguments of this Report is that the adoption of such an approach to the improvement of conditions of work is likely, far sooner than we realise, to lead to disruption and disorder in social systems on a scale quite out of proportion with the economic cost of any lucid measures to improve conditions of work taken at an earlier stage.

This argument appears valid for all countries. In under-development there is no situation in which greater attention to the working conditions of men and women and resolute action to improve these conditions are not likely to bring about beneficial results tomorrow out of proportion with the effort put in today.

ILO: *Making work more human*, Report of the Director-General, International Labour Conference, 60th Session, Geneva, 1975, op. cit., pp. 1-2.

are known to have a favourable impact on productivity (reduction of the permissible weight to be lifted by one person, shortening of excessive working hours); this aspect should be given more publicity.

A real difficulty may be encountered when the initial cost of a social improvement is beyond a company's financial means, and this may be particularly so in the case of small undertakings. Yet some very positive measures cost little or nothing to introduce, and may even produce immediate financial gain. However, cost-benefit analysis requires a more accurate assessment of factors that are often overlooked because they are considered insignificant, or of adverse circumstances that are not of an economic nature. Factors often overlooked include day-to-day incidents which, although minor, when added together may cost firms large sums of money or may be the cause of occupational accidents. The control of such incidents is therefore both a cost-saving and an accident-prevention measure. Adverse circumstances not considered to be of an economic nature include absenteeism, labour turnover, social conflict, and upheavals in social and cultural traditions. This intentionally brief list may be used to illustrate some of the shortsighted actions that too often determine choices and decisions. Social and qualitative problems have delayed "fall-out" which can be quantified economically. Many examples of this can be given for occupational safety and health: costs of disability, health-care cost increases, hours of work (health effects of long working hours or shift work, dislocation of family life), work organisation and job content (mental health disorders, distaste for specific jobs), and so on.

Chapters 2 and 8, in particular, will deal further with the cost of measures to improve working conditions and environment. Nevertheless, the economics of these measures must be assessed lucidly, and the numerous constraints that are limited in time and space should not be confused with the real issues of worker welfare that must be looked at over the medium and long term.

Aims and scope of the global approach

A diagrammatic presentation of the above concepts has been attempted in figure 3. The circle immediately outside the "situation in which the worker lives" contains a simplified grouping of the main factors comprising or contributing directly to working conditions and environment. These factors may act jointly or in combination, and the arrows schematically symbolise these inter-relationships. Each factor affects the worker's situation and simultaneously has an impact on others. The factors in the upper half of the circle are related to technology, economics and organisation; those in the lower half to man and the environment. Outside the circle, at the top and bottom, are factors forming the more general context.

Figure 3. Working conditions and environment: A complex system and its context

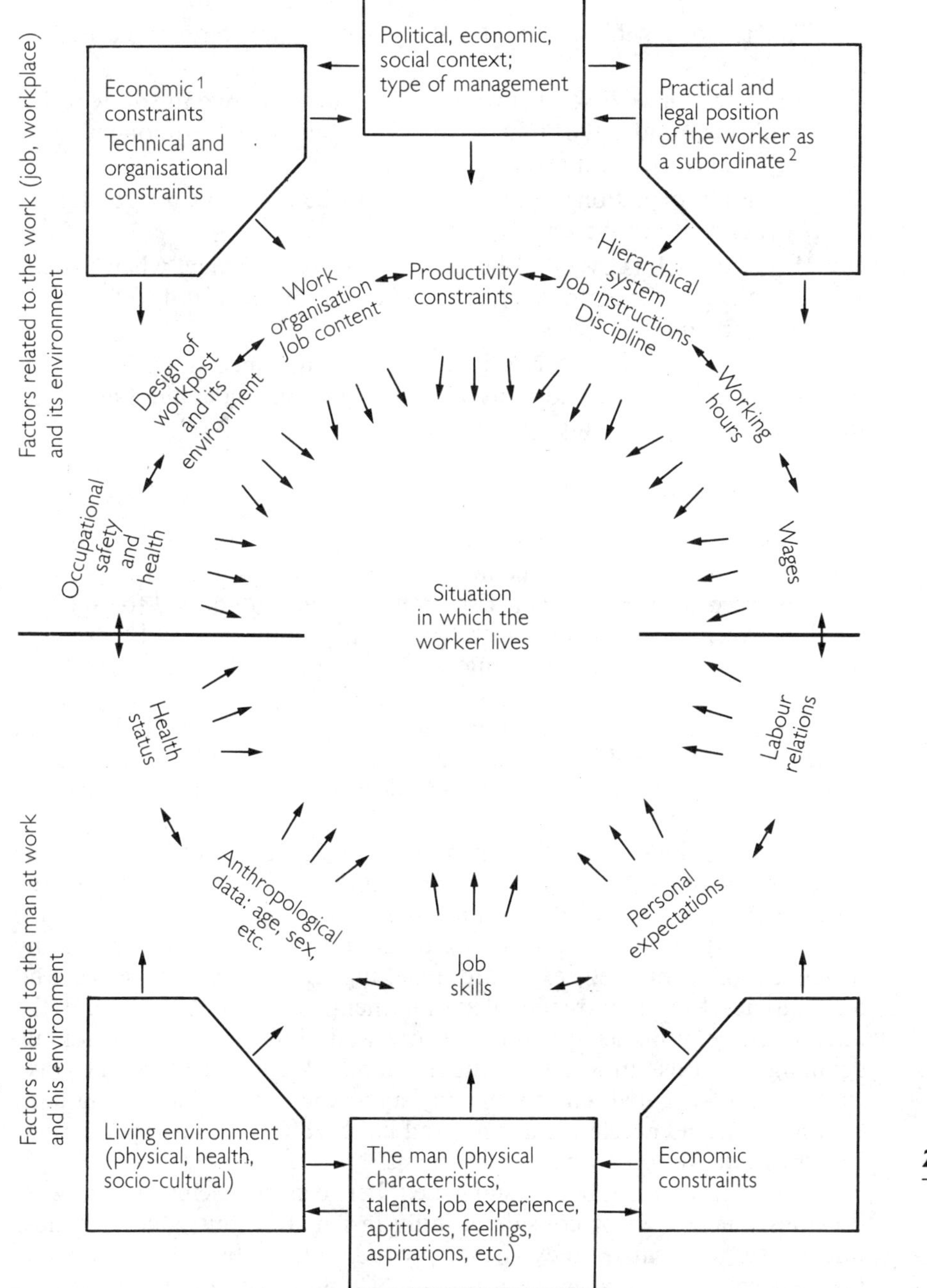

[1] This diagram shows the situation the way it is felt and presents forces rather than scientific laws. The term "constraint" should be taken in its relative sense: these constraints are very strong in many cases, especially in economic matters, but they should not be considered absolute. Techniques and organisation, in particular, are not intangible and can be changed.
[2] In the case of the wage earner, *de jure* or *de facto*. In the case of self-employed workers (in the rural or informal sectors, craft work, small family firms), this box should be replaced by another entitled "economic constraints".

Meaning and objectives of the global approach

The reasons behind this global approach may be summarised as follows:

(a) there are close links between individual components of working conditions and environment, and between these components as a whole and their context;

(b) any study or action in this area should be aimed at improving the worker's situation; and

(c) this approach is essential because there is a tendency, when delving into the problems, to isolate them from each other and to lose sight of their unity.

The approach is, moreover, a realistic one since it relates to the situation as it really is, and aims at improving our understanding of the situation and at achieving effective action.

Practical applications

There is a long-standing desire to bridge the gap between science and technology, which give the impression of being isolated and perhaps even divorced from each other. Ergonomics [15] has attempted to establish a link between the engineer and the physician or physiologist by bringing together the means for adapting work to the worker (by modifying machinery and plant or, better still, by designing them correctly in the first place). The result is usually greater safety, improved welfare and higher productivity. Unfortunately, the knowledge and application of ergonomics are still not sufficiently widespread, and only too rarely is the subject taught.

Recent interest in environmental problems has contributed to the development of a global approach, and emphasis has been given to: the absence of any division between the working and the living environment; the likelihood that risk exposure occurs 24 hours a day; [16] the increasing incidence of combined exposure resulting from new processes and technologies (for a more detailed treatment, see Chapter 2); and the interaction between occupational hazards and life-style factors (smoking, drinking, nervous tension and fatigue, etc.).[17] Medical research has also shown that repeated environmental aggression may have long-term health effects and accelerate ageing, and has highlighted the imperfections in man's adaptation to his environment.[18]

Links have also been established between problems which were previously neglected or considered only in isolation but which are now under study because they have proved intractable. For example, absenteeism and the negative attitudes of young workers have led to a consideration of inadaptation phenomena which are certainly related to industrial job content; inequalities in the richer countries are the subject of similar considerations.

All these phenomena and problems are inevitably related to job content or to the physical, legislative or social conditions under which the work is done. They raise various questions. Does industrialisation automatically remove from work both the interest and the social role or "status" that identify the worker in the social hierarchy? What are the implications of the growing number of clandestine workers who, in many countries, have no legal protection?

Implementation and impact

The distinguishing feature of the global approach is that it encompasses all components of working conditions and environment, and also their determining factors and their inter-relationships and interactions. Georges Spyropoulos [19] has stated that one of the basic concepts behind PIACT is that the problems of working conditions and environment should be viewed globally by taking into consideration all the dimensions of the problem and by attempting to relate them to those factors which are as yet still unrelated; it is essential to understand the interaction between the various components of the system of working conditions and environment if one is to control them. Adopting the global approach does not mean acting on all components at the same time, with the danger of unduly spreading one's energies or compounding the difficulties; it means being aware that, *at the moment of action, one should be attentive to the interdependence of the various factors in the work environment.*

Four final comments need to be made:

1. *The global approach does not replace specific techniques and disciplines.* The adoption of a global approach does not make specific techniques and disciplines (occupational safety and health, labour legislation, occupational sociology, etc.) redundant, nor does it limit their scope or diminish their significance. Those working as generalists or specialists on the working environment should develop a mental attitude and a working procedure by which they keep a constant eye on the overall context of the problem they are dealing with.

2. *The global approach is not an obstacle to localised action.* Localised action is not diminished in value because it derives from a global approach; on the contrary, its value may be increased, its effects may be multidirectional and it may break the vicious circle of negative chain effects. We return to this point in Chapter 8.

3. *Specific problems are caused by environmental change.* Environmental factors such as climate, health status, undernutrition, malnutrition, long commuting distances, social traditions and structures must always be considered. This is particularly important when the man-work environment relationship undergoes rapid change, such as when, for example, the worker emigrates from a rural to an urban

environment, or to another region, country or continent, with all the adaptation problems this entails. The rapid introduction of new technology implies a similar break with the environment. The installation of industrial plant, factories and machines inevitably brings with it less tangible but very real factors related to work organisation (e.g. production-line work), the pattern of working time (e.g. continuous shift work), work forms (e.g. piece-work, payment by results) and certain approaches (e.g. separation of design from execution). The transfer of plant and premises also implies transfer of life-styles, usually with inadequate thought and preparation; these transplanted material and non-material factors may have been suited to the country and to the physical and socio-cultural environment in which they developed, but not to the country that receives them; they may also be unsuited to both. In rural regions as well – and perhaps even more so – the implantation of new cultures or techniques may (if the environment is not correctly assessed) result in a major upheaval of the existing society, since work is an integral part of family and social life. This is a major problem for developing countries since the pressing nature of development problems in general, and those of nutrition, unemployment and income distribution in particular, lead rapidly to the import of industries and technologies; occupational health, working conditions and job content are relegated to a relatively low place on the list of priorities, and this creates situations which are, to a large extent, irreversible. Chapters 5 and 8 will return to this subject.

4. *The improvement of working conditions and environment affects all dimensions of man.* Work concerns man as a whole: not just muscle and nerves, but intelligence, capabilities, feelings and aspirations. "All human beings . . . have the right to pursue both their material well-being and their spiritual development in conditions of freedom and dignity", to quote from the Declaration of Philadelphia (see Panel 6). Progressing in one's work and learning something new each day, being more than just part of a mechanical process, having initiative and making use of it, profiting from human contact – these are all basic needs. This book does not, therefore, deal only with safety and health and the most flagrant problems of working time. Protecting life and safeguarding health are, of course, top priorities; but it is when men rise above material concerns and express their aspirations that they make an effective contribution to society. A society of robots would be the "end of modern society". The present Director-General of the ILO has listed the following interdependent factors: improvement of working conditions and environment; the defence and organisation of workers, in particular the most underprivileged; and the dignity of man:

For the ILO . . . employment creation is not enough. The jobs created must be good jobs. We are ready to contribute to all efforts by member States to *improve the conditions of work and life* in the secondary sector, in services and in agriculture. The success of such efforts will, we believe, largely depend on action taken or encouraged to enable workers, particularly those who have hitherto known nothing but material privation and isolation, to organise themselves freely to make their voice heard and defend their interests. . . . It is not just progress which is at stake, we feel, but *human dignity and freedom.* That is one of the fundamental goals which the international community should pursue more vigorously than ever before in its search for an economic and social order which can meet the *aspirations of mankind in this day and age.*[20]

Notes

[1] See ILO: *Making work more human*, Report of the Director-General, International Labour Conference, 60th Session, Geneva, 1975. The discussion of this report at the Conference led to the adoption of the resolution on the future action of the ILO in the field of working conditions and environment (see Panel 2).

[2] A brief description of the history and activities of the International Labour Organisation is contained in Appendix A.

[3] Hours of Work (Industry) Convention, 1919 (No. 1).

[4] See the list of these international labour Conventions and Recommendations in Appendix C.

[5] J. de Givry: "New world programme to make work safer, healthier and more human", in *ILO Information* (Geneva, ILO), 1976, No. 6, pp. 1-2.

[6] PIACT is described in the document entitled *International Programme for the Improvement of Working Conditions and Environment* (Geneva, ILO, doc. GB.200/PFA/10/8), adopted by the ILO Governing Body in June 1976. See also Appendix B.

[7] Resolution on work and its environment, adopted by the International Labour Conference on 25 June 1974.

[8] *International Programme for the Improvement of Working Conditions and Environment*, op. cit., p. 17.

[9] In particular, ILO: *Encyclopaedia of occupational health and safety* (Geneva, 3rd (revised) edition, 1983), 2 vols. See especially the articles on: Accidents, commuting; Beverages at work; Climate and meteorology; Cold and work in the cold; Heat acclimatisation; Heat and hot work; Heat disorders; Housing of workers; Mental health; Nutrition and food.

[10] N. Rao Maturu: "Nutrition and labour productivity", in *International Labour Review*, Jan.-Feb. 1979, pp. 1, 11.

[11] G. Lambert: "Ergonomics and industrialisation", in ILO: *Ergonomics in machine design*, Occupational Safety and Health Series, No. 14 (Geneva, 1969), Vol. II, pp. 1005-1006.

[12] A. E. Malyševa: "Climate and meteorology", in ILO: *Encyclopaedia of occupational health and safety*, op. cit., Vol. 1, pp. 484-485.

[13] Albert Thomas had already conceived these minimum safety rules that should be applied no matter what the circumstances.

[14] A. Thomas: *The first ten years* (Geneva, ILO, 1931), p. 12.

[15] See W. T. Singleton: "Ergonomics", in ILO: *Encyclopaedia of occupational health and safety*, op. cit., Vol. 1. This term will be defined and explained at greater length in Chapter 2.

[16] See E. Bolinder and G. Gerhardsson: "A better environment for the worker", in *International Labour Review*, June 1972, pp. 495-505.

[17] ibid. See also R. Murray: "One hundred years behind", in *World Health*, Special issue on work and health (Geneva, WHO), July-Aug. 1974, pp. 14-21.

[18] "Mankind is displaying, as it has always displayed in the past, a startling ability to survive and function in environments which are highly unnatural and hostile to its biological and physiological nature – including environments created by scientific technology and urban complexity. The long-range cost of this adaptability, however, is the emergence of new disease patterns and an impoverishment of emotional life." From R. Dubos: "Progressive degradation", in *World Health*, July 1975, pp. 8-15.

[19] G. Spyropoulos: "L'évolution des conditions et du milieu du travail au Bureau international du Travail", in *Avenirs 2000*, Special issue "Changer le travail?" (Paris), 1978.

[20] Speech by F. Blanchard, Director-General of the ILO, at the 63rd Session of the United Nations Economic and Social Council, 12 July 1977.

2 OCCUPATIONAL SAFETY AND HEALTH

Extent of the problem

Introduction

Still cause for concern

After so many years of efforts to prevent occupational accidents and diseases, and bearing in mind the often significant progress in machine guarding, personal protection and the improvement of the working environment, one may feel surprised that the effectiveness of the measures taken is still in question and that new approaches are being sought to break out of a state of affairs that remains unsatisfactory. Yet the occupational accident frequency rate curves, which were falling in most industrialised countries, have now flattened out, whilst in developing countries they are continuing to rise. Although the frequency rates of compensable occupational diseases – always difficult to assess – appear to be following a downward trend in many countries, the physical and mental fatigue induced by the subordination of workers to their machines and by certain work schedules tend to accelerate body wear and tear in workers whose physiological capacities are often subjected to stress beyond the limits of recovery.

The cost of occupational accidents and diseases is very high; yet it may not be generally realised just how high it is. It has recently been estimated that in certain industrialised countries the total cost – both direct (treatment expenses and benefits to the injured and their dependants) and indirect (damage to plant, equipment and goods, lost production, etc.) – amounts to some 4 per cent of gross national product. Nor is the financial cost the only element, for one cannot overlook the cost in human terms: the suffering of the injured, the grief of all concerned and the social, as well as economic, consequences for the families. Occupational accidents and diseases have very severe personal and social repercussions and are a major burden for the national economy.

Considerable efforts have been made

Nevertheless, in many countries, considerable efforts have been made by governments and employers' and workers' organisations and by undertakings to reduce this heavy tribute to technical progress and rapid industrialisation. Regulations have become stricter and their enforcement more thorough, and information and education at all levels have made the workers themselves more aware that the prevention of occupational accidents and the improvement of the working environment play an important role in their well-being and in industrial productivity. This increasing awareness of the dignity of labour has promoted the idea of "built-in safety" at the design stage of plant and equipment, and the consideration of physiological and psychological factors in workplace layout. However, the trend is far from universal, since it is seen mainly in countries with high levels of social and economic development, and especially in the advanced technology sectors. In vast areas of the world, the protection of the worker from occupational hazards or from work-induced, premature physical and mental ageing is a need which either is virtually unheard of or is subordinated to the demands of production. In this regard, one may observe that national progress is measured not only by levels of productivity but also by the way in which productivity is achieved and by such criteria as wage levels, the degree of economic independence that these wages provide, the cultural development they permit, and general health levels.

Although these various contributions to progress are, to a large extent, dependent on levels of remuneration, they are also related to the conditions under which work is carried out, and any direct or indirect effect of the work on the health and physical integrity of the worker may have undesirable repercussions not only on the worker himself but also on his family and on society as a whole. Many workers still suffer from stress and are badly paid, with the result that their health gradually deteriorates and that efforts to improve the socio-economic status of a large part of the population are nullified.

The rise of new concepts

New technologies, substances and work methods have brought with them new hazards which are often not widely understood but which may produce such severe long-term effects as malignancies and premature ageing.

In his fascination for technology and progress, man has always underestimated these hazards, and consequently injury, disease and physical and mental exhaustion are often considered to be the unavoidable corollaries of work in general, or at least of work in certain specific occupations. Absorbed by performance criteria, designers strive to develop plant and equipment which are more powerful, accurate, rapid and sophisticated and which, once in use, are carefully nurtured and maintained; the worker's only task is to feed these machines, keep them running and check their output – in a word, to "serve" them.

However, there is a dawning realisation of the true situation: an appreciation of the machine's hazards, its inherent power and the often stultifying work pace that it imposes; a growing understanding of fatigue phenomena and the importance of the worker's total health; and a perception of the need to care not only for the machine but also for its operator. Whereas the guarding of mechanical hazards was previously a prime concern, more and more attention is now being devoted to the substances, dusts, fumes and other harmful agents released into the working environment which, although not always an immediate threat to the worker's life, may have severe long-term effects on his health.

Legislators too are beginning to show an interest. Whereas at the start of the industrial era the machine was nearer to the focal point of the man-machine system, the emphasis has now shifted progressively towards the man, and the worker is gradually finding his true place as an essential partner in any production process. The machine is still nurtured, but ever more thought is being given to its adaptation to the worker's capacities and abilities.

Occupational injury and disease are no longer considered to be the inevitable tribute to progress: the hazards can be controlled. However, this change in the attitude of authorities, employers, workers and society to the status of the worker and the dignity of labour is clear in the highly industrialised countries only; it is nowhere near as evident in the developing countries, where the situation is scarcely better than that of today's developed countries during the period of their early industrialisation.

Attitudes to the working environment and to occupational stress have changed in line with socio-economic development. It is now accepted that a worker's physical integrity and health are an asset for both the nation and the undertaking. Bad working conditions, an unhealthy and dangerous working environment, excessive pace of work and poorly scheduled hours of work are all sources of disease, fatigue and accidents and causes of absenteeism, high labour turnover and dissatisfaction.

A safe, healthy and ergonomically acceptable working environment will have a positive effect on an undertaking's prosperity. Ergonomically designed machines, working environments in which harmful factors have been minimised, production organisation which does not lead to excessive fatigue, wages which will allow the worker's family to develop its potential in the way to which it is entitled in a developing society – all this will permit optimal use of production plant and investments and be a factor of stability and social well-being.

Optimal working conditions and environments can require the convergence of efforts in fields such as production plant and equipment, environmental factors, work organisation, management and workers' education and training.

Material factors

Machinery and plant

Machines are increasingly both slave and master to the worker: on the one hand, they relieve him of heavy tasks, spare him excessive effort and perform repetitive operations in his stead; on the other, they often subject him to their blind power and rhythm. Under the impulse of rising demand and rapid technological development, machines have become more powerful, plant more complex and, consequently, more expensive, and the pace of work more rapid. Machinery and plant call for greater vigilance, and even a fleeting loss of attention can have disastrous consequences, so that ever more sophisticated safety devices and systems are being developed and installed. Many countries have introduced legislation in line with the ILO Guarding of Machinery Convention, 1963 (No. 119), and Recommendation (No. 118), requiring manufacturers to supply machines in which the transmission mechanism and controls are adequately guarded, and this is an important step towards "built-in safety".

However, this does not always mean that the working parts of the machine are guarded or that the employer enforces the use of the guards when the machine is working. Certain provisions of Parts II (Sale, Hire, Transfer in any other Manner and Exhibition) and III (Use) of Convention No. 119, and in particular Articles 2, 6 and 7 (see Panel 8) are of especial relevance in this respect.

Harmful agents in the work environment

The pollution of the working environment by gases, vapours, fumes and dusts of all sorts is at present one of industry's severest problems. This pollution is not limited to the area of the undertaking; it is also a hazard for the surrounding neighbourhoods, and endangers the health of the general population in large industrial centres.

Industrial wastes and residues are often released into the atmosphere around the factory or dumped into watercourses or the sea without any prior treatment. They may contaminate regions at some distance from the plant in areas where the population is unaware of this, and where the relevant safety measures are consequently non-existent.

PANEL 8

Guarding of Machinery Convention, 1963 (No. 119)

Article 2

1. The sale and hire of machinery of which the dangerous parts specified in paragraphs 3 and 4 of this Article are without appropriate guards shall be prohibited by national laws or regulations or prevented by other equally effective measures.

2. The transfer in any other manner and exhibition of machinery of which the dangerous parts specified in paragraphs 3 and 4 of this Article are without appropriate guards shall, to such extent as the competent authority may determine, be prohibited by national laws or regulations or prevented by other equally effective measures: Provided that during the exhibition of machinery the temporary removal of the guards in order to demonstrate the machinery shall not be deemed to be an infringement of this provision as long as appropriate precautions to prevent danger to persons are taken.

3. All set-screws, bolts and keys and, to the extent prescribed by the competent authority, other projecting parts of any moving part of machinery liable to present danger to any person coming into contact with them when they are in motion, shall be so designed, sunk or protected as to prevent such danger.

4. All flywheels, gearing, cone and cylinder friction drives, cams, pulleys, belts, chains, pinions, worm gears, crank arms and slide blocks, and, to the extent prescribed by the competent authority, shafting (including the journal ends) and other transmission machinery also liable to present danger to any person coming into contact with them when they are in motion, shall be so designed or protected as to prevent such danger. Controls also shall be so designed or protected as to prevent danger.

. . . .

Article 6

1. The use of machinery any dangerous part of which, including the point of operation, is without appropriate guards shall be prohibited by national laws or regulations or prevented by other equally effective measures: Provided that where this prohibition cannot fully apply without preventing the use of the machinery it shall apply to the extent that the use of the machinery permits.

2. Machinery shall be so guarded as to ensure that national regulations and standards of occupational safety and hygiene are not infringed.

Article 7

The obligation to ensure compliance with the provisions of Article 6 shall rest on the employer.

The ILO Working Environment (Air, Pollution, Noise and Vibration) Convention, 1977 (No. 148), emphasises the importance of preventing hazards at their source, and stipulates in its Article 9:

As far as possible, the working environment shall be kept free from any hazards due to air pollution, noise or vibration –

(a) by technical measures applied to new plant or processes in design or installation, or added to existing plant or processes; or, where this is not possible,

(b) by supplementary organisational measures.

Modern personal protective equipment is highly effective. Respiratory protection masks are typically fitted with high-efficiency, low-resistance filters, and gloves, safety helmets, aprons, goggles, and so on, are designed to provide maximum protection with minimum hindrance and discomfort. Nevertheless, collective safety measures (i.e. those related directly to the machine, plant, products or working methods) are still the most effective and should be given preference wherever practicable, since their use does not depend on the worker's co-operation.

Noise and vibration have become major problems. Increasing mechanisation and machine power and the concentration of several machines into a limited space have raised noise levels, and the number of cases of noise-induced hearing loss has increased. Moreover, noise also has a deleterious effect on the nervous and cardiovascular systems and increases nervous fatigue. Since it hinders communication, concentration and the detection of acoustic signals, it may also play a role in accident causation. In many countries, legislation has empirically set maximum allowable sound pressure levels to limit pathological effects on hearing, reduce difficulties in verbal communication and prevent fatigue.

Fatigue

Multiple facets of fatigue

Only recently has it been appreciated that fatigue is a multifactoral problem. Previously, it was linked only to physical exertion, but the various forms of mental fatigue are now also recognised, and there is growing realisation that physical and mental fatigue are frequently linked. A typical example is the greater vigilance made necessary by poor lighting or poor workplace design, the discomfort caused by, for instance, excessive cold, heat or humidity. Workers exposed only occasionally to these disadvantages often fail to appreciate their effects; however, day-long exposure is fatiguing and may be harmful to health.

Although easy to appreciate, fatigue is difficult to define. In general, it may be described as the result of excessive stress, i.e. of effort exceeding the individual's limits of resistance and consequently demanding a period of recovery in the form of a break or, in severer cases, complete rest.

Fatigue may be physical (following dynamic or static muscular effort, for example) or mental (resulting from prolonged vigilance or concentration), and the two forms may be cumulative – as, for example, where repetitive or monotonous work calls for the maintenance of a set posture. The sensation of fatigue may be influenced by environmental factors or by the worker's motivation or lack of it.

Since fatigue lowers levels of vigilance, inhibits concentration, increases reaction times and dulls muscular reflexes, it may be a major factor in accident causation and should therefore be seriously considered in the implementation of safety and health measures. The prevention of fatigue requires a careful assessment of such factors as the type and intensity of the physical effort, the ergonomic features of machinery, air temperature and humidity, lighting, noise and vibration levels, work organisation, the psychological climate, and so forth.

Making correct allowance for these factors may not only reduce fatigue but also directly affect the output of the man-machine system (i.e. productivity and production), accident frequency and severity, worker health status, absenteeism, labour turnover and, last but not least, the functioning of the enterprise and the national economy as a whole.

The contribution of science and technology

Research on the improvement of working conditions and environment has benefited from a multidisciplinary approach (ranging from engineering through the human sciences to medicine) and has highlighted the need for co-ordinated action aiming at an improved quality of life.

Ergonomics has made a major contribution to this research. It offers a multidisciplinary approach to the adaptation of work to the worker and deals with both the biosomatic interactions between the worker and his equipment and the individual's physiological and psychological capacity for his work. Design ergonomics is a planning tool for equipment, workplaces, processes, and so on, whereas corrective ergonomics is used in defining remedies for existing unsatisfactory situations, thereby contributing to, inter alia, a reduction in fatigue.

The innovative feature of ergonomics is that it studies work problems from the point of view of the worker involved. An ergonomic approach takes into account anthropometric criteria, the worker's physiological and psychological characteristics, the position, shape and direction of movement of controls, the types of information displayed by the panels and instruments, display design, tool shape and size, seating characteristics, thermal comfort, lighting, work organisation, and so on.

Hours of work and work organisation

The daily or weekly arrangement of working hours, shift work, night work, work breaks and rest periods may all influence the physical and mental workload (for more details, see Chapter 3).

Originally, hours of work were related to hours of daylight, with the worker setting his work pace and breaks to match his physical capacity, and fatigue was essentially physical. In modern industry, machines set the work pace, control the worker's movements, and often demand unrelenting concentration. Dynamic physical effort has now been replaced by static effort (imposed postures, etc.) and, outside agriculture, the building industry and craft-work, occupational fatigue is basically the result of vigilance, concentration and responsibility, i.e. nervous fatigue which "exhausts" without "fatiguing".

Night work, now common in many industries, is in conflict with the natural biological cycle. Normally, at night, biological activity, pulse, respiration, oxidation phenomena and nervous activity are all reduced, and attention and reflex thresholds fall significantly. During the night, physical effort, concentration and capacity for critical judgement take on another dimension; exhaustion occurs earlier, and physical fatigue is more pronounced.

Even with a normal schedule of working hours and with a work pace that is not machine-dependent, the system of work organisation may itself extend the worker to the limits of his capacity by imposing production quotas and payment systems that bring him down to the level of a machine. If, under these conditions, energy consumption exceeds nutritional intake, the body burns up its reserves, and the result is chronic fatigue, decreased resistance to disease and premature physical wear. The worker's nervous resistance is affected, and imbalances occur which will have negative effects at all levels and, in particular, in the worker's relations with his superiors, his colleagues, his family or the community in which he lives.

Social problems

Nutrition

The human "machine" cannot produce more than is permitted by the energy input from nutrition. However, food must not just provide the fuel required for physical activity; it must also be correctly balanced.

Originally, diet depended on the local geography, climate and resources. In colder regions, animal protein and fats with a high calorific content predominated; in temperate regions, a wide variety of vegetables, animal and vegetable proteins was available, but fewer fats; in tropical regions, vegetables, fruits and liquids were abundant, but little existed in the way of proteins and there were virtually no fats.

The energy that the local population could expend at work was determined by the calorific input from their diet. Their activities, which were normally carried out in the open air, followed a tempo and a time-schedule stipulated by climatic conditions and the food available. With the arrival of the machine, these relationships were changed. Hours of work were determined, at least in part, by technical considerations; the work pace was not freely chosen by the worker but was imposed by the machine or the payment system. The workload became dependent on technical and economic factors, while the timing and quality of energy intake changed. Diet was adapted by increasing the protein intake for those doing hard physical work, by reducing the fat intake for sedentary workers, and by adapting the carbohydrate, vitamin and liquid intake to the requirements of the workplace microclimate. Finally, traditional diets had to be supplemented to counteract certain deficiencies.

At times, local commercial supply problems, coupled with the constraints of work-schedules and commuting, compelled enterprises to set up canteens to ensure an adequate diet at an attractive price (see Chapter 6). These initiatives were certainly praiseworthy, but they met with obstacles and their success was varied.

Rest, housing and transport

Just as the body requires an adequate and balanced diet to reconstitute the energy reserves that have been expended, it also needs adequate rest to overcome accumulated fatigue. To have its full effect, rest should be taken in an environment with noise levels and temperatures that permit total relaxation. Thus, housing is not only a fundamental factor in family and social life but also a major factor in recovery from work fatigue. This housing should incorporate a room reserved for rest, which is not permanently exposed to the sun (in warmer climates), is suitably isolated from the family living quarters, and is well ventilated and protected from external noise. This last requirement is often one of the most difficult to achieve in urban settings.

In developing countries, workers tend to live in the large, densely populated, commercial areas of towns, and are consequently exposed to both the noise of everyday family life (radios and cassette recorders, television, children's games, domestic animals, loud talking and discussions, various household noises, etc.) and that of social activities around the home, in particular where there are public establishments in the vicinity (bars, shops, cinemas, markets, etc.), not to mention traffic noise. This acoustic background may continue late into the night and be not only a serious obstacle to sleep but also a factor of nervous stress that disturbs sleep. The problem is even more acute for shift or night workers compelled to rest when family and community life is in full swing, disturbing to a critical degree not only sleep but also family life and involvement in social life.

In their search for housing at a price they can afford, workers tend to settle on the outskirts of the town, often at some distance from their workplace. Commuting time is often longer because of dense rush-hour traffic, and the journey is all the more unpleasant because it is frequently made standing up in a crowded, hot and smoke-filled vehicle.

Workers who commute in their own means of transport – a practice that is becoming more and more common in industrialised countries – find the purchase and maintenance of a vehicle a heavy economic burden, and the tension of driving in heavy traffic, with delays and accident hazards, is a cause of nervous fatigue. No matter which way workers commute, they arrive at work stressed and return home with additional fatigue. Finally, long commuting distances may expose workers to inclement weather and urban pollution, prolong the working day, cause additional fatigue, contribute to general morbidity and aggravate pre-existing diseases.

It can therefore be seen that nutritional, housing, transport and many other personal, family and social factors which ostensibly have no direct connection with work are all interdependent and may affect the worker's health, physical and mental equilibrium and, ultimately, productivity (more detailed consideration of these items is contained in Chapter 6).

Health conditions

A specific problem faced by developing countries is the general health status of the population and the lack of resources for improving the situation.

Studies on work and its organisation commonly take as a starting point a worker of medium stature, of normal intelligence, of normal physical strength and with adequate capacities of vigilance and critical judgement, i.e. a "normal" healthy person with an adequate level of education and training, rested and relaxed. However, and especially in many developing countries, this picture of the "normal" worker is incorrect as far as health status is concerned.

On many occasions, experts' reports issued by the World Health Organisation and the International Labour Office have emphasised the severity of workers' health problems in Third World countries, owing in particular to the incidence of epidemic and endemic diseases. This situation may be aggravated by malnutrition, general hygiene deficiencies, inadequate health organisation, shortage of adequate sanitary facilities, overcrowded and inadequate housing and a hot and humid climate. A high percentage of the population suffers from infectious diseases, including gastro-intestinal disorders, parasitic diseases (malaria, schistosomiasis and ankylostomiasis, etc.), respiratory tract disorders, eye diseases and chronic otitis.

Such chronic, attenuated diseases may not prevent the patient from working in some way or other; however, when combined with nutritional deficiencies, they may cause generalised debility with weight loss and signs of malnutrition, anaemia and chronic asthenia that considerably reduce working capacity and output. In recently industrialised regions, various factors favour the spread of these diseases, and the shanty towns of industrial suburbs are an illustration of this. To these structural problems should be added the adaptation difficulties of people who have recently been transplanted from a rural to an urban environment and who have moved without transition from the independent organisation of agricultural work or craft-work to the constraints of industrial timetables, work pace and discipline.

Undertakings in many countries are making efforts to deal with these problems and to improve the worker's general health status; the extent of the efforts made usually depends on factors such as the situation and effectiveness of the local public health services and the distance between the undertakings (especially mines, plantations, dam construction sites, etc.) and a large town.

Larger undertakings have set up dispensaries providing various services that range from first aid for minor injuries given by a part-time or full-time nurse to health screening by specialised medical personnel (see also Chapter 6). In this way, it is possible to offer diagnostic and therapeutic services, perhaps using specialised staff, and provide a relatively complete health service for the workers' families.

Plant medical services should be designed and organised to meet local requirements. In industrialised countries, services will concentrate on health screening, the detection of occupational diseases, monitoring adaptation to work, workplace analyses and health consultations (see "Measures involving the worker", pp. 72-77 below). In newly industrialised countries where medical personnel are scarce and health services less well equipped, plant medical services are required to devote much time to the control and treatment of endemic and communicable diseases, health education, consultations for the workers and for their families, medical prescriptions, and so on.

In all cases, however, the plant medical officer must acquire a knowledge of the technology and substances used in the plant, the characteristics of the working environment, special types of stress attributable to work organisation, and so on. Even where the plant medical service is required to function as a general health-care dispensary, its basic purposes and task are still to adapt the work to the worker's physical and mental capacities and to prevent ill effects of work and working conditions on the worker.

Occupational safety and health in agriculture

Extent and characteristics of farming communities

It is only recently that concern about occupational safety and health has been extended beyond industrial workers to agricultural workers. Yet, in view of the growing world demand and need for agricultural products and the capital investments involved, farming may definitely be considered an industry.

As farming activities essentially entail work in the open air with a characteristic overlap between living and working environments, it is difficult to draw hard-and-fast dividing lines between occupational and non-occupational accidents, injuries and diseases. Farming does not have the typical enclosed industrial working environment that is designed by man and amenable to human-induced change at any time; and, in contrast with their industrial counterparts, agricultural workers are often exposed to environment-dependent hazards as well as to specific occupational risks. In agriculture, safety and health problems are complex and much less "industry-specific" than in other occupations. Epidemiological patterns of agricultural occupational accidents and diseases differ considerably from country to country, and even from one area to another within the same country.

Risks and patterns of accidents and diseases

Farming is a high-risk occupation, and agricultural workers may be exposed directly or indirectly to the following disease or accident factors, either singly or in combination:

- inclement weather and temperature extremes;
- communicable diseases common in a rural environment;
- seasonal work-cycles with long hours of work;
- remote and isolated worksites;
- stationary and mobile machinery;
- electrical plant and equipment;
- pesticides, fertilisers and other agricultural chemicals;
- agricultural tools and, in certain cases, firearms;
- crop and plant cultivation and processing; and
- animal husbandry and handling.

On the basis of these natural and technological risks, it is possible to categorise agricultural accidents and diseases as follows:

- conventional accidents resulting in trauma, poisoning, electrocution or drowning;
- tractor accidents (these may involve children);

- accidents involving other stationary or mobile machinery;
- impairments due to machinery noise or vibration;
- pesticide poisoning which, although usually transient, may sometimes be acute or chronic;
- dermatitis resulting from contact with plants, flowers or chemicals;
- allergic pulmonary lesions following exposure to high concentrations of organic vegetable dusts; and
- zoonoses resulting from close contact with animals.

Accident and disease control

In industrialised countries

In the more developed countries, legislative, administrative and operational strategies and policies are being used to extend industrial occupational health and safety systems to cover the prevention of occupational accidents and diseases among wage-earning and self-employed agricultural workers. In this approach, due allowance must be made for the differences between the working environments, mix of workers, types of work and work processes in agriculture and industry. Regulations have been issued on the control of agricultural safety and health hazards. In addition, the dissemination of research and information on agricultural hazards has been pursued, farm inspectorates have been strengthened, and education and training programmes have been established.

This extension of conventional industrial safety and health systems to agriculture has been simplified by the existence of the following factors that are associated with advanced socio-economic structures:

- a well-established administration;
- easy mobilisation of resources;
- availability of information about agricultural hazards and their control;
- a high level of literacy among agricultural workers;
- adequate, accessible and effective public and personal health-care services;
- established veterinary services; and
- sound finances of agricultural undertakings.

Legislation is an essential part of an agricultural safety and health programme; but, even when it is backed up by a suitable labour inspectorate to ensure enforcement, it may have only a limited impact on accident and disease frequency rates. A comprehensive programme must also include a full commitment to good working practices on the part of employers, wage earners and the self-employed, and extensive advice from inspectors on recognised and potential hazards and their control.

The competent authorities in the industrialised countries have also become involved in education, training and information dissemination activities on agricultural safety and health, and have, in particular, promoted in-service educational programmes. A multidisciplinary approach has been adopted, and active support has been given to employers' and workers' organisations. As a result, accident and disease frequency rates have gradually been reduced and there has been a substantial increase in the number of farmers with a positive attitude to safety and health and the necessary control measures.

In developing countries

In developing countries, the task of extending industrial safety and health programmes to agricultural workers is significantly different from that in industrialised countries. Programmes and the inspectorates for enforcing them are basically weak, and operate only in large urban industrial undertakings which together employ only a minority of the total labour force.

Agricultural accident and disease hazards in developing countries are probably the same as those listed above, but no adequate epidemiologic data yet exist for realistically identifying the types of injury and disease that require priority control; in addition, information is lacking on successful operational experience with control measures. As a further complication, agricultural communities in developing countries are often poor and lack adequate health-care and other services, and the population may suffer from endemic communicable diseases, malnutrition and other health problems related to low standards of personal and public hygiene.

The need for a multisectoral approach. In these socio-economically deprived and geographically vast regions, it is necessary to address a multitude of economic and social problems, and an approach concentrating solely on occupational safety and health will have little impact and no lasting effect. Only by co-ordinated development programmes designed for specific community structures in all areas will it be possible to raise living and working standards. A balance must be sought between the pursuit of economic objectives and the parallel promotion of essential social services consistent with nationally determined minimum standards. Such a minimum occupational safety and health standard should at least ensure that no worker is denied the necessary knowledge and means to protect himself and his colleagues from known hazards.

General principles for occupational safety and health programmes. In the rural areas of developing countries, there are no well-defined structures for the implementation of conventional programmes, and available resources are inevitably over-stretched. The challenge is to devise and develop alternative and less formal approaches to the problem.

It is necessary to develop integrated and comprehensive programmes that make due allowance for differences in national resource levels, patterns of administrative infrastructure, the nature of the work, local occupational epidemiology and population structure and distribution, literacy and socio-cultural patterns. These programmes can be implemented, step-by-step if necessary, as an integral part of national development programmes with the commitment of the necessary resources. The enforcement of infrastructures set up should have the full support of wage earners, self-employed workers and employers, and due allowance should be made for multidisciplinary approaches so as to maximise the impact that can be achieved with limited resources.

Establishing priorities. A universal schedule of minimum requirements for an agricultural safety and health programme is not feasible; likewise, it would be difficult to design a universal blueprint for implementing the necessary measures. However, it is possible to identify needs common to many developing countries, and efforts to meet these needs can be made in similar ways without exerting excessive pressure on national resources. These include the education of workers concerning agricultural hazards and their control; the promotion of co-ordination and collaboration between various social and economic units in rural communities; the control of diseases transmitted to man from farm animals; safety engineering measures to control machinery hazards; the control of the health hazards of pesticides and airborne plant dusts; and the phased replacement of unsuitable farming tools and equipment.

Information, education and training are critical components of any occupational safety and health programme. Safety consciousness and safe behaviour are the outcome of unrelenting education and, consequently, efforts should be made to incorporate basic safety and health education into programmes of general education and vocational training.

However, before any such domestic programme can be put into action, it will be necessary to train a sufficient number of trainers with a detailed knowledge of local needs and how to meet them.

The role of legislation. Legislation has a valuable part to play in the prevention of agricultural accidents and diseases in developing countries, although its effects will be most marked in higher, more organised levels of national administrative hierarchies. In particular, well-drafted regulations may have a significant impact in safeguarding workers from the effects of machinery and pesticides that are being used in agriculture more and more. For the regulation of pesticides, it may be desirable to set up a national authority to control their registration, import, classification and use.

When considering regulations for farm machinery, it may be necessary to draw up standards specifying appropriate degrees of sophistication for locally manufactured and imported equipment, taking into account the need for machinery to be maintained and repaired locally. Standards of guarding for transmissions and other moving parts should also be specified.

Agricultural tools which are poorly adapted to the worker and his work are common problems, which should be dealt with by the phased introduction of appropriate standard specifications for tools that are ergonomically designed on the basis of national anthropometric data and of a thorough analysis of the work they are required to do.

Occupational risk control: Safety and health measures in the plant

In the first part of this chapter we attempted to paint an overall picture of the problems involved in the prevention of occupational accidents and diseases. We must now examine ways and means of reducing occupational hazards to the lowest possible level and, where circumstances and the nature of the hazard permit, of eliminating them altogether.

In most countries, occupational safety and health programmes are carried out at a number of levels. Efforts made by the undertakings themselves under economic, moral or legislative pressure are virtually always accompanied by action from public authorities. In industrialised countries in particular, employers' and workers' organisations and scientific and technical institutions may also carry out substantial programmes.

An important, although self-evident, point that should not be forgotten is that occupational accidents and diseases (as well as other types of health impairment which have not yet been officially recognised as such) have their onset at the workplace. Their origin may be traced to the design and layout of premises and workplaces; plant and equipment; machinery and tools; substances, agents and products used; production processes; work organisation or the individuals themselves; most often, several of these factors are involved.

Consequently, the most effective approach is for each undertaking to introduce its own safety and health programme and implement it methodically with imagination and perseverance. This approach is supported both by legislation in many countries and by the ILO Occupational Safety and Health Convention, 1981 (No. 155) (see Panel 9).

As emphasised in a recent major report, action must take place at the level of the undertaking itself: "The primary responsibility for doing

PANEL 9

Occupational Safety and Health Convention, 1981 (No. 155)

Article 16

1. Employers shall be required to ensure that, so far as is reasonably practicable, the workplaces, machinery, equipment and processes under their control are safe and without risk to health.

2. Employers shall be required to ensure that, so far as is reasonably practicable, the chemical, physical and biological substances and agents under their control are without risk to health when the appropriate measures of protection are taken.

3. Employers shall be required to provide, where necessary, adequate protective clothing and protective equipment to prevent, so far as is reasonably practicable, risk of accidents or of adverse effects on health.

. . . .

Article 18

Employers shall be required to provide, where necessary, for measures to deal with emergencies and accidents, including adequate first-aid arrangements.

something about the present levels of occupational accidents and diseases lies with those who create the risks and those who work with them." [1]

Hazards can be controlled only when their nature and effects have been recognised and when adequate attention is given to them. Very many occupational accidents still result from a lack of adequate appreciation of hazards that have been widely recognised and understood for years. The complexity of the problem is less of an obstacle to occupational risk control than the fact that those who are in daily contact with the hazards fail to take the necessary precautions, either because they are indifferent to them or because they have become accustomed to them.

The plan adopted for this section may appear arbitrary. Safety and health measures (see p. 53) are, of course, closely linked to risk analysis (see below), from which they naturally stem and from which they are inseparable. However, it seemed preferable to deal first with risk analysis, as a separate item and in some detail, since this basic requirement for any rational approach to occupational safety and health is too often neglected or dealt with only superficially.

Risk analysis

Management's first task should be systematically to identify and evaluate all obvious and hidden hazards that may occur during the operation of the undertaking; the term "management" here means staff to whom top management has delegated these specific tasks or clearly defined responsibilities for occupational safety and health. Wherever possible, risk analysis should be carried out before new plant and equipment are commissioned, since it is at this point that the necessary safety and health measures can be taken most effectively and at the lowest cost.

An idea of the wide range of questions to be considered can be obtained from the "technical fields of action" listed in Paragraph 3 of the ILO Occupational Safety and Health Recommendation, 1981 (No. 164) (Panel 10).

The first step in risk analysis is a methodical and critical study – at the planning stage of buildings, premises and other workplaces and of their access, egress, gangways and traffic aisles – of any operating factors that may endanger the life, limb or health of workers and other persons who may be affected by the operation of the undertaking, and, where local conditions or legislation so require, that may endanger the neighbouring environment.

The hazards of the materials and substances used should also be analysed. Highly toxic or flammable products continue to be used without the most elementary precautions being taken: for example, containers are left unlabelled or the labels affixed are inadequate (a mere trade name is not sufficient), the quantities stored on the premises greatly exceed operating requirements, waste is allowed to accumulate on working surfaces or floors, personal hygiene measures are non-existent.

Although in such situations it is not always easy to determine the part played by ignorance or negligence, neither is excusable. Ignorance is no excuse when reliable and fairly comprehensive information can readily be obtained about the hazards of most substances, products and equipment used at work and about techniques for their safe use. It is totally unacceptable that ignorance of the most elementary safety and health requirements should lead to accidents or diseases in the use of metals such as lead or mercury, solvents such as carbon disulphide or benzene, or processes such as arc or gas welding.

Thorough risk analysis should also cover working conditions and methods and the workers themselves. A worker who is ignorant of the main hazards of his day-to-day work is a danger to himself and sometimes to others. Similarly, a shop or shift foreman who looks only at production targets and fails to consider the safety and health of the persons under his direct control cannot be considered a "capable" person in the full sense of the word: a unit, shop or shift would be hard put to achieve and maintain the required production levels if it were hit by

PANEL 10

Occupational Safety and Health Recommendation, 1981 (No. 164)

3. As appropriate for different branches of economic activity and different types of work, and taking into account the principle of giving priority to eliminating hazards at their source, measures should be taken . . . in particular in the following fields:

(a) design, siting, structural features, installation, maintenance, repair and alteration of workplaces and means of access thereto and egress therefrom;
(b) lighting, ventilation, order and cleanliness of workplaces;
(c) temperature, humidity and movement of air in the workplace;
(d) design, construction, use, maintenance, testing and inspection of machinery and equipment liable to present hazards and, as appropriate, their approval and transfer;
(e) prevention of harmful physical or mental stress due to conditions of work;
(f) handling, stacking and storage of loads and materials, manually or mechanically;
(g) use of electricity;
(h) manufacture, packing, labelling, transport, storage and use of dangerous substances and agents, disposal of their wastes and residues, and, as appropriate, their replacement by other substances or agents which are not dangerous or which are less dangerous;
(i) radiation protection;
(j) prevention and control of, and protection against, occupational hazards due to noise and vibration;
(k) control of the atmosphere and other ambient factors of workplaces;
(l) prevention and control of hazards due to high and low barometric pressures;
(m) prevention of fires and explosions and measures to be taken in case of fire or explosion;
(n) design, manufacture, supply, use, maintenance and testing of personal protective equipment and protective clothing;
(o) sanitary installations, washing facilities, facilities for changing and storing clothing, supply of drinking water, and any other welfare facilities connected with occupational safety and health;
(p) first-aid treatment;
(q) establishment of emergency plants;
(r) supervision of the health of workers.

a high level of absenteeism due to accidents or diseases. The safety and health effects of the duration and organisation of work have already been emphasised, together with the role played by nutrition, housing, transport and hygiene, in the sections "Fatigue" and "Social problems related to occupational safety and health" above.

A plant manager planning a thorough hazard analysis of an undertaking should consult the competent official services or bodies and independent specialists, and also obtain information about the experience gained by similar undertakings within the country or abroad, so as to benefit from their findings and results.

In this chapter we cannot hope to cover even superficially all possible material and human risk factors, be they in agriculture, mining, construction, transport or the service industries; each branch of industry is affected by a number of general common hazards and has, in addition, its own specific risks. The same applies to the processing industries, textile mills, electrical or electronic equipment plants, mechanical engineering shops, tyre factories – each of these has its own specific hazards. Nevertheless, some general guidance may be given on the technique of risk analysis in certain of the above areas.

Installation, layout and maintenance of premises and workplaces

Faulty design or maintenance of premises, access routes and traffic aisles account year after year for a significant percentage of all occupational accidents, many of which are the direct result of poor housekeeping (cluttered traffic aisles, soiled floors, etc.). Major deficiencies and defects should be identified, but it should not be forgotten that even a slight oil spill, a small, scarcely visible object on the floor or a slight unevenness of the walking surfaces can cause a worker to trip and suffer an injury out of all proportion to its cause.

Floors should not be overloaded. Floor and wall openings, staircase wells, passenger and goods lift cages, gangways and overhead platforms should all be suitably fenced. Guard rails should be of adequate dimensions and strength, and toe-boards should be fitted on gangways passing over areas where people are working and where there is a danger of falling objects.

Suitable means of access to machinery and equipment should be provided so that maintenance teams are not required to perform dangerous or acrobatic manœuvres in their regular work. Crush, nip or shear points between moving and stationary objects should be eliminated or guarded (persons maintaining the tracks of overhead travelling cranes or their electric supply cables, for example, have suffered very serious injuries).

Emergency exits that ensure a means of escape in the event of fire should be sufficient in number, correctly positioned and signposted and free from encumbrance.

Good lighting is essential for the safe movement of persons, plant and equipment. Inadequate lighting levels may conceal obstacles or danger zones, whereas badly positioned or excessively powerful light sources may cause glare.

Plant and equipment hazards

With increasing mechanisation in the manufacturing industries, mining, agriculture, forestry, construction and transport, it is essential to analyse at the design stage the potential hazards of machinery and plant operation, maintenance and repair (see also the ILO Guarding of Machinery Convention, 1963 (No. 119), and Recommendation, 1963 (No. 118); some of the more important provisions are reproduced in Panel 8). As the power, speed, size, weight and complexity of machinery and plant increase, the hazards become greater while the innate abilities of the operators themselves remain virtually unchanged. Some machines are now so large that they require several operators and special communications equipment; others are so sophisticated that their instrumentation and automation systems are more expensive than the rest of the installation.

Most machine builders take great care to design and market equipment that meets current occupational safety and health requirements, and this is particularly true in countries where legislation requires that new equipment be tested and approved before being marketed; yet one still finds equipment which does not meet the most elementary requirements.

The situation is particularly serious in the developing countries, where items of equipment, acquired second-hand, are often in poor condition. The purchaser, either through ignorance, to save costs or for some other reason, may not worry unduly whether the equipment he buys is fitted with the necessary safety devices; manufacturers and retailers who supply inadequately guarded plant and equipment are consciously or unconsciously assuming a tremendous responsibility. The situation is one which must be remedied, since the firms importing plant and equipment are also, often unwittingly, importing serious hazards. Sometimes the more or less adequate safety devices fitted by the manufacturer have been removed, or are no longer operating properly because of inadequate maintenance. Such devices may be more dangerous when present than when absent, since they give only the illusion of safety.

In view of the extreme diversity of machinery available, it is scarcely possible to give hereafter more than a very summary review of the points that should be covered by risk analysis.

Managers should first assess the nature and relative severity of the potential hazards of any plant, machinery, equipment and tools they are thinking of acquiring, and then select the equipment which, in addition to being suitable for the job, offers the best guarantee of safety. Ergonomic design data, which are available from numerous relevant publications, should form part of this assessment.

Where legislation or good practice require formal approval, the order should not be placed until it has been verified that the equipment meets its specifications; if it does not, the manufacturer should be requested to make the modifications necessary for use under specific operating conditions (work in explosive atmospheres, exposure to bad weather, heat, etc.).

Particular attention should be paid to the potential hazards of transmissions, in-running nips of wheels, gears, belts and pulleys, shafts and other drive components. Modern machine drive systems, with individual motors built into the machine housing, have a considerably simplified transmission guarding, without however making it entirely unnecessary.

Protrusions on rotating or reciprocating parts are a major hazard since they may cause lacerations and catch loose clothing. A similar hazard exists with smooth shafting (e.g. drill bits) which may catch long hair or a flapping sleeve. Workers have been killed when their aprons or blouses, made from particularly strong synthetic fibres, got caught by moving machines; cotton clothing would in many cases have given way and avoided a severe accident.

Stock and finished-product feed and delivery points should be guarded to protect the worker's hands from mutilation. Guarding the points of operation is usually the most difficult problem. Squeeze, crush or shear hazards or in-running nips that may injure hands or other parts of the body, the possibility of breakages or flying particles during machining, and potential electrical or chemical hazards should be carefully analysed both by the user and by the manufacturer. The latter may not always be able to guard the operating zone in the most effective way if he does not know the exact conditions under which the machine is to be used; it is the responsibility of the user, where necessary, to improve the initial guarding or provide new or supplementary guards in order to meet actual production requirements.

The section on "Machinery guarding" below gives examples of common machinery hazards and describes a number of suitable safety devices. The modern appearance or the streamlined housing of a machine may be deceptive and give the impression that the machine is properly guarded. When mechanical risks are analysed, the machine should also be studied when it is operating, since certain major hazards may not be apparent when it is at a standstill.

Operating controls (for starting and stopping, etc.) will seldom cause injury themselves, but poorly designed controls (e.g. wrongly placed push-buttons, or levers which have a direction of travel which does not correspond with that of the machine parts they operate) may be a major hazard. Where equipment requires an emergency stop device, this should be preferably fitted by the manufacturer.

Any other machine and plant hazards (noise, heat, etc.) should be identified, and particular attention should be given to the detection of potential leakages of flammable, explosive, toxic, radioactive or irritant substances.

Although perhaps less spectacular, the hazards that occur in the transport and handling of goods and materials are still very real. Falls of heavy objects during manual handling still cause numerous hand and foot injuries, while puncture and laceration wounds caused by pointed or sharp objects (e.g. the sharp edges of metal containers) are far from rare. The manual lifting and carrying of heavy loads may cause back injuries if incorrect handling techniques are used.

Serious accidents are still common in mechanical handling but are more often the result of incorrect working methods than of the equipment itself. When mechanical handling equipment collides with a fixed structure, a mobile obstacle or a pedestrian, when loads fall to the floor or the equipment overturns, it is often because incorrect operating procedures have been used, perhaps as the result of inadequate training or supervision of drivers.

Electrical hazards are becoming more and more important as electrification spreads throughout the world; they are also all the more dangerous because they are less apparent and are not properly understood by most workers other than electricians. This is particularly the case with rural workers, who have only minimal industrial experience. Electrical safety is based on relatively advanced technical knowledge, and plant management should call on a specialist to examine electrical installations (preferably at the planning stage) and to detect electrical hazards that may occur during normal operation, maintenance and repair work. Systems should be devised to prevent plant or equipment from being switched on during repair or maintenance work and to ensure that unskilled workers do not carry out stop-gap repairs and the like. The "permit-to-work" system, which is described in greater detail on p. 69, may be of considerable assistance in preventing accidents of this type.

Whenever management does not feel sufficiently competent or experienced to make a complete and reliable analysis of material hazards, assistance should be obtained from the competent authority (labour inspectorate, etc.).

Dangerous substances

Management should check that the raw materials, intermediates, substances, agents and products used or manufactured by the undertaking are suitable for their proposed use, and should also determine their specific hazards and the relevant precautions.

Management can seldom undertake research on these hazards itself; this does not, however, absolve it from making every effort to avail itself of the knowledge and experience of others.

Any statutory prohibition on the use of specific substances should be strictly observed. Materials and products should be selected with reference to their safety and health hazards, and it may be necessary to forgo the use of particularly hazardous or harmful substances such as benzene, asbestos, etc., which, when used even under the most favourable conditions, may still entail particularly severe risks for the exposed worker. Storage and transport hazards should also be considered. Where potentially hazardous products are manufactured or supplied, there is a moral obligation (and often a legal requirement) to inform the end-user of the products' characteristics and hazards and the corresponding precautions.

Use should be made of the numerous information sources available and of the literature issued by manufacturers, suppliers, official services, research institutes and specialised bodies in industrial countries, and various international organisations. In assessing the hazards, allowance should be made for aggravating factors such as the high temperatures in tropical countries, which may, for example, accelerate the vaporisation of dangerous solvents.

The properties of flammable, explosive or toxic substances should be listed with particular reference to those that may present hazards under normal or exceptional circumstances: flashpoints of flammable liquids, explosive limits of gases, exposure limits of toxic substances (threshold limit values, ceiling levels, emergency exposure limits, etc.). Factors that may increase or reduce these hazards should also be considered (storage quantities and conditions, suitable containers correctly marked or labelled, the provision of measures for the retention or elimination of spillage or effluents in the event of a dangerous occurrence or a system failure, sources of ignition, workers' information, personal protective equipment, personal hygiene, etc.).

It is not possible to identify the potential hazards of substances once and for all, and it will be necessary to keep abreast of research that may reveal new hazards or define known hazards more clearly. If toxic substances with long-term effects are used, management should co-operate with the occupational health service to keep records on conditions of work, exposure levels and medical and biological monitoring findings of all workers exposed to them (see the section "Measures involving the worker" below).

Hazards related to working procedures and work organisation

Defects in working procedures and work organisation may also constitute hazards; these deserve greater attention, since the human factors involved are usually more difficult to identify and deal with than are purely material factors.

It will be necessary to list and analyse: individual job operations and movements; job interfaces; work instructions (where available); supervision by immediate supervisors; and sanctions (if any) if safety and health precautions are deliberately or repeatedly ignored (removal of guards, tampering with safety devices in order to make them inoperative, failure to use personal protective equipment provided by the employer, etc.).

Although certain hazards inherent in working procedures and work organisation may be recognised and taken into consideration at the design stage, most of them do not become apparent until afterwards. Consequently, risk analysis should start on the drawing-board but should continue in the field, i.e. in manufacturing shops, on worksites, and so on. It is only in this way that it will be possible to detect unsuspected hazards when – for example – equipment is used incorrectly, products are handled dangerously or work has to be carried out in exceptional or emergency situations.

Sometimes risks can be eliminated or significantly reduced by relatively simple measures embodied in the job instructions. In other instances more complex measures may be required, such as for work in confined spaces, plant and workers entering and moving around a container terminal, safe sequencing of operations by means of permit-to-work systems, etc. (see also the description of these systems on p. 69).

It is not possible to offer here general guide-lines for all the situations that may occur. Plant managers should work with line management, workers and, where necessary, specialists to identify dangerous practices and develop measures to eliminate them. Analyses of accidents or dangerous occurrences in the plant can be of great value in this respect. In practice, a dangerous occurrence differs from a minor or serious accident only in its consequences, and these are often merely the result of chance. A sound approach is to study factors that have led to a dangerous occurrence even though it has resulted in no more than material damage and local disruption of work.

Even in high-technology industries, the basic causation of accidents is more a question of organisation than one of technical factors.

Risks related to the individual

There are numerous risk factors that are directly worker dependent, but these are the most difficult to identify, assess and control.

Employers have a duty to analyse the type of safety and health problems in their undertakings and their material and human dimensions. They are in a key position to order safety measures to be taken, to enforce their application and to guide and stimulate action by subordinates. Their responsibility is considerable and, if it is not handled satisfactorily with the aid of managers and specialists and if management and worker compliance and participation are not obtained, the hazards that exist will eventually cause human injury or material damage. Whether the number of employees is small or large is immaterial; any difference in responsibility will be quantitative and not qualitative.[2]

Managerial staff at all levels play a key role in occupational risk control since they are in direct contact with operating personnel and have been delegated authority by the employer. They need knowledge, experience and resources to carry out this safety and health role, their authority must be recognised and they must be capable of establishing and maintaining good communication with and between the members of the departments or teams for which they have responsibility.

If management visibly lacks interest, technical skills or personal commitment in safety and health, this will have direct repercussions on the workers' behaviour. Shortcomings in training and management skills will, sooner or later, affect accident frequency and severity rates and the occurrence of dangerous incidents.

The workers themselves are exposed to the hazards that others have left uncontrolled and to those which they themselves may create or aggravate. It cannot be denied that workers' negligence or thoughtlessness play a decisive role in the causation of many occupational accidents; but, in many cases, it is the lack of training, information or supervision that is to blame (see also the section "Fatigue" above). Finally, accidents may also be caused by excessive zeal and an aggressive will to achieve.

A thorough risk analysis should also include consultation of the workers concerned, so as to make use of their experience, since their day-to-day contact with sources of danger makes them particularly knowledgeable about conditions under which hazards may occur and, in many cases, about the best way of eliminating or reducing them. Their involvement in safety and health endeavours is essential and will help in achieving and consolidating the desired results.

No risk analysis should overlook temporary or permanent deficiencies in the worker's adaptation to his job. These deficiencies may be physical, mental or even social, and be related to the individual's personality.

Pregnant women, young, elderly and migrant workers and those with physical or mental handicaps all form categories of workers who require special attention (see also Chapter 8, "Specific categories of workers and their problems"). However, there are also predisposing factors such as stress and fatigue due to housing conditions (overcrowding, poor sanitation, noise, etc.), inadequate diet, unsatisfactory commuting conditions and alcoholism.

Other risk factors

Comprehensive risk analysis should also cover certain outside factors which may be involved in occupational accident and disease causation, such as the lack of safety and health training for engineers and technicians or the fact that labour inspectors do not always possess the knowledge and experience required. These are cases of inadequate training and can be corrected only by efforts at another level. This aspect will be dealt with in the section "Safety and health measures" below.

The interaction of the above factors may be held responsible for most, if not all, dangerous occurrences, accidents and disasters. Indifference to and ignorance of these factors explain the occurrence in both the developed and the developing countries of accidents which may be described as having been "programmed", i.e. built into an undertaking even before it starts operation.

Everyone in an undertaking has a role to play in occupational risk control and can make a contribution to risk analysis, risk control and the undertaking's prosperity.

Employers and all levels of management should be inquisitive and imaginative, seek to broaden their knowledge but recognise their own limitations, and be willing to learn from others in the analysis of occupational risks.

Safety and health measures

Limitations of space permit only a general survey and some examples of occupational safety and health measures.

It is essential to remember that the imposing array of measures and systems currently in use is the outcome of many years of endeavours, and that these endeavours must continue. The task was started in the nineteenth century by enlightened employers and legislators in an attempt to eliminate the most flagrant abuses of the early industrial era, and has been continued by the authorities, by employers and workers and their organisations, and by many official or private institutions.

In-plant safety and health should be aimed at eliminating the unsafe or unhealthy working conditions and dangerous acts which account for nearly all occupational accidents and diseases.

In accident investigation reports a contrast is sometimes made between material factors and human factors, even though there is no value symmetry between these two apparently homogeneous terms. The role of human factors in accident causation cannot be denied, but a reference to human factors in 60, 80 or even 90 per cent of all accidents is, in itself, of little significance. Human beings are involved in nearly all production work, and the presence of the human element necessarily implies a degree of variability, unpredictability and, at times, irrationality. Moreover, not only shop-floor workers but also top and middle management are subject to human factors.

Figure 4. The four basic methods of controlling occupational hazards classified by decreasing order of effectiveness

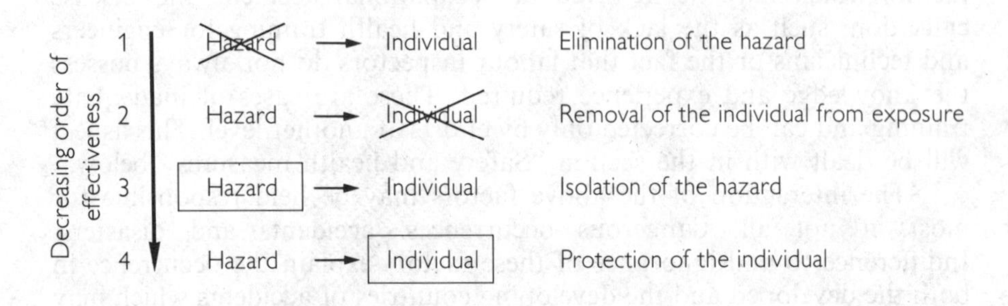

Source. Adapted from E. Gniza: "Zur Theorie der Wege der Unfallverhütung", in *Arbeitsökonomik und Arbeitsschutz*, Vol. 1, 1957, No. 1.

The human factor will, however, lead to an accident only if dangerous conditions are also present. Therefore, the consequences of human-factor failure must be prevented or neutralised by eliminating working conditions or methods that are intrinsically dangerous. Every effort should be made to promote safety consciousness and safe behaviour, even if it is widely accepted that long-lasting behavioural change is more difficult to achieve than improved material safety.

Figure 4 clearly illustrates the relative effectiveness of the main prevention or protection methods; these of course can often be combined.

1. The most radical and therefore the most effective solution is to eliminate the hazard, for instance by prohibiting silica sand in abrasive blasting or by transferring to floor level work that has previously been done at a height, so as to eliminate the fall hazard. This method is often the most difficult to apply; nevertheless, even when a hazard cannot be eliminated, it may still be possible to reduce its extent and severity, for example by selecting high-flashpoint (i.e. low-flammability) hydraulic fluids for hydraulic transmission systems.

2. When it is not feasible to eliminate the hazard or lower its severity to an acceptable level, workers and the public should be prevented from entering the danger zone; or else the dangerous installations should be located at a distance from occupied zones or traffic routes. This is the method of choice, for instance, for radioactive sources used in industrial radiography and gammagraphy, or particularly hazardous processes, such as the manufacture of explosives. Operations may also be dispersed in a number of small workshops instead of being concentrated in a single large building.

3. In most cases, however, it will be necessary to install "barriers" (not necessarily physical) between the worker and the hazard, for example by enclosing dangerous machine parts or live electrical conductors, by hermetically enclosing the processing of toxic or dangerous substances or by guarding the danger zone of a mechanical press with a photoelectric barrier.

4. Finally, when none of the above approaches is feasible, or when the degree of safety achieved is considered inadequate, the only solution is to provide exposed persons with suitable personal protective equipment. This is the last line of defence and should be used only as a last resort, since it entails reliance on active co-operation and compliance by the workers. Moreover, such equipment may be heavy, cumbersome and uncomfortable, and restrict movement.

Having outlined the general framework for plant-specific measures, we must now look at some concrete measures for: the workplace; the plant and equipment; the substances and agents being processed; the operating methods and work organisation; and, finally, the workers in the production process. First, however, we shall briefly review the main occupational safety and health responsibilities of the social partners.

Obligations of employers and workers

Panels 11 and 12 list the possible obligations of employers, as listed in the ILO Occupational Safety and Health Recommendation, 1981 (No. 164), Paragraph 10 of which supplements Articles 16 and 18 of the Occupational Safety and Health Convention, 1981 (No. 155), specifying the employer's major responsibilities.

The obligations listed in Panel 11 are supplemented by the provisions of Paragraph 15 of the same Recommendation (Panel 12).

A counterpart to these employers' obligations is the list of workers' obligations in Article 19 of Convention No. 155 (Panel 13), and in Paragraph 16 of Recommendation No. 164 (Panel 14).

A specific example of prevention and protection measures will now be given for each category and studied in some detail to illustrate the recommended approach.

Buildings, workplaces and means of access

Detailed regulations usually exist for industrial and commercial buildings and premises. These may be supplemented by standards or guide-lines from either specialist bodies or official services, dealing with matters such as the minimum cubic space and workplace dimensions, floor construction, the layout of traffic aisles and means of access (stairs, ladders, platforms, gangways, ramps), guard rails and toe-boards for floor and wall apertures, and natural and artificial lighting.

PANEL 11

Occupational Safety and Health Recommendation, 1981 (No. 164)

10. The obligations placed upon employers . . . might include, as appropriate for different branches of economic activity and different types of work, the following:

(a) to provide and maintain workplaces, machinery and equipment, and use work methods, which are as safe and without risk to health as is reasonably practical;

(b) to give necessary instructions and training, taking account of the functions and capacities of different categories of workers;

(c) to provide adequate supervision of work, of work practices and of application and use of occupational safety and health measures;

(d) to institute organisational arrangements regarding occupational safety and health and the working environment adapted to the size of the undertaking and the nature of its activities;

(e) to provide, without any cost to the worker, adequate personal protective clothing and equipment which are reasonably necessary when hazards cannot be otherwise prevented or controlled;

(f) to ensure that work organisation, particularly with respect to hours of work and rest breaks, does not adversely affect occupational safety and health;

(g) to take all reasonably practical measures with a view to eliminating excessive physical and mental fatigue;

(h) to undertake studies and research or otherwise keep abreast of the scientific and technical knowledge necessary to comply with the foregoing clauses.

PANEL 12

Occupational Safety and Health Recommendation, 1981 (No. 164)

15. (1) Employers should be required to verify the implementation of applicable standards on occupational safety and health regularly, for instance by environmental monitoring, and to undertake systematic safety audits from time to time.

(2) Employers should be required to keep such records relevant to occupational safety and health and the working environment as are considered necessary to the competent authority or authorities; these might include records of notifiable occupational accidents and injuries to health which arise in the course of or in connection with work, records of authorisations and exemptions under laws or regulations in the field and any conditions to which they may be subject, certificates relating to supervision of the health of workers in the undertaking, and data concerning exposure to specified substances and agents.

PANEL 13

Occupational Safety and Health Convention, 1981 (No. 155)

Article 19

There shall be arrangements at the level of the undertaking under which —

(a) workers, in the course of performing their work, co-operate in the fulfilment by their employer of the obligations placed upon him;

(b) representatives of workers in the undertaking co-operate with the employer in the field of occupational safety and health;

. . . .

(f) a worker reports forthwith to his immediate supervisor any situation which he has reasonable justification to believe presents an imminent and serious danger to his life or health; . . .

PANEL 14

Occupational Safety and Health Recommendation, 1981 (No. 164)

16. The arrangements provided for in Article 19 of the Convention should aim at ensuring that workers –

(a) take reasonable care for their own safety and that of other persons who may be affected by their acts or omissions at work;

(b) comply with instructions given for their own safety and health and those of others and with safety and health procedures;

(c) use safety devices and protective equipment correctly and do not render them inoperative;

(d) report forthwith to their immediate supervisor any situation which they have reason to believe could present a hazard and which they cannot themselves correct;

(e) report any accident or injury to health which arises in the course of or in connection with work.

Example: Loading bays

These installations are widely found in industrial and commercial premises but, owing to design and equipment deficiencies, they are often not safe enough. They are used for vehicle loading and unloading and for the transit or storage of goods. The following data are taken from a publication by the Institut national de recherche et de sécurité (French National Research and Safety Institute – INRS).[3]

Studies show that accidents in road or railway loading bays are usually attributable to design and construction faults. The necessary safety measures would have been easy to build in at the design stage, had they been foreseen in due time, but may be difficult to implement on existing bays.

The above-mentioned data sheet starts by listing the hazards of the movement of persons and vehicles (excluding the hazards of plant or machinery, the transported goods or electricity):

(a) falls are the major hazard when persons are moving around. Falls on the level may be caused by flooring which is in bad repair, cluttered or slippery, or by stationary obstacles; workers may fall from a height from the top of a bay, from vehicles or transfer points, or from access ladders or staircases;

(b) moving vehicles seldom cause accidents in loading bays; however, these accidents may have very serious consequences.

Pedestrians may be hit (mainly when vehicles are reversing) or may be crushed between a vehicle and, usually, a stationary obstacle (such as a bay wall, posts, palletised goods or containers).

The data sheet then examines bay layout (which depends on vehicle parking and manœuvring space, the width of façades, the type and quantities of goods being trans-shipped; the length and width of the bays (their extension is always desirable); the shape of the bays (in a straight line if a sufficiently large façade is available, indented or star-shaped if the façade is limited in size); the bay height (the ideal height is that of the loading platform of the delivery vehicle, but variations in truck size make it difficult to stipulate a single height and it is necessary to use special devices to bring the vehicle loading platform to bay level); the kerbing of the bays (non-slip and resistant to vehicle impact); bay flooring (ability to withstand heavy concentrated loads and dynamics stresses, absence of holes, protruding parts and other obstacles, flat, uniform surface made of a material which is not likely to become slippery after wear, and with an anti-slip covering at points where there is a slip hazard, etc.); and access to the bay and protection of workers against bad weather.

The second part of the data sheet deals with the equipment and emphasises means of access, lighting and signals. Access is by stairs, ramps, fixed ladders and steps which should be close enough to each other to dissuade workers from jumping down from the bays rather than using the correct means of access (distances between means of access should not be greater than 10 m). Stairways are preferable to ramps or steps.

It is not possible to reproduce here all the practical advice offered to builders and users, and reference will be made only to the value of installing along the bays an inset or external stairway (depending on the shape of the bay) which will provide protection for pedestrians by giving them a refuge point. The data sheet also shows the structure, dimensions and layout of these stairs. Figure 5 illustrates these various points.

The data sheet also illustrates hand signals for controlling the movement of vehicles and pedestrians. Road and warning signs for vehicles should indicate the direction of traffic movement and delineate "no entry" and "no parking" areas; warn of the danger of collision with stationary obstacles (pillars, etc.); guide truck drivers (direction is indicated by continuous strips painted on the ground and distance by indelible markings on the ground). Even with these signs, a banksman should guide the driver in his manœuvres. Signs for pedestrians should warn of danger and prohibit entry into particularly dangerous zones; correct interpretation will be helped if standard road signs are used.

Figure 5. Stairway for a road haulage bay

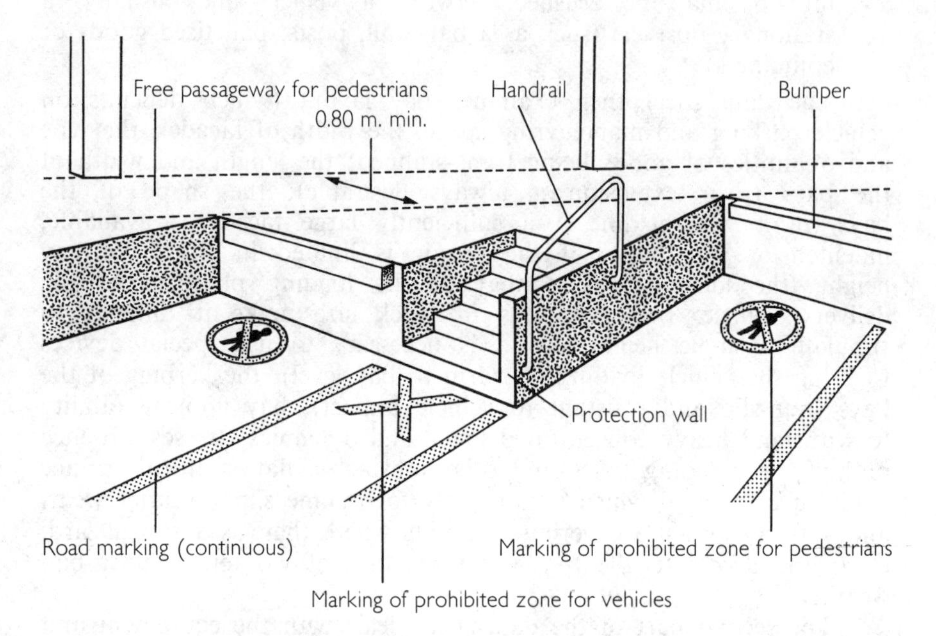

Example of a small access stairway on an indented bay with markings on the courtyard pavement
A bay access stairway provides a refuge for pedestrians

Source. Based on plans included in Comité central de coordination: "Les quais et la sécurité", Note technique n° 15, 27 June 1973, in *Cahiers de notes documentaires – Sécurité et hygiène du travail,* n° 76, Note 921-76-74 (Paris, INRS, 1974).

Machinery guarding

A wide range of examples could be given under this subheading, since each type of mechanical plant can cause specific accidents. Hand injuries on vertical spindle-moulding machines, for instance, are typical of the accidents on this type of machine.

The main machinery hazards (transmissions, protruding parts on rotating or reciprocating components, stock feed and delivery points, operating zones and controls), have been reviewed in the section on "Plant and equipment hazards" (pp. 47-49). The implementation of suitable safety measures will be illustrated by a detailed examination of the circular saw, a tool which is widely used in workshops and on building sites; this will be done on the basis of an information sheet published by the CIS in 1962.[4]

Example: Circular saws

Numerous versions are available, ranging from the simple bench saw with a single fixed blade, through pendulum saws and automatic saws with roller feeds to combination machines with multiple blades. Listed below are the hazards of the common type of bench circular saw, together with corresponding safety measures.

Hazard factors on these machines (figure 6) are mainly related to:

- the material characteristics of wood (wood is far from being a homogeneous material and may include knots, frost cracks, heart-shapes, soft parts adjoining harder areas, splits and other defects);
- contact between a part of the worker's body and the saw teeth (at the end of the cut when the worker's hands pushing the stock are close to the saw teeth; during sawing when a sudden change in the feeding pressure may cause the hands to move suddenly towards the blade, etc.); contact with the section of the blade not engaged in the stock (when the operator removes a piece of wood from behind the saw, etc.) or with the lower portion of the blade under the saw table (when the operator is removing accumulated sawdust from under the saw table, etc.);
- kickback of the stock (this will be considered further below, since it is a characteristic hazard of the circular ripping saw);
- contact between a part of the worker's body and the saw transmission;
- hands being drawn into the automatic feed device;
- inadequacy of the guards (intrinsic defect of the guard, guard incorrectly adjusted, etc.);
- bad working practices (loose clothing, sawing long stock with an overhang and without a horse or guide roller, etc.);
- noise;
- the stock being sawn.

Over 70 per cent of saw accidents occur as a result of contact with the saw blade during sawing. The great majority of reported accidents come about because the blade guard is not in place, or because it is inadequate.

Kickback occurs mainly when stock is being ripped, and is usually caused because the kerf is not kept open and the sawn stock pinches the saw blade; in other cases, it may be caused by a resin deposit on the blade sticking to and lifting the stock, by the wood jamming between the blade and the saw guide if it extends beyond the blade shaft, or by a movement which places the wood on the rising teeth of the blade.

Kickbacks cause over 10 per cent of circular saw accidents, and the resultant injuries may be very serious. Workers 15 m from the saw have been killed by the stock thrown out by a saw blade rotating at a peripheral speed of 40-70 m/s (i.e. approximately 150-250 km/h).

Figure 6. Dangerous parts of a circular saw

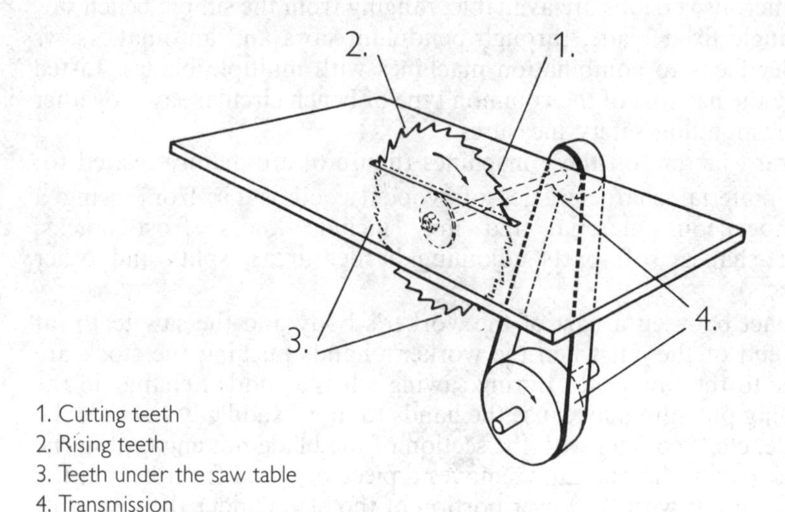

1. Cutting teeth
2. Rising teeth
3. Teeth under the saw table
4. Transmission

Source. A. Chavanel: *Circular saws*, Information Sheet No. 7 (Geneva, ILO/International Occupational Safety and Health Information Centre (CIS), 1962).

The most important safety devices for a circular saw are the blade guard and the riving knife (see figure 7) which protect the non-cutting part of the blade protruding above the table.

Early models of blade guards were often fragile and unsuitable for stock of different height. However, significant improvements have been made and the latest models are more robust, completely cover the blade at rest, are instantaneously adjustable to the stock height, can be easily adjusted to blade diameters, do not hinder adjustment of the riving knife (which is not a part of the guard but fixed to the table) and cannot come in contact with the blade even if the retaining screw shakes loose.

Modern blade guards may be fixed to the saw frame, saw table, ceiling, wall or post anchored to the floor; they are mounted on levers and rods to permit instant adjustment of the blade guard to blade diameter and stock height.

The riving knife holds the saw kerf open to prevent blade-jamming and kickbacks, and guards against contact between the operator's body and the teeth at the rear of the saw. To ensure safety against kickbacks, the thickness of the riving knife must be between that of the blade and the kerf (for non-set teeth, the riving knife will be the same thickness as the blade). To prevent blade contact, the riving knife must closely follow the blade periphery and, consequently, be adjustable both horizontally and vertically (see figure 8); if blades of different diameters are used consecutively, a set of riving knives of different sizes will be required. In

Figure 7. Guard for a single-blade circular saw

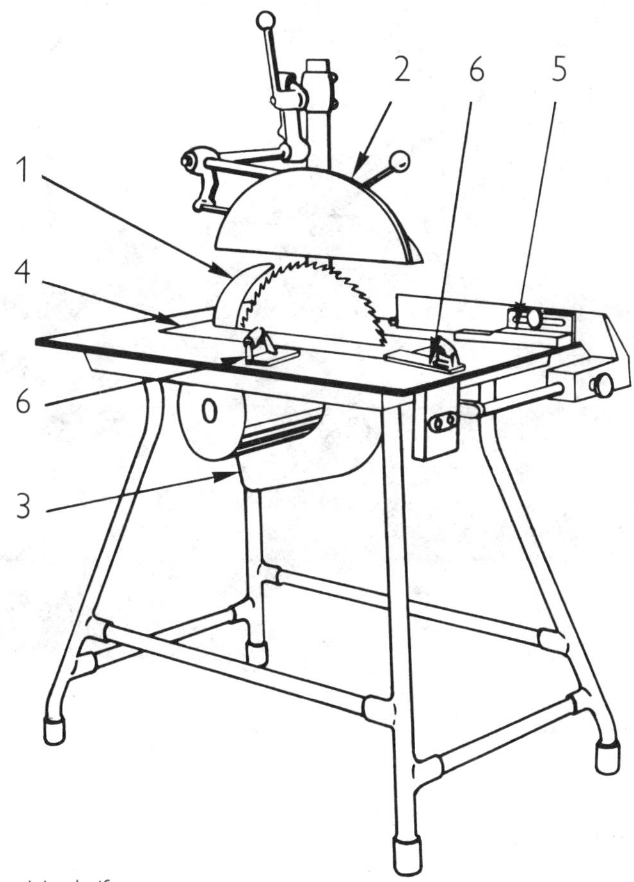

1. Adjustable riving knife
2. Blade guard that can be rapidly adjusted vertically and horizontally
3. Under-table blade guard
4. Easily replaceable plate which fills the table aperture
5. Intermediate guide for stock
6. Stock pushers

Source. Caisse nationale suisse d'assurance en cas d'accidents (CNA), Lucerne, Switzerland.

all cases, the riving knife must be mounted rigidly on the table but still be rapidly adjustable.

Kickback caused by jamming between the blade and the saw guide can be avoided by using an adjustable intermediate guide which should be an integral part of any well-designed machine.

Figure 8. Riving knife correctly adjusted for the smallest blade diameter

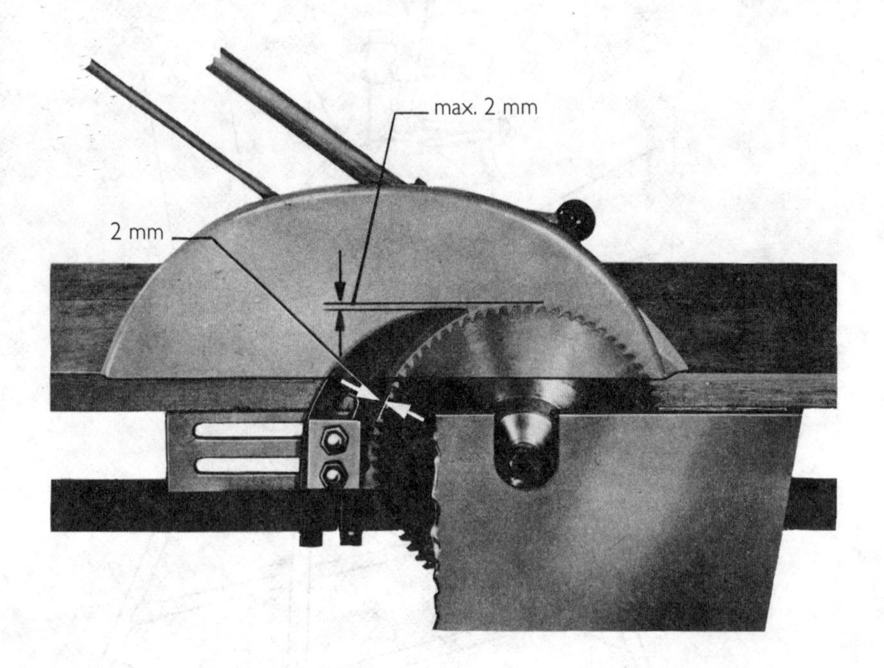

Source. Caisse nationale suisse d'assurance en cas d'accidents (CNA), Lucerne, Switzerland.

Drive mechanisms must, of course, also be guarded. The area around the saw must be kept clean and tidy, and adequate glare-free lighting should be provided; serious accidents have occurred because the floor was cluttered with waste or slippery with oil or grease, or because the worker did not have sufficient space around the machine. The start and stop controls should be within easy reach of the operator; the start button should be inset or surrounded by a fixed collar to avoid accidental starting, while the stop button should protrude and be easily accessible. Electrical hazards should be eliminated in accordance with standard electrical practices.

Before sawing, stock should be checked to detect nails, bolts or other inclusions which could act as missiles if projected during sawing.

A rigid frame, solidly anchored to the floor, will help to reduce noise and vibration. The housing of modern saws guards the underside of the saw table, eliminating the need for guard panels which may vibrate and cause high noise levels. Studies are currently under way with special toothing to reduce circular saw noise levels – which may exceed 100 dB (decibels) under load and 110 dB under non-load conditions, i.e. higher than the usual permissible limit of 85-90 dB.

Fine sawdust may present an explosion hazard in the use of circular saws, and fire prevention and protection measures are required. Exhaust ventilation devices on the saws and the regular removal of combustible waste from the workshop will control the hazard. Local exhaust ventilation devices will also contribute to accident prevention on circular saws by ensuring better visibility of the stock. When certain exotic woods are being sawn on a more or less regular basis, exhaust ventilation may be essential.

Mention should also be made of stock push-sticks which eliminate contact between the blade and the operator's hands.

The range of hazards encountered with a circular saw of simple design makes it easy to imagine the complex risk analysis required for the more sophisticated types of machine used nowadays.

Dangerous substances and agents

The hazards encountered during the manufacture and use of dangerous substances and agents are:

- ignition (acetone, petrol (gasoline), ether, carbon disulphide, magnesium, cotton, sawdust, etc.); [5]
- explosion [6] (acetylene, kerosene, liquefied petroleum gas, hydrogen, fine organic particles – wood, coal, cork, starch, grain, sugar, cocoa, etc. – or certain metals – magnesium, titanium, etc.);
- poisoning (benzene, trichloroethylene, aniline, carbon disulphide, hydrogen sulphide, phosgene, cyanides, lead, mercury, arsenic, pesticides, etc.); under this heading may also be included those diseases caused by mineral dusts (asbestos, silica, talc, etc.);
- skin and mucous membrane irritation (respiratory irritation – sulphur dioxide, nitrous oxides, chlorine, ozone, etc.; skin irritation – solvents, pickling agents, detergents, phenols, etc.); chemical or caustic burns (acids, bases, etc.). Many chemicals act as allergens and cause skin sensitisation, etc. (chromium, nickel, turpentine, synthetic resins and hardeners, solvents, paints and varnishes, isocyanates, antibiotics, etc.);
- biological action of ionising radiation (X-ray apparatus, particle accelerators, natural or artificial radioactive substances);
- pathological effects of other physical agents (noise, vibrations, ultrasound, ultraviolet radiation, radio-frequency radiation, etc.);
- climatic factors (heat, cold, humidity, etc.) and microclimatic factors (dry air, humidity, use of water at the workplace, etc.).

Many chemicals have hazards which fall into more than one of the above categories. Safety and health measures are based on a number of major principles which may often be combined: substitution of a dangerous substance for a less dangerous substance (replacement of benzene by xylene or toluene, trichloroethylene by perchloroethylene, silica sand by shot or olivine, ammonia by freon, silica by pumice, white lead by zinc oxide, etc.); isolation by distance (ionising radiation, explosives, etc.) or by the enclosure of operations (hermetically sealed chemical installations, pneumatic conveying of toxic particulates, etc.); ventilation (general ventilation should usually be supplemented by local exhaust ventilation); wet processes (where it is technically feasible to use moisture to reduce the release of dangerous dusts or vapours into the atmosphere, as is the case with the hydraulic fettling of castings); sampling and analysis of toxic substances in the atmosphere; medical supervision of exposed workers; use of suitable personal protective equipment (respirators, goggles, gloves, aprons, boots, etc.); education and information of workers.

Example: Dangerous substances in garages

In motor-vehicle maintenance and repair shops and garages, running engines and the presence or use of flammable products involve hazards which are often aggravated by confined and overcrowded premises and by the use of processes without the minimum essential precautions (spray painting, arc welding, etc.).

A brief review will be made of two common hazards in these installations, with particular reference to a publication on poisoning and explosion hazards issued in 1974 by the Caisse nationale suisse d'assurance en cas d'accidents.[7]

Poisoning may be caused by engine exhaust gases and the numerous products used to clean engines and car bodies (trichloroethylene, carbon tetrachloride, etc.); some of these products are also corrosive or irritant to the skin.

Exhaust gases contain mainly carbon dioxide, nitrogen oxides, water vapour and carbon monoxide which is odourless, highly toxic and, therefore, particularly dangerous. The widely accepted exposure limit for carbon monoxide is around 50 ppm (parts per million), i.e. around 55 mg/m^3 air, for exposure not exceeding eight hours per day and 40 hours per week. Petrol-engine exhaust gases have a much higher carbon monoxide content than diesel-engine exhaust gases.

The carbon monoxide content of the atmosphere at the workplace must be accurately determined. The simplest instrument for this purpose is a small hand-pump which draws a specific quantity of the air through a reagent-filled detector tube; the carbon monoxide content is measured by comparing the colour of the reagent with that of a calibrated standard. Other more sophisticated instruments exist, including some that continuously monitor and record the carbon monoxide concentration and automatically give a warning and switch on the ventilation when the critical concentration is reached.

Fires and explosions are caused mainly by the flammable products used in garages and the presence of explosive mixtures of petrol vapours and air either in the vehicle fuel tanks or in confined spaces, together with such ignition sources as naked flames, unprotected heating equipment, electric arcs caused by short circuits or a welding unit, or sparks produced by static electricity or grinding. Prohibition on smoking must be strictly enforced. Working clothes impregnated with petrol, oil and grease are an additional risk in the event of fire. Electrical installations – especially those left on during work-breaks or which are switched on when entering the premises – must be explosion proof if they are less than 1 m above ground level (petrol vapours, which are three to four times heavier than air, accumulate at floor level).

Artificial ventilation is essential to evacuate toxic gases and vapours and to dilute explosive mixtures; it should be designed by a competent technician and correctly maintained. The contaminated air drawn off must be balanced by an equivalent intake of fresh air, and ventilation inlet and outlet apertures should be arranged so that the whole premises are ventilated. Air should flow through the garage in a specific direction without producing vortexes; this can be verified by walking round the premises with a piece of thread hanging from the hand.

In calculating the rate of air flow, allowance should be made for the density of traffic in the garage and the type of work being carried out. For example, a medium-sized motor car entering or leaving a garage will produce around 400 l of exhaust gas containing around 20 l of carbon monoxide per minute. Normal manœuvring to enter or leave the garage may therefore produce around 50 l of carbon monoxide; if the permissible volumetric concentration of 0.005 per cent is not to be exceeded, these 50 l of carbon monoxide should be diluted 20,000 times, which requires a supply of 1,000 m^3 of fresh air. A repair shop in which workers are continuously present and where ten motor cars enter or leave per hour therefore requires a ventilation air flow of 10,000 m^3/h, i.e. some 3 m^3/s.

When it is necessary to have the vehicle engine running, exhaust gases should be collected directly from the exhaust-pipe and led into the open without contaminating the workshop atmosphere. Since, in this case, exhaust gases are released in surges, the air flow in the exhaust take-off should be higher than for steady-state exhaust; each take-off should exhaust around 3 m^3 of gas per minute.

Special precautions should be taken when fuel tanks are being welded, since they may contain explosive mixtures of petrol vapour and air. Before welding, the tank should be emptied and filled with water, inert gas (nitrogen or carbon dioxide), steam or dry ice. Vehicle inspection pits (see figure 9) in garage floors make inspection, cleaning and repair work underneath vehicles considerably easier, but they may

Figure 9. Diagram of an inspection pit in a motor-vehicle repair and maintenance shop (all items are not shown)

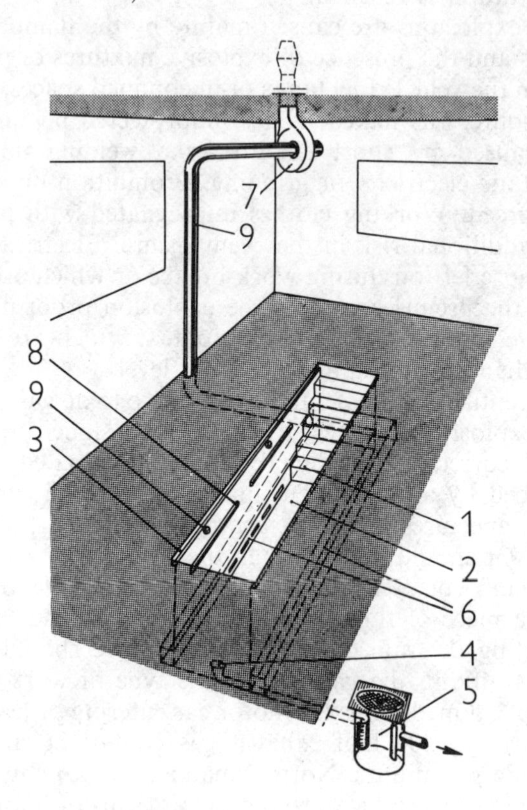

1. Ladder
2. Smooth, fire-resistant, joint-free floor
3. Slots for planks or grills covering the pit
4. Drain with an incline to the collector
5. Collector with oil and fuel separator
6. Exhaust ventilation duct
7. Non-sparking exhaust ventilation fan (the exhausted air is released into the atmosphere through the roof)
8. Electrical installations (e.g. lighting) sunk into the pit wall and linked to the pit ventilation
9. Electrical installations on the edge of the pit not linked to the ventilation

Source. Caisse nationale suisse d'assurance en cas d'accidents (CNA), Lucerne, Switzerland.

be the site of accidents unless precautions are taken (the pit should be covered or guarded when not in service, accumulations of explosive gas mixtures should be exhausted, explosion-proof lamps and electrical equipment should be used, etc.). Note that the use of "non-sparking" tools does not ensure total safety. Since petrol vapours are heavier than air, even an open pit will not ventilate itself. The air must be evacuated from the pit bottom, and the exhaust air flow must be so calculated as to ensure that the pit air is renewed at least 25 times per hour.

Finally, workers should practise good personal hygiene so as to prevent skin disorders caused by solvents, oils and cleaning products.

Operating methods and work organisation

Each operating method, work process, form of work organisation or technology introduces specific hazards but may at the same time eliminate or reduce others.

Automation has often eliminated accident and health hazards by making human intervention unnecessary; the robotisation of such tasks as welding and spray painting has a similar effect. However, new technologies may introduce new hazards due to the monotony of work, the increased pace of work, the isolation of the worker, and so forth.

Example: The permit-to-work system

This section will be illustrated not by the hazards of a method of work but by a method of work organisation – the "permit-to-work" system – which has made a significant contribution to improved occupational safety and health.

The permit-to-work system has notably increased the safety of persons working in confined spaces or in tanks which have contained flammable, explosive or toxic substances, on electrical installations, on very large machines, in container terminals, and so on. Under the system, work cannot be started until the responsible supervisor has delivered a written permit signed by everyone involved in the work; this permit must be returned when the job has been completed.

The permit specifies, where appropriate, the correct sequence of operations and the relevant precautions, necessary prior actions (ventilation, locking-out, etc.), equipment and installations to be closed down, checks to be made, and any personal protective equipment to be used. The permit is delivered by the responsible supervisor only after he has ascertained that the workers concerned understand the hazards and the necessary precautions. The system is most effective and should be used more widely; it is similar to the check-lists developed by a number of industries (airlines, nuclear plants, etc.).

The use of the system will be illustrated by way of examples described by the United Kingdom Factory Inspectorate,[8] which review specific risks of repair and maintenance work on machinery and plant when normal operational safety measures no longer function. In such cases, the supervisor should identify foreseeable risks and devise a working system that will effectively prevent or eliminate them. The system should also provide for hazards during emergency work and routine maintenance; in particular, it should ensure the necessary liaison and co-ordination between different groups of workers (production and maintenance services, or teams belonging to subcontractors and the main company).

The permit-to-work system aims above all at avoiding the common danger that verbal instructions, requests and simple assurances are poorly understood, incorrectly interpreted or forgotten and therefore are not a basis for reliable action on which human lives depend. Safety can and should be ensured by using a system of written authorisations.

The permit-to-work is a document listing the job phases and the safety precautions for each. It specifies a working procedure that covers all foreseeable risks and safeguards. A permit-to-work should:

– contain accurate, precise and detailed information about the plant or equipment, the safety measures to be taken and the job to be done;
– indicate that it will remain valid until the job has been completed or until it has been cancelled by the person who issued it;
– have priority over all other instructions that might be given;
– prohibit the entry of non-specified persons into the danger area (this also applies to management and supervisors);
– prohibit work not specified in the permit; any changes found necessary during the work require the cancellation of the original permit and the issue of a new one;
– in cases of divided responsibility, be signed by all responsible persons;
– be in several copies, one of which will be posted in the immediate vicinity of the worksite.

It is not possible to list here all the situations in which a permit-to-work system would increase safety. However, typical accidents that such a system would prevent include: workers crushed in machines which were started up whilst they were working on them or inside them; workers electrocuted by someone unwittingly switching on electrical equipment; maintenance workers crushed by a vehicle or an overhead travelling crane moving along a section of track that had not been locked out; errors or inadequacies in verbal instructions given to maintenance teams; workers asphyxiated in tanks which had not previously been checked and certified as safe to enter, etc.

The three drawings in figure 10 show the various stages of a fatal accident in which a worker was asphyxiated by solvent vapours when inspecting the inside of a mixer tank which had just been cleaned with a solvent. A permit-to-work would have prevented this accident by: defining the procedure to be followed; specifying the precautions to ensure that the atmosphere inside the tank was harmless and that there was an adequate supply of oxygen; making it impossible to start up the mixer accidentally; and specifying the need for proper supervision of the operation.

Figure 10. A fatal accident which occurred when a solitary worker was asphyxiated by solvent vapours in a mixer tank

Source. Reproduced from *Accidents at factories, offices, shops, docks and construction sites* (London, British Health and Safety Executive). Reproduced with the permission of the Controller of Her Britannic Majesty's Stationery Office.

Measures involving the worker

A word should be said about safety measures relating to the human factor, i.e. measures aimed at adapting man to work and the working environment.

These involve: occupational physiology, psychology and sociology; personal characteristics (such as age, occupational skills and experience, marital and socio-professional status, motivation, attention, perception, motor ability, emotivity, susceptibility to fatigue, reaction time, etc.); adaptation to the working environment, including vocational guidance, selection and training, supervision, medical surveillance, etc.[9] (Chapters 1, 7 and 8 also deal with these aspects).

An undertaking is a system of interacting material and human factors, and occupational accidents and injuries to health indicate that the system is not functioning properly. Human dysfunction may occur at any time; either this must be prevented or the deleterious consequences must be limited.

Occupational safety and health organisation should therefore make provision for variations from "average" or "normal" behaviour during work and, where necessary, introduce a "safety" factor to compensate for this possible deviation from "standard" in the same way that engineers provide for safety factors in their calculations on the behaviour of materials.

Example: The contribution of occupational health

Occupational health is the application of medicine and related sciences to industrial, social and economic factors. As occupational health services have grown and expanded, they have gradually become integrated into the production units in which they have responsibility for worker health.

The ILO Occupational Health Services Recommendation, 1959 (No. 112), states that these services have the tasks shown in Panel 15.

Other important stipulations of the Recommendation are reproduced in Panel 16.

The functions of occupational health services are reproduced in Panel 17.

The implementation of these functions is dealt with in the Paragraphs of the Recommendation reproduced as Panel 18.

The Recommendation concludes with a set of provisions on the personnel and equipment of these services.

Although occupational health services are concerned with a number of material factors (job analysis, monitoring of sanitary installations, etc.), much of their activity is related more directly to the human factor, and the following paragraphs will deal in more detail with the important field of medical examinations.

PANEL 15

Occupational Health Services Recommendation, 1959 (No. 112)

1. For the purposes of this Recommendation the expression "occupational health service" means a service established in or near a place of employment for the purposes of —

(a) protecting the workers against any health hazard which may arise out of their work or the conditions in which it is carried on;

(b) contributing towards the workers' physical and mental adjustment, in particular by the adaptation of the work to the workers and their assignment to jobs for which they are suited; and

(c) contributing to the establishment and maintenance of the highest possible degree of physical and mental well-being of the workers.

Medical examinations may be divided into pre-employment and periodic examinations.

Pre-employment examinations are usually carried out when the employee starts work and are designed to determine whether the worker is suitable for the job in question.

Periodic examinations take place at set intervals, for the early detection of health impairment caused by the work or which may affect fitness for work.

Both types of examination are closely related and may be supplemented, where necessary, by special examinations. Pre-employment medical examinations define the worker's health status on engagement, highlight items for periodic follow-up and provide a baseline for the subsequent assessment of his health. Periodic examinations should detect early signs and symptoms of disease (whether or not of occupational origin), flare-up of pre-existing pathological conditions or deterioration in general health status, and provide the necessary medical opinion. The interval between examinations will depend on the type of work and its health hazards, and the results of prior examinations.

PANEL 16

Occupational Health Services Recommendation, 1959 (No. 112)

2. Having regard to the diversity of national circumstances and practices, occupational health services may be provided, as conditions require —

(a) by virtue of laws or regulations;

(b) by virtue of collective agreement or as otherwise agreed upon by the employers and workers concerned; or

(c) in any other manner approved by the competent authority after consultation with employers' and workers' organisations.

3. Depending on the circumstances and the applicable standards, occupational health services —

(a) should either be organised by the undertakings themselves or be attached to an outside body;

(b) should be organised —

   (i) as a separate service within a single undertaking; or

   (ii) as a service common to a number of undertakings.

4. In order to extend occupational health facilities to all workers, occupational health services should be set up for industrial, non-industrial and agricultural undertakings and for public services: Provided that where occupational health services cannot immediately be set up for all undertakings, such services should be established in the first instance —

(a) for undertakings where the health risks appear greatest;

(b) for undertakings where the workers are exposed to special health hazards;

(c) for undertakings which employ more than a prescribed minimum number of workers.

. . . .

6. The role of occupational health services should be essentially preventive.

7. Occupational health services should not be required to verify the justification of absence on grounds of sickness; they should not be precluded from ascertaining the conditions which may have led to a worker's absence on sick leave and obtaining information about the progress of the worker's illness, so that they will be better able to evaluate their preventive programme, discover occupational hazards, and recommend the suitable placement of workers for rehabilitation purposes.

PANEL 17

Occupational Health Services Recommendation, 1959 (No. 112)

8. The functions of occupational health services should be progressively developed, in accordance with the circumstances and having regard to the extent to which one or more of these functions are adequately discharged in accordance with national law or practice by other appropriate services, so that they will include in particular the following:

(a) surveillance within the undertaking of all factors which may affect the health of the workers and advice in this respect to management and to workers or their representatives in the undertaking;

(b) job analysis or participation therein in the light of hygienic, physiological and psychological considerations and advice to management and workers on the best possible adaptation of the job to the worker having regard to these considerations;

(c) participation, with the other appropriate departments and bodies in the undertaking, in the prevention of accidents and occupational diseases and in the supervision of personal protective equipment and of its use, and advice to management and workers in this respect;

(d) surveillance of the hygiene of sanitary installations and all other facilities for the welfare of the workers of the undertaking, such as kitchens, canteens, day nurseries and rest homes and, as necessary, surveillance of any dietetic arrangements made for the workers;

(e) pre-employment, periodic and special medical examinations — including, where necessary, biological and radiological examinations — prescribed by national laws or regulations, or by agreements between the parties or organisations concerned, or considered advisable for preventive purposes by the industrial physician; such examinations should ensure particular surveillance over certain classes of workers, such as women, young persons, workers exposed to special risks and handicapped persons;

(f) surveillance of the adaptation of jobs to workers, in particular handicapped workers, in accordance with their physical abilities, participation in the rehabilitation and retraining of such workers and advice in this respect;

(g) advice to management and workers on the occasion of the placing or reassignment of workers;

(h) advice to individual workers at their request regarding any disorders that may occur or be aggravated in the course of work; ▷

(*i*) emergency treatment in case of accident or indisposition, and also, in certain circumstances and in agreement with those concerned (including the worker's own physician), ambulatory treatment of workers who have not been absent from work or who have returned after absence;

(*j*) initial and regular subsequent training of first-aid personnel, and supervision and maintenance of first-aid equipment in co-operation, where appropriate, with other departments and bodies concerned;

(*k*) education of the personnel of the undertaking in health and hygiene;

(*l*) compilation and periodic review of statistics concerning health conditions in the undertaking;

(*m*) research in occupational health or participation in such research in association with specialised services or institutions.

PANEL 18

Occupational Health Services Recommendation, 1959 (No. 112)

9. Where one or more of the functions enumerated in the preceding Paragraph are carried out, in accordance with national law or practice, by appropriate services other than occupational health services, these should provide the industrial physician with any relevant information he may wish to request.

10. Occupational health services should maintain close contact with the other departments and bodies in the undertaking concerned with questions of the workers' health, safety or welfare, and particularly the welfare department, the safety department, the personnel department, the trade union organs in the undertaking, safety and health committees and any other committee or any person in the undertaking dealing with health or welfare questions.

11. Occupational health services should also maintain relations with external services and bodies dealing with questions of the health, safety, retraining, rehabilitation, reassignment and welfare of the workers.

Special examinations will be at the physician's discretion, for instance on return to work following an accident or disease, or on exposure to new hazards. High-risk workers require particular attention: for example, workers with earlier exposure to carcinogenic substances or ionising radiation should be carefully monitored. Medical examinations should not be used to exclude from employment persons with physical or mental limitations, but rather to guide them into suitable jobs and help to remedy these deficiencies.

In occupational health practice, a study of pre-clinical findings is particularly useful in examining man-work adaptation. As stated by R. Elias, "recording symptoms of mental and somatic dysfunction unaccompanied by clinical disability is of major practical significance. In contrast with other medical practitioners who do not deal with 'normal' people, the industrial physician can make long-term studies of industrial cohorts and observe the good and bad effects of new technologies or new forms of work organisation. In this way, the industrial physician is at the heart of occupational safety and health problems." [10]

In developing countries and isolated regions of industrialised countries, the occupational health service may be the only health-care delivery system for workers and their families, and the work may be increased and complicated by the occasional need to act as a public health service for infectious, parasitic or epidemic disease.

Overall occupational safety and health organisation

The control of occupational hazards through risk identification and analysis, safety and health measures and the evaluation of success, calls for adequate organisation.

No organisational structure is good or bad in itself, and each approach should be judged by its results. In planning organisational structures for occupational safety and health, moderation is essential, especially in a large undertaking, since one may be tempted to devote more effort to designing an impressive organisational chart than to selecting high-quality resources: an organisational chart, in itself, is no guarantee of effective organisation. Undertakings of different sizes may achieve good results by very different organisational routes.

Over-ambitiousness at the start may be dangerous, and a "step-by-step" policy may prove better than a "headlong" approach. No matter how large the organisation, it cannot maintain a continuing offensive on all fronts.

It is first necessary to establish priorities by assessing the main factors and the hazards with the severest consequences (i.e. endangering the largest number of workers). High priority may further be allocated to hazards in dealing with which rapid results may be expected and which can be used as examples of success (stricter discipline as regards good housekeeping, installation of guard rails, improvement of lighting, etc.).

A priority objective should not be dropped until results have been achieved, since a setback cannot be concealed for long and may have a disastrous effect on the involvement and participation of workers.

Selecting the means of action

The extent of the measures shown to be necessary by the size and resources of the undertaking will determine the size, management and staffing of an occupational safety and health unit. Large undertakings or groups of undertakings may have a safety service, an occupational or industrial hygiene service and an occupational health service as well, whereas smaller undertakings may combine all these activities under a single person. In a small undertaking with no special hazards, safety may be handled by the head of personnel, the maintenance service or the employer himself.

The personnel service usually deals with recruitment, promotion, discipline, wages, working hours, working conditions in general and the recording of absences. There may be a special service to assess suitability for different jobs and carry out psycho-technical testing where necessary for special skills or responsibilities, and also a social service responsible for canteens, nurseries, leisure activities, and so on.

The safety service [11] is responsible for the engineering aspects of accident prevention; and also, in some cases, for the fire service and general security (guard duties).

The occupational health service, which may cover a group of undertakings, deals mainly with the medical prevention of ill health among workers; its activities have already been described in the example on the contribution of occupational health on pp. 72 ff.

The occupational hygiene service is responsible for health engineering, for instance through ventilation, lighting, noise control, sanitary engineering, and so on.

No standard rules govern the relations between these services and their position in the undertaking's hierarchy. However, safety, occupational health and occupational hygiene services are basically – if not exclusively – advisory, and it is widely agreed that direct reporting to management is usually the most satisfactory arrangement (this is in line with the statements made earlier in this chapter on the subject of employer responsibility).

A risk analysis may often overlook or underestimate health hazards, and this may explain the still low number of occupational hygienists and the shortage of training courses at university level for these "health engineers", who have an intermediate position between safety engineers and industrial physicians.[12]

In figure 11, Guillemin uses Swiss experience to explain current widespread ignorance as to the true extent of occupational injuries to health.[13] Whereas in 1977 there were 215,000 occupational accidents in

Figure 11. Reported cases of occupational diseases are no more than the tip of the iceberg (Swiss statistics, 1977)

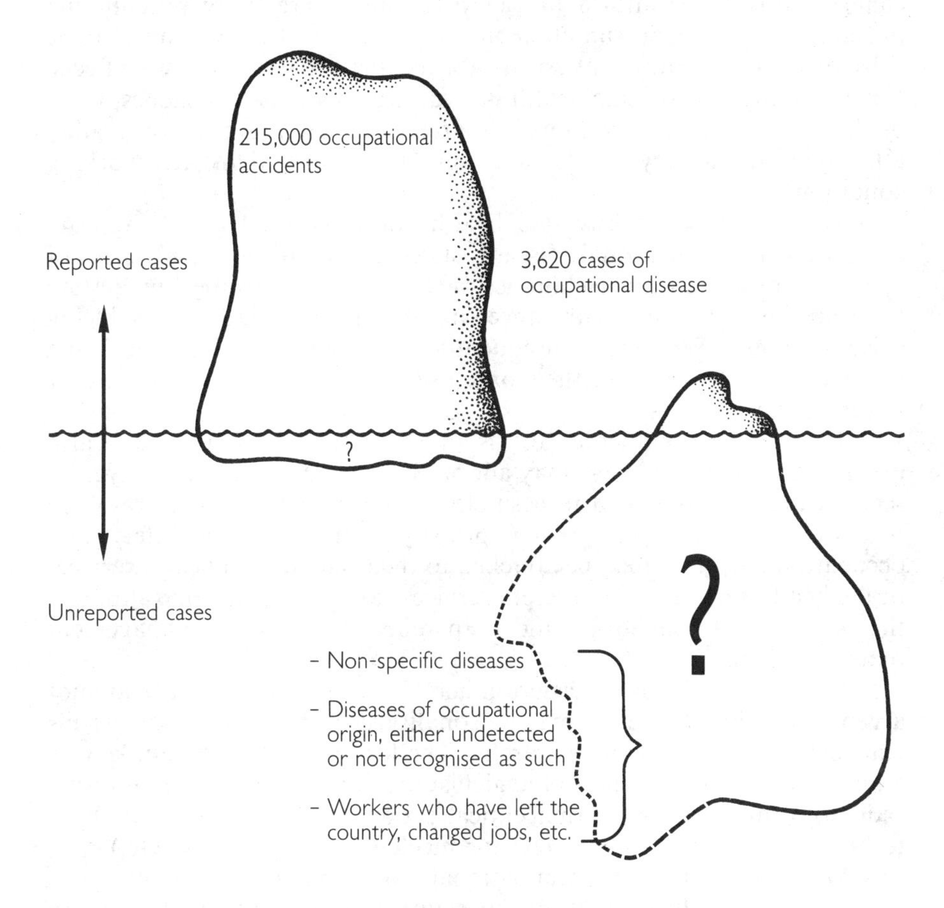

Switzerland, only 3,620 cases of compensable occupational diseases were reported; that is, occupational accidents accounted for 98.3 per cent and occupational diseases for only 1.7 per cent of all occupational injuries. It is tempting to conclude that efforts for occupational accident and disease prevention should have the same percentage distribution; but this of course would be erroneous.

First, the average compensation cost is proportionally some four times higher for occupational diseases than for occupational accidents. Furthermore, the schedule of compensable occupational diseases is usually limited (e.g. only 19 items and 119 substances for Switzerland in 1977). Many health effects develop only slowly (long latency period), and their course is often insidious and their detection difficult. They may not always be reported in official statistics, whereas accidents are more likely to be notified. An accident is a sudden occurrence, while occupational diseases develop over a period of time and are often not recorded – especially in the case of migrant workers or persons who have changed jobs. Little information exists about synergetic or potentiating phenomena in which simultaneous exposure to two or more toxic substances or harmful physical agents may have additive effects. Finally, many non-specific health disorders or diseases (headaches, renal or hepatic disorders, pulmonary carcinoma, etc.) are more readily attributed to life-style (alcohol, smoking, etc.) than to working conditions.

The extent of work-related health effects is greater than appears from official occupational disease statistics. Ignorance and lack of adequate measurement techniques account for difficulties in setting exposure limits for toxic substances in air, and in periodically reducing existing limits. Workers consequently feel that they are not being adequately protected, and their organisations are now demanding ever lower exposure limits.

There are a number of forms of occupational safety and health organisation that employers may adopt. Safety and occupational hygiene services may be combined in a single unit (although this may result in one sector being given greater priority than another); safety and occupational hygiene may be attached to the occupational health service; or the health service and the safety service may be independent (although they should still collaborate and both report to the same management representative).

Efficiency is the prime criterion here, and dogmatic attitudes are not acceptable. Thus the director's personality is just as important as his competence or level in the hierarchy. The leading role may be taken by a physician or an occupational hygienist (as, for example, where ionising radiations are used or pharmaceutical products manufactured) or by a technician or engineer (as in civil or mechanical engineering, etc.).

Whichever is the case, occupational safety and occupational hygiene cover such wide fields that no specialist can be expected to have a complete mastery of them. A multidisciplinary approach is essential in research, risk assessment and hazard control. Berry observes that the original "team" approach to occupational hygiene must be expanded beyond the traditional, basic involvement of the physician, the nurse, the safety engineer and the occupational hygienist, to include (depending on the circumstances) such disciplines as toxicology, statistics and epidemiology, among others.[14] The existence of a safety, occupational health or occupational hygiene service in no way prevents the undertaking from asking outside specialists to advise on specific problems or to check the implementation of the necessary measures. For example, an undertaking may not have the equipment needed for noise measurement or audiometric examinations, and an official service or specialised occupational safety and health organisation may be called in to do the work.

PANEL 19

Occupational Safety and Health Recommendation, 1981 (No. 164)

14. Employers should, where the nature of the operations in their undertakings warrants it, be required to set out in writing their policy and arrangements in the field of occupational safety and health, and the various responsibilities exercised under these arrangements, and to bring this information to the notice of every worker, in a language or medium the worker readily understands.

All these services have an important safety and health training function, through designing suitable programmes, co-ordinating programmes organised by other services, incorporating safety and health information into existing programmes or checking the results obtained. The staff should therefore be well versed in up-to-date didactic methodology, and capable of using it profitably.

Workers' participation

However important the employer's role, that of the worker is no less essential; however, obtaining workers' participation is not always easy. The employer should establish a safety and health policy, inform the workers and obtain their acceptance of its objectives. The ILO Occupational Safety and Health Recommendation, 1981 (No. 164), contains a formal requirement on this point (see Panel 19).

Joint employer-employee action on safety and health may also contribute to the establishment and maintenance of a good social climate and to the achievement of wider objectives.

The workers' duties in risk control have as their counterpart the recognition of certain basic rights (see the section "Obligations of employers and workers", p. 55). These are dealt with in Convention No. 155, and in more detail in Recommendation No. 164 (see Panel 20).

Over the past decade many workers have been granted specific safety and health rights, such as that to refuse to carry out or continue dangerous work, and protection from sanctions in exploiting these rights in good faith. The stipulations of Convention No. 155 and Recommendation No. 164 in this regard are reproduced in Panel 21.

PANEL 20

Occupational Safety and Health Recommendation, 1981 (No. 164)

12. (1) The measures taken to facilitate the co-operation [of employers and workers] should include, where appropriate and necessary, the appointment, in accordance with national practice, of workers' safety delegates, of workers' safety and health committees, and/or of joint safety and health committees; in joint safety and health committees workers should have at least equal representation with employers' representatives.

(2) Workers' safety delegates, workers' safety and health committees, and joint safety and health committees or, as appropriate, other workers' representatives should —

(a) be given adequate information on safety and health matters, enabled to examine factors affecting safety and health, and encouraged to propose measures on the subject;

(b) be consulted when major new safety and health measures are envisaged and before they are carried out, and seek to obtain the support of the workers for such measures;

(c) be consulted in planning alterations of work processes, work content or organisation of work, which may have safety or health implications for the workers;

(d) be given protection from dismissal and other measures prejudicial to them while exercising their functions in the field of occupational safety and health as workers' representatives or as members of safety and health committees;

(e) be able to contribute to the decision-making process at the level of the undertaking regarding matters of safety and health;

(f) have access to all parts of the workplace and be able to communicate with the workers on safety and health matters during working hours at the workplace;

(g) be free to contact labour inspectors;

(h) be able to contribute to negotiations in the undertaking on occupational safety and health matters;

(i) have reasonable time during paid working hours to exercise their safety and health functions and to receive training related to these functions;

(j) have recourse to specialists to advise on particular safety and health problems.

PANEL 21

Occupational Safety and Health Convention, 1981 (No. 155)

Article 13

A worker who has removed himself from a work situation which he has reasonable justification to believe presents an imminent and serious danger to his life or health shall be protected from undue consequences in accordance with national conditions and practice.

Occupational Safety and Health Recommendation, 1981 (No. 164)

17. No measures prejudicial to a worker should be taken by reference to the fact that, in good faith, he complained of what he considered to be a breach of statutory requirements or a serious inadequacy in the measures taken by the employer in respect of occupational safety and health and the working environment.

Joint safety and health committees and workers' safety delegates are now common practice, but their effectiveness in promoting and achieving active workers' involvement in safety and health work is not unanimously accepted. It is interesting in this connection to note that Convention No. 155 and Recommendation No. 164 do not impose them systematically. However, the effectiveness of safety delegates is generally recognised in following shop-floor operations and safety and health conditions and in introducing corrective measures where necessary.

Joint safety and health committees provide a valuable framework for discussion and concerted action to improve safety and health. It has been recommended that they should meet at least once a month and should periodically inspect the workplace. They usually have the following tasks:

- to investigate accidents, recommend corrective measures and study the plant's accident statistics;
- periodically to inspect plant and equipment and make recommendations on the improvement of safety and health conditions;
- to enforce official safety and health regulations, instructions, and so on, monitor the implementation of safety and health measures and check their effectiveness;
- to obtain workers' participation in the improvement of safety and health, and see that newly employed workers receive safety instructions, training and counselling;

- to participate in the drafting of plant safety rules; and
- to co-operate with the safety, occupational health and occupational hygiene services, where these exist, and with the competent official services.

Safety committees or joint safety and health committees have already been set up in larger undertakings in a number of countries, and smaller undertakings may have regional safety and health committees for each branch of activity. In some cases, they are making a decisive contribution to the improvement of safety and health conditions, while in others they have not had the expected results. It is often claimed that this situation is attributable to a lack of training and experience in safety and health matters.

Many undertakings with remarkably successful safety and health programmes give more importance to informal education and the consultation of workers and their representatives than to setting up formal structures. The most encouraging results seem to have come when management has concentrated on increasing workers' awareness of their important role in safety and health and encouraging them to assume their responsibilities more fully.

The larger an undertaking becomes, the more difficult it is to ensure effective in-house communication. This is a problem that has often not been fully resolved. A prime condition for a significant, positive contribution by workers and their representatives to occupational risk control is their access to better information. Workers should be able to obtain the necessary assistance from their trade union organisations, which have a legitimate right to be involved in anything that concerns the protection of their members' health, life and limb. It is to be hoped that workers' organisations will, in future, devote as much attention to safety and health as they have done to working hours and wages. Their growing involvement in national or regional joint safety and health committees – usually within the framework of a given industry – is evidence of their increasing interest in occupational safety and health.

Evaluation of results

Evaluation is an essential step in safety and health endeavours, but it is not an end in itself. Inspectors often find that management does not know the number, type or severity of the undertaking's accidents, and the information has to be obtained from the physician, nurse, head of personnel or social security institute. In such situations, management can do little to promote safety and health.

Statistical data are not always sufficient to guide safety and health efforts since, although they are relatively accurate about the type of injury and its effects on working capacity, they are less informative about the causes of accidents or diseases. Therefore, data that may help to identify or assess hazards or develop preventive measures should be collected on the spot.

Unfortunately, far more time and effort is put into compiling accident and disease statistics than in making use of them for preventive action. Yet well-compiled statistics can be most valuable in identifying new risks, developing new protective measures for existing hazards, identifying dangerous procedures or equipment, highlighting information and training needs, deciding on problems to be studied and where the advice of outside specialists is needed, and so forth.

Too much weight can, however, be given to numerical data in general and statistics in particular, and caution is essential in comparing statistics from different undertakings, industries and countries. The risk of error is less within an undertaking, since those who know the material and human context of the data are in a better position to interpret them.

Accident severity rates may be affected by the effectiveness of first-aid and medical care, and by the worker's own decision on resuming work after a minor accident. Statistics on fatal and serious accidents are therefore more reliable.

Statistics may also be affected by the occurrence of a group accident or the amalgamation of individual accidents. Wherever possible, statistical analyses should have a sufficiently long time-basis, or employ other means, in order to reduce the effect of particularly severe events, and the previous year's experience should be used to specify the limits of "normal" monthly frequency-rate variations. A distinction can then be made between random fluctuations, actual changes in trends and the appearance of a new factor calling for rapid action.

The law of "diminishing returns" (an outcome which is not proportional to the effect or the resources involved) acquires its full significance in safety and health (see figure 12). Following the fall from A to B on the curve, it will be equally difficult to reduce the accident frequency rate further to C. Once C has been reached, a similar level of effort will be required merely for consolidation. This is a cause of frustration among safety and health workers, especially as it is not usually possible to give indisputable proof that the application of a given safety measure has led to the prevention of an accident.

Variations in accident frequency and severity rates over time should not be the only object of evaluation. The extent to which hazards have been eliminated or reduced to an acceptable level, the effective implementation of the undertaking's safety and health policy and the rate of progress towards long-term policy objectives should also be assessed. Incidents as well as accidents should be investigated, since incidents indicate a transient or permanent perturbation of one or more components of the man-work system which, under other circumstances which cannot always be foreseen, might have led to serious injuries or a large-scale disaster.

Figure 12. Illustration of the law of diminishing returns in occupational safety and health

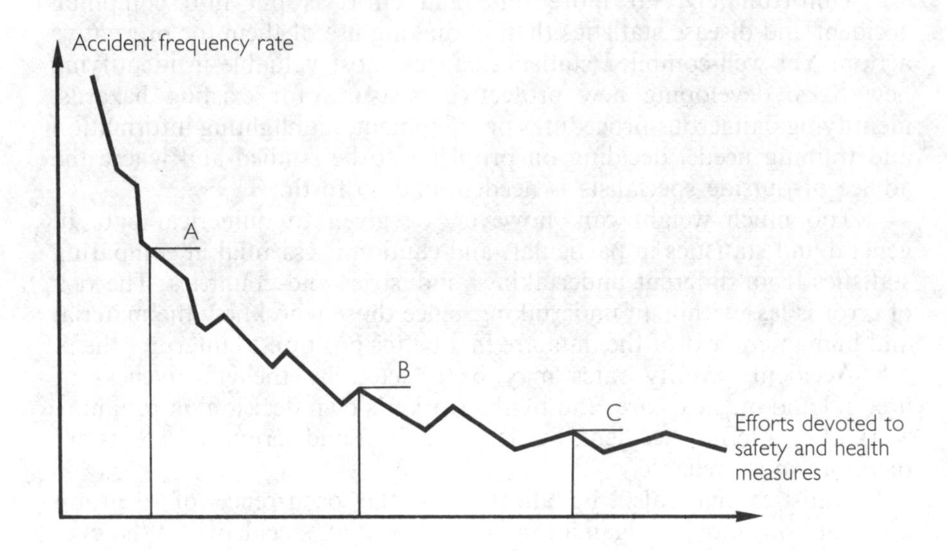

If a programme proves a failure, either partial or total, this will usually be attributable to one or more of the following factors:

(a) inadequate or superficial definition of objectives;

(b) uncritical acceptance of traditional preconceptions of occupational safety and health by management or specialists;

(c) acceptance of the phrase "Our safety problems are completely different";

(d) lack of in-depth safety and health analysis of jobs and working methods;

(e) exaggerated confidence in the firm's in-house structures (safety and health committees, etc.); and/or

(f) insufficient emphasis on good communication between employer and worker.

Safety and health qualities and achievements should be taken into consideration when the work of managerial and supervisory staff is being evaluated and decisions on promotions are being taken.

Where an undertaking has a number of departments, each with very different hazards, statistical data should be collected department by department, or shop by shop, to identify any safety and health problem areas. The general situation should first be ascertained, trends should be analysed and the necessary data should be compared with those of similar undertakings in the same branch.

Occupational risk control: Safety and health measures at the national level

At the start of the section "Occupational risk control: Safety and health measures in the plant" above, the importance of action at the level of the undertaking was repeatedly underscored. However, public authorities, professional organisations and scientific or technical institutions that work at the national level also have an essential contribution to make towards occupational risk control at the level of the undertaking.

The following paragraphs also apply to federal States in which occupational safety and health may be basically the responsibility of the federated units, and in which the federal authorities have only a co-ordinating or consultancy role.

The sequence in the section "Need for a coherent policy" below is arbitrary, and it would have been equally possible to examine the role of the public authorities, employers and workers and their organisations, the specialised institutes and teaching establishments in that order. However, to do so would have involved much repetition, since each of the "actors" plays or may play a part in the drafting of legislation, standards and practical guide-lines, in education and training, in research, in the dissemination of information, and so on. The plan adopted does not eliminate repetition but at least minimises it.

Need for a coherent policy

If satisfactory, durable results are to be achieved in safety and health, a coherent policy is essential. This is emphasised in Convention No. 155 (see Panel 22).

Recommendation No. 164 gives the additional details reproduced in Panel 23.

Occupational safety and health is a dynamic process, even though the objectives are long-term and the implementation of any well-thought-out programme may be expected to extend over several years. Significant developments or phenomena should be identified and the necessary action taken at the level of the undertaking and at the national level before a disaster occurs. This factor is also considered in Convention No. 155 (see Panel 24).

The general policy referred to in Article 4 of Convention No. 155, reproduced in Panel 22, should be set down unambiguously in writing; and it will have been noticed that Article 6 of the Convention insists on a clear definition of the responsibilities of all the bodies and individuals involved.

The competent national authority, like the employer, must identify problems and determine priorities before setting its policy goals. The relative urgency of individual problems, the difficulties that have to be

PANEL 22

Occupational Safety and Health Convention, 1981 (No. 155)

Article 4

1. Each Member shall, in the light of national conditions and practice, and in consultation with the most representative organisations of employers and workers, formulate, implement and periodically review a coherent national policy on occupational safety, occupational health and the working environment.

2. The aim of the policy shall be to prevent accidents and injury to health arising out of, linked with or occurring in the course of work, by minimising, so far as is reasonably practicable, the causes of hazards inherent in the working environment.

Article 5

The policy referred to in Article 4 of this Convention shall take account of the following main spheres of action in so far as they affect occupational safety and health and the working environment:

(a) design, testing, choice, substitution, installation, arrangement, use and maintenance of the material elements of work (workplaces, working environment, tools, machinery and equipment, chemical, physical and biological substances and agents, work processes);

(b) relationships between the material elements of work and the persons who carry out or supervise the work, and adaptation of machinery, equipment, working time, organisation of work and work processes to the physical and mental capacities of the workers;

(c) training, including necessary further training, qualifications and motivations of persons involved, in one capacity or another, in the achievement of adequate levels of safety and health;

(d) communication and co-operation at the levels of the working group and the undertaking and at all other appropriate levels up to and including the national level;

(e) the protection of workers and their representatives from disciplinary measures as a result of actions properly taken by them in conformity with the policy referred to in Article 4 of the Convention.

Article 6

The formulation of policy referred to in Article 4 of this Convention shall indicate the respective functions and responsibilities in respect of occupational safety and health and the

▷

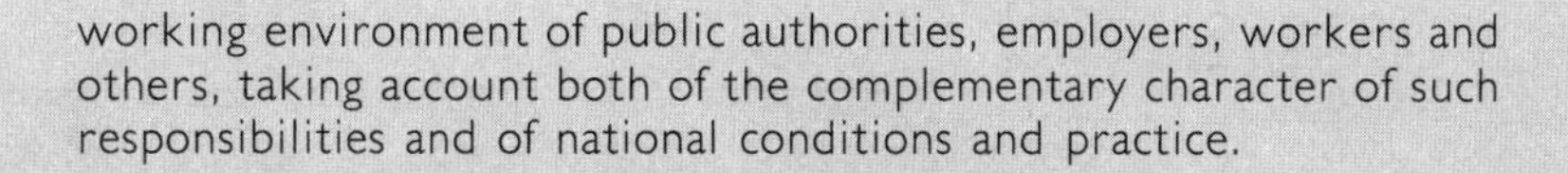

working environment of public authorities, employers, workers and others, taking account both of the complementary character of such responsibilities and of national conditions and practice.

. . . .

Article 15

1. With a view to ensuring the coherence of the policy referred to in Article 4 of the Convention and of measures for its application, each Member shall, after consultation at the earliest possible stage with the most representative organisations of employers and workers, and with other bodies as appropriate, make arrangements appropriate to national conditions and practice to ensure the necessary co-ordination between various authorities and bodies called upon to give effect to Parts II and III of this Convention.

2. Whenever circumstances so require and national conditions and practice permit, these arrangements shall include the establishment of a central body.

PANEL 23

Occupational Safety and Health Recommendation, 1981 (No. 164)

7. The main purposes of the arrangements referred to in Article 15 of the Convention should be to –

(a) implement the requirements of Articles 4 and 7 of the Convention;

(b) co-ordinate the exercise of the functions assigned to the competent authority or authorities in pursuance of Article 11 of the Convention and Paragraph 4 of this Recommendation;

(c) co-ordinate activities in the field of occupational safety and health and the working environment which are exercised nationally, regionally or locally, by public authorities, by employers and their organisations, by workers' organisations and representatives, and by other persons or bodies concerned;

(d) promote exchanges of views, information and experience at the national level, at the level of an industry or that of a branch of economic activity.

8. There should be close co-operation between public authorities and representative employers' and workers' organisations, as well as other bodies concerned in measures for the formulation and application of the policy referred to in Article 4 of the Convention.

PANEL 24

Occupational Safety and Health Convention, 1981 (No. 155)

Article 7

The situation regarding occupational safety and health and the working environment shall be reviewed at appropriate intervals, either overall or in respect of particular areas, with a view to identifying major problems, evolving effective methods for dealing with them and priorities of action, and evaluating results.

overcome and the resources available should be considered realistically; nothing is more discouraging than to be assigned objectives which are over-ambitious and unachievable.

However, policy is not all: it is also necessary to have the resources to implement it, both of personnel and of equipment. These resources, even though small in comparison with the insurance premiums paid by employers for occupational injuries, still involve large sums of money. The rational allocation of limited resources can be a major task for national authorities.

Simply appointing a national co-ordinating body is no more a policy than the establishment of complex structures without powers of decision. The competent authorities must establish, in accordance with the provisions of Convention No. 155 and Recommendation No. 164 quoted above, the main lines of national safety and health policy and take all measures necessary to ensure success.

These measures may vary, depending on the degree of technological, economic and social development and the type and extent of the resources available. In general, they will include the drafting of laws and regulations (with the necessary administrative provisions for their enforcement), education and training, studies and research, and the dissemination of information. Each of these facets of national action will be examined briefly below.

Legislation, standards and directives

Occupational safety and health legislation and regulations have developed somewhat erratically since the beginning of the twentieth century and have not, in most cases, been the result of coherent planning. However, over the past ten years, concerted efforts have been made towards simplification and unification; and it seems that the belief that occupational hazards can be eliminated or limited by legislation has lost a good deal of ground. Countries with the best present-day safety records recognise that it is more effective to stipulate the duties of those with prime responsibility for occupational safety and health measures in general terms rather than to attempt to regulate a multitude of hazards in minute detail.

Indeed, the imposing mass of laws and regulations may have given employers and workers the impression that the legislator had taken over responsibility for safety and health and that their own responsibilities were fulfilled merely by complying with statutory provisions.

The trend in many industrialised countries is to promote occupational safety and health rather than compensation for injury, to limit the volume of legislation and to encourage public authorities and specialised professional organisations to develop standards and practical guide-lines which do not have the force of law but which can be more rapidly adapted to technological change. In its most advanced form, this legislation insists on the general duties of employers, workers, manufacturers, suppliers, and so on, although specific regulations may be required to deal with serious but previously unrecognised hazards such as those caused by asbestos and vinyl chloride.

The more rapid the pace of technological progress, the more difficult it becomes to control it by the enactment of new instruments. Older legislation contained precise specifications, and its application could usually be checked by simple procedures; however, recent texts set general objectives for employers, and it is for them to achieve these objectives by the techniques they deem appropriate. Furthermore, it is now widely accepted that employers' and workers' organisations should be consulted during the drafting of laws and regulations.

Convention No. 155 deals with the scope of action of the public authorities, as indicated in Panel 25.

This is supplemented by the provisions of Recommendation No. 164 reproduced in Panel 26.

These provisions are sufficiently explicit and require no explanatory comment. Standards, specifications and codes of practice issued by national standards organisations or professional or specialised institutions have, in some cases, been given the force of law by the competent authority. This practice, which is more common in countries where such organisations and institutions are public, considerably lightens the legislator's task but increases the burden on the occupational safety and health administrations responsible for enforcing these standards, specifications and codes of practice.

Enforcement of regulations and advisory activities

Regulations are valid only to the extent that they are enforced, and Convention No. 155 points out the two approaches to enforcement (see Panel 27).

PANEL 25

Occupational Safety and Health Convention, 1981 (No. 155)

Article 11

To give effect to the policy referred to in Article 4 of this Convention, the competent authority or authorities shall ensure that the following functions are progressively carried out:

(*a*) the determination, where the nature and degree of hazards so require, of conditions governing the design, construction and layout of undertakings, the commencement of their operations, major alterations affecting them and changes in their purposes, the safety of technical equipment used at work, as well as the application of procedures defined by the competent authorities;

(*b*) the determination of work processes and of substances and agents the exposure to which is to be prohibited, limited or made subject to authorisation or control by the competent authority or authorities; health hazards due to the simultaneous exposure to several substances or agents shall be taken into consideration;

(*c*) the establishment and application of procedures for the notification of occupational accidents and diseases, by employers and, when appropriate, insurance institutions and others directly concerned, and the production of annual statistics on occupational accidents and diseases;

(*d*) the holding of inquiries, where cases of occupational accidents, occupational diseases or any other injuries to health which arise in the course of or in connection with work appear to reflect situations which are serious;

(*e*) the publication, annually, of information on measures taken in pursuance of the policy referred to in Article 4 of this Convention and on occupational accidents, occupational diseases and other injuries to health which arise in the course of or in connection with work;

(*f*) the introduction or extension of systems, taking into account national conditions and possibilities, to examine chemical, physical and biological agents in respect of the risk to the health of workers.

Article 12

Measures shall be taken, in accordance with national law and practice, with a view to ensuring that those who design, manufacture, import, provide or transfer machinery, equipment or substances for occupational use —

(*a*) satisfy themselves that, so far as is reasonably practicable, the machinery, equipment or substance does not entail dangers for the safety and health of those using it correctly; ▷

(*b*) make available information concerning the correct installation and use of machinery and equipment and the correct use of substances, and information on hazards of machinery and equipment and dangerous properties of chemical substances and physical and biological agents or products, as well as instructions on how hazards are to be avoided;

(*c*) undertake studies and research or otherwise keep abreast of the scientific and technical knowledge necessary to comply with subparagraphs (*a*) and (*b*) of this Article.

PANEL 26

Occupational Safety and Health Recommendation, 1981 (No. 164)

III. ACTION AT THE NATIONAL LEVEL

4. With a view to giving effect to the policy referred to in Article 4 of the Convention, and taking account of the technical fields of action listed in Paragraph 3 of this Recommendation, the competent authority or authorities in each country should —

(*a*) issue or approve regulations, codes of practice or other suitable provisions on occupational safety and health and the working environment, account being taken of the links existing between safety and health, on the one hand, and hours of work and rest breaks, on the other;

(*b*) from time to time review legislative enactments concerning occupational safety and health and the working environment, and provisions issued or approved in pursuance of clause (*a*) of this Paragraph, in the light of experience and advances in science and technology;

(*c*) undertake or promote studies and research to identify hazards and find means of overcoming them;

(*d*) provide information and advice, in an appropriate manner, to employers and workers and promote or facilitate co-operation between them and their organisations, with a view to eliminating hazards or reducing them as far as practicable; where appropriate, a special training programme for migrant workers in their mother tongue should be provided;

(*e*) provide specific measures to prevent catastrophes, and to co-ordinate and make coherent the actions to be taken at different levels, particularly in industrial zones where undertakings with high potential risks for workers and the surrounding population are situated; . . .

PANEL 27

Occupational Safety and Health Convention, 1981 (No. 155)

Article 9

1. The enforcement of laws and regulations concerning occupational safety and health and the working environment shall be secured by an adequate and appropriate system of inspection.

2. The enforcement system shall provide for adequate penalties for violations of the laws and regulations.

Article 10

Measures shall be taken to provide guidance to employers and workers so as to help them to comply with legal obligations.

The main function of official services entrusted with the enforcement of safety laws and regulations (labour, mine and public health inspectors, social security engineers and physicians, etc.) is an executive one, but they must also give employers and employees practical advice on how to achieve compliance.

In countries that are more advanced in occupational risk control, the inspection services are tending to expand their advisory role, whilst not abandoning their traditional inspection function. This double function may give rise to practical difficulties that have, in some cases, been tackled by creating additional specialised advisory services attached to the social security fund or to technical or professional organisations. Although this solution resolves one difficulty, it may create others since it is not always easy, in practice, to establish an effective relationship between parallel services carrying out their activities in the same area.

Nevertheless, the advisory function of inspection services or specialised bodies is an essential one in achieving safety and health conditions that are above the basic minima required by the law, and in instituting measures to cover cases for which legislative provision has not been made – an important consideration in view of the accelerating pace of technological progress. This emphasis on the inspector's advisory role does not reduce his enforcement function nor the employer's responsibility for compliance.

On the other hand, inspectors who devote too much time to undertakings which frequently call for their advice may neglect less demanding firms with poorer working conditions and where energetic enforcement is the only possible avenue for progress. Employers should not and must not be allowed to transfer their responsibilities to inspectors.

However, inspectors can do much to help undertakings to help themselves. Visiting a wide variety of undertakings is a distinct advantage, since the inspectors can be informed of new hazards and of new measures to deal with existing hazards. The labour inspectorate has a special responsibility with regard to small undertakings which do not have easy access to the advice of a specialist. In the developing countries, the labour inspectorate (or the authority which assumes these functions) may often be the only source of information available to the employer, the workers and their trade union organisations.

Much could be said about the methods used by inspection services in their enforcement activities. In view of their limited resources, inspectors must make empirical decisions as to the undertakings or parts of undertakings which should be visited and the methods of enforcement. By experience, each service has developed its own methods for routine inspections (intended to cover the majority of the undertakings during a given period) and *a posteriori* inspections (following an accident, a complaint, etc.) (see also Chapter 8, section on "Public institutions").[15]

Education and training

These overlapping terms cover all approaches intended to provide the individual with the theoretical and practical knowledge he needs for his integration into the working environment.

All those involved in the production cycle should receive occupational safety and health education and training appropriate to their technical background (see also Chapter 8, section on "Education and training").[16]

Reference has already been made to the role of designers in the safe planning of premises, plant, equipment and processes, the responsibility of the employers, the role of line management, the contribution of the workers and the functions of specialists (safety engineers, industrial physicians, occupational hygienists and labour inspectors) in safety and health. Each of these must also receive adequate training. At present this is far from being the case, even in the most advanced countries.

Employers, plant managers, engineers and supervisors will usually have received a technical, economic, legal or administrative education. Technical education seldom covers occupational safety, and medical education seldom deals with occupational health. Considerable efforts are therefore required to heighten the awareness of those whose collaboration is needed, and these efforts are not always successful. It is extremely difficult to persuade employers or top managers to attend special occupational safety and health courses.

The workers' own vocational training, whether in the plant or at school, often leaves them poorly prepared to deal with the hazards of their trade. When they have learnt to work with defective or badly guarded machines and tools, it would be surprising if they were later to

PANEL 28

Occupational Safety and Health Convention, 1981 (No. 155)

Article 14

Measures shall be taken with a view to promoting, in a manner appropriate to national conditions and practice, the inclusion of questions of occupational safety and health and the working environment at all levels of education and training, including higher technical, medical and professional education, in a manner meeting the training needs of all workers.

be much concerned about safety; if, on leaving school, they are unaware of the importance of good personal hygiene, they are scarcely likely to practise it in the workshop. If they are to be taught to earn their living, would it not be reasonable to teach them how to protect their lives?

Many occupational safety and health "specialists" acquired their training on the job or late in life, since no comprehensive specialised training was available as part of their academic studies.

Convention No. 155 makes the stipulation reproduced in Panel 28. This provision aims at incorporating occupational safety and health concepts – related to the student's needs – into the teaching of all trades and professions with a part to play in occupational safety and health. In general, the individual has great difficulty in modifying acquired habits or abandoning ingrained gestures and reflexes. Schooling or apprenticeship should therefore inculcate safe working methods and behaviour. The difficulties involved are no valid reason for not making the attempt.

In their own training, employers should also be taught how to gain the confidence of their workers and motivate them; this aspect is as important as the technical content of the training.

In occupational safety and health training, the functions of the public authorities and of industry are complementary; the public authorities' task usually finishes when the individual starts work, although adult permanent education is now very much in vogue. Industry (in the widest meaning of the term) seems to have the major responsibility for the safety training of the persons it employs. This is in line with industry's own interests and the fact that industry is in the front line of the battle against occupational hazards. The public authorities have the responsibility of promoting training and of acting as a catalyst by providing the necessary resources and specialised personnel where necessary; such support is essential in developing countries.

The achievements of undertakings that have obtained outstanding results in occupational risk control deserve closer study; they contribute valuable "case histories" and should assist in educating or supplementing the education, as the case may be, of undertakings which are lagging behind. Lessons from the methods used (in particular in the practical organisation of safety and health) will probably have more impact on the management of other undertakings than courses or formal lectures.

Where exemplary achievements are lacking, the public authorities might collaborate with management and workers to select a small number of representative undertakings and provide them with resources (personnel and equipment) with which to establish a reasonably extensive general safety and health programme and to launch specific programmes in priority sectors. This would demonstrate what can be done with relatively modest, but carefully husbanded, resources to correct hazardous situations. Immediate and spectacular results should not be expected, and a final review of the achievements of such an experiment should not be made until at least three to five years have elapsed.

Safety training is a long-term task, and one which is never completely finished. Initial training, even under the best of conditions, cannot cover all foreseeable and unforeseeable situations. Regulations and warning signs will not prevent dangerous behaviour – either because the worker is unaware of them, thinks safety measures valueless or does not understand them; or because the payment system compels him to choose between safety and a reasonable income. The immediate supervisor has the essential and indispensable task of instructing new recruits, keeping them under close supervision and ensuring that they have understood the precautions and their significance.

In conclusion, a word should be said about information and educational films and posters. Posters seem of doubtful value; at most, they can serve as a reminder of certain risks or safety and health measures. Workers quickly become used to them and no longer notice them. The message they carry is necessarily brief, which may limit its scope and comprehensibility.

Films are a useful accompaniment to a lecture and can make a strong impression on an audience. However, they are relatively expensive to produce, and can seldom be used outside the country or branch of industry in which they were originally filmed, owing to the differences in situations, conditions and mentalities. For this reason, preference is given to simpler audio-visual aids (slides and filmstrips) which can be more easily adapted to local conditions and circumstances.

Research

The contribution of research to occupational risk control is obvious and its value is clear. Research by undertakings with the necessary resources and by scientific institutes, professional organisations, social security bodies and the public authorities has supplied the knowledge and experience needed to prevent and effectively control occupational hazards.

Information

The wide range and diversity of occupational safety and health problems are such that no one can claim to have a thorough knowledge of all subjects. Moreover, some problems have become so complex that a multidisciplinary approach may be needed for their solution. On the other hand, simple common sense dictates some basic safety and health measures, such as preventing falls on the level or from a height. Common sense may also be used to prevent air pollution resulting from deficient ventilation systems and maintenance, or to prevent dermatitis caused by the abusive use of solvents for hand cleaning.

Information is a vital need for safety and health workers. The latest scientific and technical findings may be of particular interest in occupational safety and health; but older tried and tested solutions should not be neglected. Symposia and other national or international meetings are often a convenient way of obtaining information and exchanging experience.

The ILO's *Encyclopaedia of occupational health and safety*, to which reference has frequently been made in these pages, is the most extensive source of information available to those working in the field of safety and health. Some 2,500 pages in length, this work gives an overall picture of technological hazards at the end of the twentieth century and the precautions they call for. It is therefore of particular interest both to developed and to developing countries.

Such a mass of information is now available that specialised services have been created to scan, "digest" and disseminate it in a suitable form. These services [17] offer safety and health workers a means of exploiting the resources that modern science and technology have developed for detecting, assessing and controlling occupational hazards. Management should be aware of the importance of easy and regular access to the information they may one day require. The most efficient information services not only provide details of new publications but also respond, on request, to questions and even attempt to anticipate their users' needs.

Personal contacts, especially amongst specialists in a given branch of science or industry, continue to be an effective means of keeping abreast of advancing knowledge; it is obviously important to know where to obtain the right information at the right time.

Official services often disseminate information through their publications and advisory activities, as do also many professional bodies and specialised libraries. The dissemination of information is sometimes backed up by demonstrations or exhibitions of safety devices and film shows.

Any effort to obtain information should be preceded by an inventory of the most reliable and accessible data sources. Where local or national sources are non-existent or inadequate, the competent international institutions may be approached.

Public authorities and specialised institutions in many countries are making outstanding efforts to provide information for workers and their organisations with a view to enabling them to bargain for better occupational safety and health conditions in collective agreements. This emphasis on workers' information should eventually facilitate the task of official services, which may then need to intervene only when employers and employees have not been able to overcome the problems at the plant or industry level. This objective of better-informed workers and workers' representatives is fully compatible with the more extensive rights on occupational safety and health being accorded to them in many countries, including access to the results of measurements of harmful substances or agents at the workplace, and information about and consultation on the introduction of new equipment and processes and on any changes that may have a significant impact on occupational safety and health.

Conclusions

It cannot be stressed too strongly that the control of occupational hazards is a long-term task, and that the closer it gets to the shop-floor level, the greater its effects will be; the further the distance from the undertaking, the more difficult it becomes to influence occupational safety and health conditions. But this in no way reduces the significance of a coherent national policy drawn up in close consultation with the employers and workers concerned and of continuing co-ordination of safety and health efforts at all levels.

Occupational risk control is more a matter of determination than of knowledge. Many undertakings in reputedly dangerous industries operate year after year with very low accident rates. The great majority of occupational accidents and a substantial proportion of occupational health injuries occurring today could have been avoided by the implementation of simple measures that have been common knowledge for many years. These measures have long been applied by undertakings with the best occupational safety and health records, and these are also the undertakings that are well managed and have built up a harmonious worker-management relationship.

If, in a given branch of economic activity, each undertaking were able to achieve and maintain the safety and health record of the "best" undertaking in this regard, the number of occupational injuries would fall to a quarter or a fifth of what it is today. This is in no way a utopian objective: it is perfectly attainable by those determined enough and willing to pay the cost.

Notes

[1] HMSO: *Safety and health at work: Report of the Committee 1970-72* (London, 1972), p. 7. Commonly known as the "Robens Report", from the name of its chairman, Lord Robens.

[2] See also idem: *Managing safety*, Health and Safety Executive Occasional Paper Series, OP3 (London, June 1981).

[3] Comité centrale de coordination: "Les quais et la sécurité", Note technique n° 15, 27 June 1973, in *Cahiers de notes documentaires*, No. 76, Note 921-76-74 (Paris, INRS, 1974).

[4] A. Chavanel: *Circular saws*, Information Sheet No. 7 (Geneva, ILO/International Occupational Safety and Health Information Centre (CIS), 1962).

[5] Under this heading may be included substances which give off oxygen and which, although not flammable, can cause or promote ignition and combustion. They are called oxidising substances.

[6] Explosive substances are also flammable.

[7] W. R. Burri: "Risques d'intoxication et d'explosion dans les ateliers de réparation d'automobiles, dans les garages et dans les tunnels de lavage", in *Cahiers suisses de la sécurité du travail* (Lucerne, CNA), No. 114, Mar. 1974.

[8] *Accidents: How they are caused and how to prevent them* (London, HMSO), No. 78, Jan. 1969, pp. 3-28, and No. 94, Jan. 1976, pp. 19-26.

[9] *Human factors and safety*, Information Sheet No. 15 (Geneva, ILO International Occupational Safety and Health Information Centre (CIS), 1967).

[10] R. Elias: "Une approche médico-biologique de l'étude de la charge de travail", *Cahiers de notes documentaires — Sécurité et hygiène du travail*, No. 91, Note 118-91-78 (Paris, INRS, 1978).

[11] The Occupational Medical Service Recommendation, 1959 (No. 112), does not have a parallel recommendation on safety services. However, the Appendix to Council of Europe Resolution 76 (1) gives comprehensive guide-lines for safety services.

[12] See C. M. Berry: "Occupational hygiene", in ILO: *Encyclopaedia of occupational health and safety*, op. cit., Vol. 2, pp. 1511-1512.

[13] M. Guillemin: "Why is industrial hygiene lagging behind?", in ILO: *Education and Training Policies in Occupational Safety and Health and Ergonomics: International Symposium (Sandefjord, Norway, 16-19 August 1981)* (Geneva, 1983), pp. 59-63.

[14] loc. cit.

[15] For a more detailed treatment, see *Second Congress of the International Association of Labour Inspection* (Geneva, June 1977) and, in particular, Theme A: Choice of priority objectives in the action of the labour inspectorate for the prevention of occupational risks – Definition of the methods for carrying out priority activities (pp. 21-55). (In French)

[16] For a more detailed treatment of these aspects, see ILO: *Education and Training Policies in Occupational Safety and Health and Ergonomics . . .*, op. cit.

[17] In 1959 the ILO founded the International Occupational Safety and Health Information Centre (CIS) whose task it is to collect, select and disseminate information from all countries.

WORKING TIME

3

The importance of working time

Working time is perhaps second only to remuneration as the aspect of conditions of work that has the most direct, most perceptible impact on the day-to-day lives of workers.

The level of hours of work and the way those hours are organised can significantly affect not only the quality of working life but also the quality of life in general. They can influence the health of the worker, safety at work, the degree of strain and fatigue, the level of earnings, the amount of free time available, the extent to which that free time represents usable leisure, the family and social life of the worker – in short, many of the elements that determine a worker's well-being.

For the undertaking, too, working time can be a vital issue. Hours of work can be a factor – in some circumstances, a major factor – determining output, cost of production and ultimately the profitability or viability of the undertaking. Regulations and practices concerning shift work and other ways of organising working time can be crucial elements in the organisation of work, the optimal utilisation of equipment, the level of manning, the continuity of services and hence the efficiency of operation.

For the economy and society as a whole, decisions on working time can have wide-ranging consequences, often going well beyond the immediate interests of a particular undertaking or group of workers. They can have repercussions on the health of the economy, the competitiveness of industry, the levels of employment and unemployment, the need for transport and other facilities and the organisation of public services. Through its legislative and regulatory powers on hours of work and related matters, a government may have an important means of applying its policies and priorities for the economy, the resolution of social problems and the protection of the health, safety and well-being of workers.

Given these multiple dimensions, it is easy to understand why questions relating to working time have long been a preoccupation not only of employers and trade unions but also of governments and indeed of the international community.

Collective bargaining

Union demands for shorter normal hours of work, better conditions for shift workers, higher overtime or night differentials, longer annual holidays and other improvements are often at the forefront of collective bargaining. Such demands may be almost as important as wage claims – and when, for example, economic conditions may preclude real wage increases, this type of improvement in working conditions may have even greater practical importance.

Collective bargaining may be a particularly suitable way of dealing with these questions. It enables the parties directly concerned to fix the priorities and set the pace for action. It allows account to be taken of variations in economic conditions and operational requirements from industry to industry or even from undertaking to undertaking, and the need for gradual rather than rapid change so that the necessary adjustments can be made. It can provide, when successful, a firm foundation of agreement on which legislation can build. In many countries – particularly industrialised market economy countries – collective bargaining has initiated much of the progress in this field, whereas legislation has often served mainly to consolidate it and extend it to other workers.

Legislation

The need for at least some legislation to regulate working time is now universally recognised. This was not always so. In the nineteenth and well into the twentieth century, governmental intervention to limit hours of work was a subject of bitter controversy. Early efforts concentrating on the protection of women and children were sharply contested, and legislation covering adult males was fiercely opposed. For trade unions, the ten-hour day and then the eight-hour day were more than just claims: they were articles of faith. The consecration of these goals through legislative action was a landmark of social progress.

At present, virtually all countries have laws, regulations, wages orders, arbitration awards or comparable texts on hours of work. Some texts simply establish a framework of minimum standards and leave employers and trade unions to determine the specific conditions. Others prescribe detailed rules on all relevant points. The approach adopted depends on the industrial relations system and traditions of the country: the essential point is the recognition of the responsibility of government to set certain limits.

A government may operate other than through direct statutory action. Through its policies, its influence on the social partners and its public sector measures, a government can affect the course of negotiations between employers and trade unions. If it considers a reduction in hours without loss of pay to be inadvisable for the economy at a particular time, it can discourage employers from conceding on the issue. If it favours a reduction as a social objective or as a means of promoting employment, it can create a climate of opinion conducive to agreement on a reduction. A government will not necessarily see its views prevail, but it is bound to make its position felt if the result is likely to have national implications.

International action

In the early debates on governmental regulation of hours of work, it was widely argued that a country which did limit working hours by law would put its industry at a competitive disadvantage against countries which did not. As a response, internationally co-ordinated and harmonised action on this and certain other labour problems became a recurring theme in trade union and other social thought in the years before the First World War and was a fundamental force in the founding of the International Labour Organisation in 1919. The prominence of hours of work can be seen from the Preamble to the ILO Constitution, in which the list of measures urgently required to improve conditions of labour starts with "the regulation of the hours of work, including the establishment of a maximum working day and week".

International labour standards

Appropriately, the first international labour standard adopted by the International Labour Conference was the Hours of Work (Industry) Convention, 1919 (No. 1). While this was a complex instrument containing detailed provisions on various aspects of hours of work, its essential purpose was to establish the eight-hour day and the 48-hour week – still the basic standard in much of the world today. Many other instruments on hours of work have been adopted over the years. The following are some of the most important:

- Hours of Work (Commerce and Offices) Convention, 1930 (No. 30), which extended the eight-hour day/48-hour week standard to non-industrial workers;
- Forty-Hour Week Convention, 1935 (No. 47), which advocated – rather unsuccessfully – a reduction of working hours as a means of fighting unemployment;
- Reduction of Hours of Work Recommendation, 1962 (No. 116), which reaffirmed the 48-hour week as a basic standard and called for progressive reduction towards the social objective of 40 hours.

Additional instruments have dealt with hours of work in particular industries, and numerous international labour standards have been adopted on related matters such as weekly rest, annual holidays with pay and paid educational leave.

Historically, the main aim has been to provide at least minimal protection and then gradually to improve conditions as circumstances permitted. The essential objectives have been quantitative: limiting hours of work and then further reducing them; guaranteeing a certain amount of weekly rest; providing a minimum annual holiday with pay and then increasing its duration. For most of the world, these remain the priority. In many developing countries, large numbers of workers do not in practice receive even minimal protection and most others are lucky to enjoy even basic standards. In many industrialised countries, however, while problems of protection may persist and while quantitative improvements are no less important, more and more attention is being paid to the qualitative aspects of working time: not just how many hours are worked but how those hours are organised; the questioning of established practices such as shift work and night work; the introduction of new patterns such as flexible hours.

This chapter will discuss the main elements contributing to the effective regulation and organisation of working time, and attempt to clarify the essential points: normal hours of work, overtime, excessive hours in poorly regulated sectors, shift work and annual holidays. Certain newer ideas and practices on the arrangement of working time will also be described, although these may be more relevant to industrialised than to developing countries. One fundamental issue – the economic effects of reduced hours of work (e.g. on employment) – will be considered but, given its complexity, only in a brief and indicative way. Throughout the chapter, an effort will be made to define key terms, explain the basic concepts, identify common problems, outline the advantages and disadvantages of possible solutions and suggest questions for further reflection.

Finally, a word of caution: working time is a notoriously controversial subject. Any discussion of hours of work, shift work, night work and similar topics is bound to provoke disagreement. It is extremely difficult to simplify and to generalise and yet still to take into account all the different points of view and to do justice to all the different interests at stake.

Hours of work

The key concept is that of normal hours. A convenient definition of this term is given in Recommendation No. 116:

Normal hours of work shall mean, for the purpose of this Recommendation, the number of hours fixed in each country by or in pursuance of laws or regulations, collective agreements or arbitration awards, or, where not so fixed, the number of hours in excess of which any time worked is remunerated at overtime rates or forms an exception to the recognised rules or custom of the establishment or of the process concerned.[1]

Time worked beyond normal hours constitutes overtime or exceptions. Weekly hours that are temporarily below normal hours (with a corresponding reduction in pay and usually because of adverse business conditions) constitute short-time working or "partial unemployment". Weekly hours that are substantially below normal hours and are worked on a regular and voluntary basis can be considered part-time work.

Policy debates and negotiations on the reduction or rearrangement of working time may focus on such specific problems as overtime, short-time working, part-time work, rest periods, work schedules and shift work. But the reference point is the level of normal hours and the fundamental question is what that level should be.

Normal hours of work

Normal hours can be fixed in terms of the day or of the week: usually a combination of the two is retained. Virtually all countries have laws or regulations setting upper limits to the number of normal hours in both a day and a week. Statutory provisions on maximum normal hours should not be confused with provisions limiting the maximum total hours (normal hours plus overtime) that may be worked in any given period. Provisions of the latter type are found in some countries and are discussed below in the section on overtime. When "legal hours", "maximum hours", "legal maxima" or similar terms are used, it is essential to know whether they refer to normal hours or to some other concept.

Legal limits

The traditional legal limits on normal hours, and still the most widespread in developing countries, are eight in a day and 48 in a week. These are the standards laid down in Convention No. 1 and restated as basic standards in Recommendation No. 116 (see Panel 29). Many countries, both industrialised and developing, have prescribed lower limits. Most have legal maximum normal hours between 40 and 48 a week, but some have adopted a normal working week of 40 hours. This latter standard was first set forth in Convention No. 47 and was subsequently reaffirmed as a social objective in Recommendation No. 116.

Variations in legal standards within a country are common. Some countries promulgate specific laws, regulations, orders or other statutory instruments for particular sectors or industries. These may provide for different levels of normal hours, especially when the fixing of hours is

PANEL 29

Reduction of Hours of Work Recommendation, 1962 (No. 116)

I. GENERAL PRINCIPLES

1. Each Member should formulate and pursue a national policy designed to promote by methods appropriate to national conditions and practice and to conditions in each industry the adoption of the principle of the progressive reduction of normal hours of work in conformity with Paragraph 4.

. . . .

4. Normal hours of work should be progressively reduced, when appropriate, with a view to attaining the social standard indicated in the Preamble of this Recommendation without any reduction in the wages of the workers as at the time hours of work are reduced.

5. Where the duration of the normal working week exceeds forty-eight hours, immediate steps should be taken to bring it down to this level without any reduction in the wages of the workers as at the time hours of work are reduced.

6. Where normal weekly hours of work are either forty-eight or less, measures for the progressive reduction of hours of work in accordance with Paragraph 4 should be worked out and implemented in a manner suited to the particular national circumstances and the conditions in each sector of economic activity.

. . . .

D. *Overtime*

16. All hours worked in excess of the normal hours should be deemed to be overtime, unless they are taken into account in fixing remuneration in accordance with custom.

17. Except for cases of *force majeure*, limits to the total number of hours of overtime which can be worked during a specified period should be determined by the competent authority or body in each country.

18. In arranging overtime, due consideration should be given to the special circumstances of young persons under 18 years of age, of pregnant women and nursing mothers and of handicapped persons.

19. (1) Overtime work should be remunerated at a higher rate or rates than normal hours of work.

. . . .

linked to wage determination. Other countries have labour codes or hours-of-work legislation laying down a general standard, but make special provision for certain industries or occupations. Under one such system, normal hours are also fixed for occupations considered to include substantial periods of mere presence (night-watchmen, hairdressers) on the basis of "equivalences", that is, a higher number of hours deemed equivalent to the general level of normal hours. But both the theory and the actual equivalences are increasingly disputed. In a number of countries, shorter hours are found for certain workers, such as those in arduous or hazardous jobs, those working under severe climatic conditions and, sometimes, young workers and older workers.

In several industrialised countries, particularly in Western Europe and the Pacific regions, there has recently been growing pressure for the general level of normal hours to be statutorily reduced towards a target of 35 a week.

In many countries, large groups of workers enjoy normal hours below the legal limit by virtue of collective agreements or other arrangements. When comparing levels of hours of work in different countries, one should remember that statutory provisions on normal hours in most cases merely prescribe upper limits: the prevailing standard may correspond to those limits or may be lower. The prevailing level of normal hours should again not be confused with actual hours, which take account of overtime, short-time working, and so on. Reductions in normal hours, for example through collective agreements and often industry by industry, have often led ultimately to reduction in the legal limits. In some countries, normal hours of below 40 have been achieved by substantial categories of worker such as public servants, office employees and other non-manual workers; however, these categories traditionally receive little or no compensation for overtime, though that is beginning to change.

Even where the prevailing standard is below the legal limit, the latter may remain important as a point of reference in applying other legal provisions. For example, in some countries, certain overtime regulations apply only when the legal limits on normal hours have been exceeded. To take a hypothetical case, normal hours by law are 44 and, by collective agreement, 40: hours worked between 40 and 44 are remunerated as overtime under the collective agreement but only the hours exceeding 44 are subject to legal overtime constraints such as a need for prior authorisation or a limit on the additional hours permitted. It is thus necessary to take full account of both the legal standard and the standard prevailing in practice in order to appreciate the true situation.

Because of its importance to the regulation of working time in general and its economic implications in particular, the determination of normal hours of work is a crucial issue for governments, employers and trade unions. Usually it is debated in terms of the normal working week – the generally accepted measure of the length of working hours.

Legislation and collective agreements refer mainly to weekly hours, and most statistics are kept on the basis of the week. Demands for, and negotiations on, the reduction of normal hours generally focus on the number of hours per week.

Length of the working day

However, many countries have recently given renewed attention to the length of the working day. Normal daily hours of eight or below are now so well established in the great majority of countries that the only serious policy questions have related either to special circumstances (e.g. shift work or activities with a long daily spreadover, such as hotels and restaurants, or those alternating relatively long rest periods, such as transport) or to the manner in which reductions in weekly hours should be distributed over the working days.

The introduction of new ways of arranging working time has to some extent re-opened the debate on the eight-hour day. Even in the past, it was found necessary to breach the eight-hour limit to permit the five-and-a-half-day week when weekly hours were around 48, and the five-day week when weekly hours exceeded 40. For example, a five-day week of 42 hours would bc feasible only if more than eight hours were worked on at least some of the days. Convention No. 1 took similar possibilities into account by providing that, subject to certain conditions, where the hours on one or more days of the week were below eight, the limit of eight could be exceeded on the other days but not by more than one hour: that is, an upper limit of nine. Thus the qualitative aim of arranging working time in a more convenient, more attractive and perhaps more efficient way – over five days rather than six – already entailed certain quantitative concessions on the eight-hour day.

With the advent of such practices as the compressed working week and flexible hours, the question of limits on the working day must be posed in new terms. These practices will be discussed further towards the end of this chapter, but their implications for the working day are relevant to the fixing of normal hours.

The compressed working week typically means a 40-hour week distributed over four days with each normal working day being of ten hours, not counting any overtime. Is this acceptable? Does the prospect of a three-day weekend justify a regular ten-hour day? Unions have usually said no, arguing that the eight-hour day remains a valid social principle and that a substantially longer day would be detrimental to the worker's health.

Flexible hours schemes

Similar problems arise with flexible hours schemes, which permit workers to start, finish and take meal breaks at times of their own choosing, subject to certain conditions. Under a typical scheme, workers may start any time between 7 a.m. and 9.30 a.m., finish any time between 4.30 p.m. and 7 p.m., and take from 30 minutes to two hours for lunch. They could therefore work on any one day as much as 11.5 hours, and the extra hours would not be considered overtime. Such long days would not be worked regularly and they would be balanced by short days – theoretically, as short as five hours – or by compensatory time off. Flexible hours are therefore not comparable to the compressed week, but they do permit a departure from the eight-hour standard.

Is this an acceptable price to pay for the greater degree of individual discretion afforded by flexible hours? The eight-hour day was for many years a rallying cry for workers and their unions: its emergence was hailed as a great achievement. Should it now be considered outmoded as an absolute standard? These are difficult questions.

Flexible hours have other advantages and disadvantages, but unions have had reservations about them for several reasons – of which respect for the eight-hour day is only one. Nevertheless, flexible working is now widespread in some industrialised countries, and it is therefore necessary to examine the policy questions it raises. Many countries have laws and regulations, often modelled on Convention No. 1, prescribing upper limits to the number of normal hours permitted in a day or even limits to the total number of hours. Flexible hours schemes may entail violating, or at best bending, these regulations. Should the legal provisions be amended? Should the daily limit be set at 11 hours, 12 hours or even higher? Should there be a legal limit on normal daily hours at all? When does flexibility become dangerous and when does protection become interference?

Changes in technology, work content and organisation, labour force composition, economic structure and living conditions may force a re-thinking of well-established patterns and raise questions that pose real dilemmas for employers, workers and governments.

Overtime

The regulation or reduction of normal hours will have little practical effect on the actual level of working hours unless overtime is kept within reasonable limits. Some overtime work is essential to give enterprises the flexibility to deal with unusual pressure of work, fluctuations in demand or other special circumstances. However, frequent and substantial recourse to overtime can virtually negate statutory or negotiated provisions on normal hours, and may lead to levels of actual hours that are prejudicial to workers' health, safety and well-being.

Definitions

The terms in which overtime is defined and regulated may vary and may sometimes lead to confusion.

Reference may be made, for example, to "permanent exceptions", "temporary exceptions" and "periodical exceptions" to normal hours of work. Special rules may apply to intermittent work (i.e. jobs involving long periods of mere presence). A system of equivalences may be established (thus, in specified types of work, a given number of hours higher than the usual level of normal hours is considered equivalent to the latter). Provision may be made for preparatory and complementary work (e.g. maintenance and cleaning, starting-up, change-overs between shifts) and for making up time lost owing to shortages of supplies, bad weather, power failures, and so on. Grounds for exceptions may include abnormal pressure of work, urgent work on machinery or plant, accidents and *force majeure*. Different categories of exceptions may be subject to different conditions as regards whether or not prior authorisation is required, a maximum number of additional hours is prescribed or premium rates of pay are obligatory.

While it is important to be aware of the existence of such varied provisions, the basic problems relating to overtime can be examined without going into so much detail.

For this purpose, overtime can be defined as all hours worked in excess of the normal hours fixed by laws, regulations, collective agreements or other methods. Normal hours, in this context, do not necessarily mean normal weekly hours: hours worked on any day in excess of the usual daily limit are, in some cases, regarded as overtime even if the normal weekly hours are not exceeded.

Overtime usually means not only longer hours but also higher rates of pay, although salaried employees, in particular, have traditionally not been entitled to extra pay (but in some countries there is even a trend towards paying this category of overtime as well). For hourly- or daily-paid workers, premium rates for overtime are a well-established practice. The minimum overtime rate laid down in international standards is one and one-quarter times the regular rate.[2] Rates higher than this are common, often increasing further for work at night or on weekly rest days or holidays.

Reasons for overtime work

Overtime work may be needed for technical or operational reasons or for economic reasons, and, although this distinction is not clear cut, it does help to clarify some of the problems. Technical or operational reasons include preparatory and complementary work, accidents, *force majeure*, making up of lost time, and such periodical or seasonal work as processing perishable goods, taking inventories or preparing balance sheets. Overtime for these reasons is not usually substantial, nor does it pose serious problems.

Overtime for economic reasons is more important. When the specific reason is genuinely abnormal pressure of work, recourse to overtime is authorised by international labour standards and by most national legislation. Few people would deny that overtime must sometimes be worked in order to cope with exceptionally heavy workloads, cover absences, fill output requirements or provide adequate services during busy periods, take advantage of favourable market conditions, and so forth. Unduly rigid regulations which prevent such arrangements might run counter to the interests of both the workers and the undertaking.

Problems arise when overtime becomes the rule rather than the exception. The use of overtime may be considered institutionalised in some countries: the standard working week is explicitly longer than the theoretical normal hours and includes a fixed amount of overtime (e.g. the theoretical normal hours are 40 but the standard hours are 44 with the extra four being remunerated as overtime), and exceptional pressure of work is covered by still further overtime. The normal hours of work are little more than a reference point for wage determination. In other countries, substantial overtime – recognised and paid for as such – forms part of the accepted pattern of work. Opportunities for overtime are highlighted as advantages in job advertisements; a minimum amount of regular overtime may even be guaranteed.

Why is such extensive use made of overtime? How are workers affected? What are some of the measures that may be used to control overtime?

The attraction is partly that of its flexibility, since actual hours can be increased or reduced to fit changing economic conditions or other circumstances which do not justify taking on or laying off staff. However, this is not an explanation for systematic or regular overtime since, when overtime is embedded into working patterns, the element of flexibility becomes less important.

Labour shortages – in particular, of skilled or specialised workers – may be another reason, and systematic overtime may be more convenient than the setting up of adequate programmes to recruit and train suitable staff or to improve productivity by rationalisation. Existing workers also have the advantage of being familiar with the undertaking, its tasks and its procedures.

Additional workers may also mean higher social charges, fringe benefits or other non-wage costs such as insurance or tax payments for each worker; pension fund or other contributions linked to the basic wage; and holiday or other entitlements unrelated to overtime pay. Even premium overtime rates may be less expensive than these charges and than the increased overhead costs entailed by the additional space, equipment and welfare facilities needed for more workers.

The difficulty of discharging or laying off workers may encourage the use of overtime instead of the engagement of additional workers. Employers may be reluctant to increase staffing levels unless they are certain that the increase in consumer demand will be lasting, and such reluctance may be all the stronger if the discharge of workers is subject to tight legal restrictions or expensive severance pay, redundancy payments, compensation awards, and the like.

Advantages and disadvantages

Although the employer has many reasons for favouring extensive overtime, the worker's motivation is solely financial, since substantial overtime at premium rates means substantial additions to wages. Where overtime is virtually institutionalised, such payments form, in effect, part of the regular wage. Where overtime is substantial and recurrent or frequent, the payments represent an important, routinely expected component of earnings, and workers count on them. In either case, a sharp cutback in overtime would mean an appreciable loss of income; therefore, workers – and in particular lower-paid workers – have a material stake in the maintenance of overtime.

Thus the immediate advantage of higher earnings can overshadow the potentially detrimental consequences of extensive overtime for workers.

The gravest consequences can be for workers' health and safety. The original fundamental reason for regulating hours of work was to prevent fatigue and stress which might be deleterious to health and increase accident risks. If actual hours are excessive because of overtime, this objective is defeated. "Excessive" is not, of course, an absolute, readily quantifiable term. A given amount of overtime will have a less serious effect where normal hours are 40 a week than where normal hours are 48. When actual hours regularly and substantially exceed 48 a week, overtime may well be considered excessive.

The reduction of hours of work is also meant to promote the well-being of workers by giving them more leisure; although extensive overtime increases earnings, it does so at the expense of leisure. It may be argued that the choice between income and leisure is made, implicitly or explicitly, by the workers themselves, even if the choice is conditioned by circumstances – in particular, wage levels. However, the social objective behind the reduction of hours is clearly lost, and an apparent reduction in working time becomes, in reality, a disguised wage increase.

While this result may, at least in the short run, suit the workers affected, it carries a potential risk. Overtime earnings are not necessarily stable: in an economic downturn they can be reduced or even eliminated. Workers who had become dependent on them, as if they were truly part of the regular wage, may suffer considerable financial deprivation.

To discuss overtime in terms of a choice between income and leisure is to assume that a choice exists. Although, in many cases, workers may indeed willingly accept overtime work for the sake of the extra pay, in other cases overtime may be imposed upon them. Workers may be obliged to work overtime or risk adverse consequences, ranging from being ill regarded by the employer to losing their employment. There is perhaps less danger where overtime is demanded of a whole plant, or a clearly identifiable group, since the union, works committee, shop steward or labour inspectorate can act as a check. There is greater risk, however, where the work, including overtime, is allocated individually: a refusal would then be an individual decision and could have direct repercussions. To prevent abuses on either side – unreasonable demands or unreasonable refusals – clear rules, whether statutory or negotiated, on permissible levels of compulsory overtime are desirable.

Even if overtime may have some appeal to the individual worker, the interests of the economy or the society as a whole may be different. Is the extensive use of overtime legitimate where there is widespread unemployment? In most developing countries, unemployment and underemployment are heavy and endemic: yet some groups of workers have long, if not excessive, hours. In many industrialised countries with severe unemployment problems owing to economic recession or to structural difficulties, extensive overtime is nevertheless common. Are such practices logical and justifiable?

Whether the reduction and restriction of overtime will contribute to the reduction of unemployment depends on many factors. First, if overtime is deeply entrenched, efforts to reduce or eliminate it may well encounter resistance – not least from the workers. Where, for policy reasons, public authorities and trade unions have sought to reduce overtime, the reaction at enterprise or shop-floor level has not always been positive. Second, even if overtime were reduced, employment would not necessarily increase. In a particular situation, vacancies and jobseekers may not match. Skill requirements, geographical locations, wage rates, other working conditions and the attractiveness of the work itself can determine whether jobs will be filled and candidates placed. Moreover, different firms may react differently to a reduction in hours, whether these be normal hours or overtime. Some may take on additional workers, but others may reorganise and rationalise their processes so as to obviate the need for a larger workforce.

In spite of this, the reduction of overtime may be a useful element in a policy of reducing hours of work with a view to increasing the number of jobs available. Whether such a policy would be effective in the first place is a controversial subject which is discussed elsewhere in this chapter. The effects of measures to reduce overtime will depend on many factors: the level of normal hours, the amount of overtime usually worked, the way overtime is used, the way work is organised, the type of regulatory measures taken, the form of compensation required, and so forth. It is important to consider these factors in the specific context of each country and of each branch of economic activity.

Measures to reduce or limit overtime

Various approaches to the reduction or limitation of overtime have been tried in different countries, with varying degrees of success. No universally applicable formula can be given, and each of the measures outlined below has its advantages and disadvantages.

The traditional and most widely used measure is the imposition of premium pay rates for overtime – called "penalty rates" – designed to deter the employer. The advantage of this measure is that it is simple, flexible, easy to apply and relatively easy to verify and enforce. The disadvantage is that, as indicated above, even very high premium rates may be less costly than taking on additional workers, and the substantial extra earnings may attract workers to accept or even seek overtime. As a matter of equity, premium rates are not only justified but necessary: those who work beyond normal hours should be adequately compensated for the extra time and effort. However, as a deterrent to the use of overtime, premium rates by themselves can be of limited effectiveness only.

A second possible measure (which, although not common, does exist in some countries) is the stipulation of compensatory rest instead of, or in addition to, extra payment. It has the advantage that, when granted shortly after overtime work, it affords an opportunity for rest and recuperation and thus mitigates the negative effects of overtime. The difficulty is that its correct application may not always be feasible: an undertaking may not be able to grant time off soon enough after the overtime for it to be truly an opportunity for recuperation; workers may prefer cash payment in some circumstances; operational problems or cost factors may inhibit undertakings from employing replacements for workers receiving compensatory rest. Because of these difficulties and, in particular, the need to employ replacements, it has been argued that provision for compensatory rest to be granted in the form of whole days off within a prescribed period would be an essential part of a policy of reducing hours of work with a view to increasing employment.

One of the more direct means of limiting overtime is to specify by law or regulation the circumstances under which overtime is permitted. Convention No. 1 and many national legislations embody this approach, but such provisions do pose a dilemma. If the conditions are defined too precisely, the result may be undue rigidity; if they are left too general, the provisions may be ineffectual. Practical application may also be difficult since, where judgement must be exercised, there is room for disagreement between the employer, the union and the labour inspectorate.

Provisions by which the labour inspectorate or similar body must give prior authorisation for overtime may be effective, but they must be applied with care. Cumbersome procedures which make authorisations too difficult to obtain will inhibit the flexibility needed by undertakings and make the regulations a hindrance to efficiency. However, if the procedure becomes a mere formality and blanket authorisations are granted, the provisions will have little practical effect. One approach is to require authorisation only for overtime in excess of a specified number of hours in a specified period. This allows the undertaking a degree of flexibility while minimising unjustified overtime.

Probably the most effective method of limiting overtime is to prescribe the maximum number of overtime hours. Controversial in some countries, this method has been successfully applied in others. The ILO's Committee of Experts on the Application of Conventions and Recommendations has pointed out that "The setting of an absolute maximum number of hours constitutes a simple but effective form of protection against undue recourse to overtime, a practice which in many branches and indeed in many countries leads to the normal or statutory working week being little more than a theoretical standard." [3]

Exactly how many hours should be permitted and what should be the period of reference are questions to which no general answer can be given. National provisions vary widely and can be extremely complex. The permissible number of overtime hours in a particular context (country, industry, occupation) will depend on such factors as the length of the normal working week, the pay and compensatory rest requirements; other legal constraints on overtime; and the general context (i.e. climate). The reference period may be the day (e.g. not more than two hours of overtime on any day), the week, the month, an average over a given period, a number of weeks in a year, or even the whole year. A very short period, such as a day, means strict regulation but minimal flexibility; a very long period, such as a year, means great flexibility but perhaps inadequate protection. A common solution is to combine a maximum number of hours over a relatively short period with a limit over a longer period: for example, so many hours in a week but not exceeding so many in a year, or so many in a week during not more than so many weeks in a year.

The control of overtime may be a key element in the effective regulation, limitation or reduction of hours of work. The need for further control and the choice of appropriate measures are for the social partners and the public authorities in each country to decide. If the policy objective is to reduce or limit hours of work in an area where systematic or extensive overtime is prevalent, stricter regulation may be desirable. If overtime is not a serious problem, existing provisions or even premium rates of pay may suffice. What is essential is the right balance between worker protection and operational flexibility without the interests of society as a whole being lost sight of.

Excessive hours in poorly regulated sectors

Much of the above discussion on normal hours and overtime applies mainly to the better-organised and better-regulated sectors of the economy and presupposes effective laws, regulations, enforcement machinery, trade unions, collective agreements and grievance procedures.

Even then, some problems of policy implementation, supervision and enforcement may arise.

However, the gravest problems are found in activities that escape regulation or in which enforcement is ineffective. Actual hours of work that substantially exceed even the most basic standards remain common in certain sectors or types of undertakings, especially in developing countries.

Long hours are frequently found in construction, transport (especially road transport) and hotels and restaurants. These sectors often have characteristics which make them particularly vulnerable: work concentrated in a limited season to take advantage of weather or market conditions; a long daily spreadover imposed by the work rhythm; a labour force largely composed of temporary or casual workers anxious to earn and unlikely to protest; a predominance of small, owner-operated undertakings with few non-family employees; the difficulty of control, verification and enforcement. Regulations which reflect the special needs of each activity and emphasise record-keeping and inspection are essential if hours of work are to be kept within bounds. Road transport, for example, is the subject of an international labour Convention and Recommendation [4] which focus on such specific points as maximum total driving time, rest breaks, daily spreadover, daily and weekly rest and supervisory measures (including, where appropriate, the installation of tachographs).

Long hours may also be a general problem in small undertakings and in the informal sector. Much of the non-agricultural labour force in developing countries works in small industrial, commercial or service undertakings or in the "informal sector", which includes very small workshops, sales stalls, family or quasi-family establishments and street trades. These workers often have practically no protection against excessive hours. Laws and regulations may exclude from their scope undertakings such as those employing fewer than a specified number of workers or not using power-driven equipment, or engaging only members of the same family. They may also exclude homeworkers and domestic workers. Even where the law is in theory applicable, its enforcement in small undertakings is often ineffectual: understaffed, inadequately trained, ill-equipped labour inspectorates cannot easily cope with large numbers of widely dispersed undertakings with a fluctuating workforce and rudimentary record-keeping and in which the trade union may be very weak or non-existent. Workers may not know how, or may be afraid, to file complaints about excessive hours, non-payment of overtime or other abuses. The economic pressures to work long hours are heavy: employment is precarious, wages are low and earnings would not otherwise cover even minimal living expenses. In the informal sector, where income derives directly from sales or services rather than from regular wages, survival means working as long as production requires or customers are available.

The negative effects of long hours may be compounded by extremes of climate, bad working conditions (e.g. inadequate lighting, ventilation, hygiene, safety precautions), unsatisfactory nutrition, poor general health, lack of public social services, long commuting distances and overburdened transport facilities.

Where undertakings in these sectors are covered by laws and regulations, efforts must be made to improve enforcement by strengthening the labour inspectorate (by more staff, training in both advisory and inspection capacities and improved transport and other equipment).

In many developing countries, labour administrations and related services or institutions may need to develop their data collection systems, expand their statistical services and conduct more in-depth studies and training courses. Both employers and workers may need clear guidance and simple information on legal provisions, methods of implementation, their respective rights and obligations and the means available for obtaining redress.

It may be worth re-examining existing laws and regulations, with a view to extending their coverage to types of undertakings now excluded or exempt. Any such steps must be taken carefully and realistically: comprehensive but unenforceable legislation may be worse than none. Perhaps more important, a simplification of existing provisions may be desirable. In many developing countries, laws and regulations on hours of work were enacted at different times, evolved over the years and were often inherited almost without change from a previous colonial Power; they now form a tangle of provisions which may specify unrealistic general standards or different rules for different types of undertaking, or even different rules for different categories of worker in the same undertaking. They may also impose impracticable requirements for authorisations, exceptions and record-keeping. Both employers and workers may have difficulty in understanding, interpreting and applying the provisions. By harmonising or consolidating different texts, simplifying administrative requirements and streamlining procedures without weakening substantive protection, it may be possible to secure more effective implementation.

However, even the best regulations may prove inapplicable in practice to the informal sector, where most workers are self-employed or work with family members. Legislation on the opening hours of shops and other establishments and on the compulsory observance of weekly rest days may provide some protection, but its effectiveness is likely to be limited.

More tangible results will probably be achieved through practical action to enhance the level and regularity of earnings, provide better community welfare services, make available tools and equipment that are more efficient and less tiring or less hazardous to use, offer opportunities for training and for improving occupational safety and health and other physical working conditions, and encourage the development of co-operatives and other types of organisation.

Agriculture is another major sector which has notoriously long working hours and which, owing to its seasonal nature and other peculiarities, cannot be dealt with in the same terms as other industries or services. Realistic regulations are difficult to frame, and enforcement is hampered by the dispersion of agricultural undertakings, often over enormous areas, by the wide use of temporary or seasonal workers, by unsatisfactory record-keeping, and so on. A further complication is the frequent use of piece-rate systems of remuneration which push workers to seek the maximum possible earnings – especially when their employment may be limited to a few months in the year.

In agriculture, a distinction must be made between wage earners and self-employed or family workers. Smallholdings operated by the farmer and the family members with only occasional hired labour account for much of the agricultural sector, and any attempt to regulate hours of work here would be futile. A more promising indirect approach would be to ease the work processes and thereby lighten the workload, perhaps by improving the design of agricultural tools and equipment so as to lessen the need for long hours of heavy work, reduce strain and fatigue, and increase productivity. A better balance between nutrition and energy expenditure would also be an important contribution. Practical improvements in working conditions in agriculture could certainly be achieved by a multidisciplinary approach – ergonomics, nutrition, physiology and agronomy – to this still inadequately researched problem.

Rest periods and breaks

The complement to work is rest, and the complement to working time is time for rest. For workers' safety, health and well-being, it is essential to arrange hours of work so as to provide adequate periods of rest: short breaks during working hours, longer breaks for meals, daily or nightly rest and weekly rest.

The need for short breaks during working hours is ever more widely recognised and, although not usually prescribed by law, such breaks are more and more the subject of collective agreements. They are particularly important in jobs requiring a fast pace of work or a high degree of vigilance. Where the individual or small groups determine the work pace, informal arrangements may suffice; but where the machine sets the pace or workers are highly interdependent, clear rules on the number, timing and duration of breaks are advisable. Such breaks are usually paid.

Meal breaks are always provided, and are sometimes regulated by law. A 30-minute meal break is usually considered indispensable in a working day of eight hours or more. Collective agreement or practice usually determine whether or not these breaks are paid, but more often they are unpaid.

Traditionally, two different patterns have been followed: with the first, a long break of one-and-a-half to two hours or more allows most workers to go home for a substantial meal and effectively divides the working day into two parts; with the second, there is a break of 30 minutes to one hour, and most workers have their meal at or near the workplace. Which pattern predominates depends on the conditions and customs of the country: however, the latter – often known as the "continuous" or "unbroken" day – is becoming more and more common. Although the change usually results from the difficulty of commuting in urban areas or the employment of more married women outside the home, it can offer some other advantages: savings in overhead costs and lost time for the undertaking, shorter daily spreadover, more usable free time and less commuting strain for workers.

The unbroken working day creates certain needs: the operation of canteens, restaurants and similar facilities at the workplace or, where sufficient outside restaurants are available, the provision of subsidies such as luncheon vouchers. The organisation of other welfare facilities for rest, child care and recreation may also be desirable.

Variations on these basic patterns are sometimes found. For instance, special provision is often made for shiftworkers, particularly those on the night shift. To avoid having too long a shift, and to facilitate the rotation system, the meal break may last only 20-30 minutes – and may be counted as paid working time. Again, during the summer months in very hot climates, the working day starts early in the morning, is interrupted by a short break and finishes in the early afternoon. Sometimes the normal hours are slightly reduced for the period.

In finding an appropriate solution, those concerned should take account of climate, customs and other local conditions, as well as the operational requirements of the undertaking. If a change from a well-established pattern is to be made, all the implications should be studied and the widest possible agreement sought among the workers so as to prevent discord.

The limitation of daily hours of work implies at least a minimum period of daily or nightly rest. This does not, as a rule, pose any problems except possibly in connection with rotating shift work, which will be discussed in the next section.

The guarantee of weekly rest is fundamental to the protection of workers. It was the subject of early international labour standards – in particular, the Weekly Rest (Industry) Convention, 1921 (No. 14) – and is prescribed by laws or regulations almost everywhere. The minimum amount of weekly rest required by Convention No. 14 and by the Weekly Rest (Commerce and Offices) Convention, 1957 (No. 106), is 24 consecutive hours in any seven-day period. Recommendation No. 103, accompanying the latter instrument, calls for at least 36 hours of weekly rest which, wherever practicable, should be an uninterrupted period. The Conventions further provide that, wherever possible, the weekly rest should be granted simultaneously to all persons concerned in each establishment and that it should coincide with the customary rest day in the country or the district. Convention No. 106 also calls for respect of the traditions and customs of religious minorities as far as possible.

Almost all countries legally require a minimum of 24 hours of weekly rest, and this amount is very frequently exceeded in practice. The widespread adoption of the five-day week has made two weekly rest days common.

The main problems arising in connection with weekly rest are of two sorts.

First, in continuous shift work, the weekly rest day cannot always coincide with the customary rest day, e.g. Friday, Saturday or Sunday. More seriously, under certain archaic schemes, even the provision of 24 consecutive hours of weekly rest for some workers means an unacceptably long stretch of work for others. These problems will be discussed in the next section.

Second, and more generally, everything said above about excessive hours of work in poorly regulated sectors also applies to weekly rest. In many developing countries, workers in, for instance, small workshops, stores and the street trades commonly work seven days a week. Legislation prohibiting opening on Fridays, Saturdays, Sundays or other customary rest days may provide some measure of protection. But the problem is largely similar to that of limiting hours of work in such activities.

Shift and night work

Shift work is a widespread and long-established method of organising working time, and it exemplifies a classic dilemma: economic advantages versus social disadvantages. Because it permits the fuller utilisation of productive capacity, it can have definite economic benefits for the undertaking. Yet, because it disrupts normal living patterns, it can pose equally definite social problems for the workers. Which way the balance should tilt and how to profit from the economic possibilities while minimising the adverse effects on workers are difficult questions for governments, employers and trade unions. Some elements to be considered in formulating responses are outlined below. The problem of night work is a closely related, though not identical, one and will also be discussed in this section.

Shift systems

The concept of shift work is a generally familiar one and may be defined as follows: "A method of work organisation under which groups or 'crews' of workers succeed each other at the same work stations to perform the same operations, each crew working a certain schedule or 'shift' so that the undertaking can operate longer than the stipulated weekly hours for any worker."

The main systems are discontinuous, semi-continuous and continuous shift work:

(a) discontinuous shift work: the undertaking operates less than 24 hours a day with a daily break and usually a weekend break; since there are typically two shifts a day, this is often called the "two-shift system";

(b) semi-continuous shift work: the undertaking operates 24 hours a day: that is, without a daily break, but with a break at the weekend; and

(c) continuous shift work: the undertaking operates 24 hours a day, seven days a week: that is, without a daily break or a break at weekends or on public holidays.

Within the framework of these systems, the crews can be assigned to shifts according to the following two basic patterns:

(a) fixed (or permanent) shifts in which each worker belongs to a crew which is permanently assigned to a given shift (this is essentially used in the discontinuous, or two-shift, system with crews being permanently assigned to the day shift or the night shift); and

(b) rotating (or alternating) shifts in which each worker belongs to a crew which alternates between the day and the night shift or rotates between the morning, the afternoon and the night shift (this is used in all three systems).

In the case of rotating or alternating shifts, there are two other important variables:

(a) frequency of rotation: crews may change shifts every week (the most common practice) or at shorter or longer intervals; and

(b) the length of the rotation cycle (that is, the period necessary for a worker to get back to the same point and resume the sequence of days of work and rest over a number of weeks), which will depend, in a continuous shift system, on the frequency of rotation and the number of crews being used.

Familiarity with these terms is essential to any discussion of shift work, and the practices they denote form the basic elements of shift-work arrangements. However, under each system, variations in specific practices are possible.

The two-shift system may, for example, use a day shift and a night shift, well separated from each other, or a double day shift, that is to say, a morning shift followed closely by an afternoon shift. The specific starting and stopping times and the length of the meal break in the course of a shift may vary, and a separate part-time shift may be added in the evening. The shifts may be permanent or may alternate through the use of different rotation cycles in which a crew may change shifts at intervals of a week, fortnight or more.

The semi-continuous system, usually based on three shifts a day, may use a varying number of shifts per week depending on how Saturdays (or equivalent days) are treated. The frequency of rotation (again usually once a week) and the direction of rotation may vary: each crew may move from the morning shift to the afternoon shift to the night shift and back to the morning shift, or vice versa.

Numerous variants of the continuous shift are possible, and the frequency of rotation, length of the rotation cycle, the direction of rotation and the number of crews are the key variables. Their combinations and permutations offer many different schemes, allowing an undertaking to operate over the whole 168 hours in a week with different levels of normal weekly hours of work, different patterns of rest and different numbers of free Sundays (or equivalent days) in a cycle. The continuous system is the most complex and the most difficult to manage. It is also the one most often criticised as having negative effects on workers.

Reasons for shift work

Shift work may be used to meet technical or operational requirements and/or for economic benefits. A decision to introduce or to expand shift work may be based on both considerations, which, though inter-related, may still be regarded as distinct.

Shift work in the modern sense was first introduced for technical reasons. Certain industrial processes had to be carried out continuously, and to shut down at nights or weekends would have been technically unfeasible or prohibitively expensive. The classic example was the continuous firing of the blast furnace. Technical requirements are now given as the justification for shift work, even though continuous operation of the main processes may not be strictly necessary but is strongly desirable for the smooth organisation of production. The concept might be stretched still further to include advanced automation technology (e.g. in the chemical industry), in which operations are programmed in long cycles with relatively small numbers of workers performing what are essentially control and maintenance functions.

Operational requirements are typically invoked where services or facilities are provided to the public round the clock, or at least well into the night. Some of these may be considered matters of necessity: post and telecommunications, public transport, power and water supply, hospitals and ambulances, police and firemen, radio and television. Others are more matters of convenience: cinemas, restaurants, stores, full-service hotels. Shift work in industry tends to receive the most attention, but it should not be forgotten that large numbers of workers are employed on rotating shifts or fixed night shifts in service activities. In some industrialised countries, the increasing unwillingness of workers to accept "unsocial hours" makes the provision of night, weekend and holiday services ever more difficult.

The most important reason for shift work, apart from the maintenance of essential public services, is economic benefit. Shift work is a means of increasing the productive capacity of a plant, since it allows machinery and equipment to be used more intensively, output to be increased, overhead costs per unit of output to be reduced and investment to be amortised more rapidly. This does not mean – and the point needs emphasising – that shift work is always economically advantageous. The economics of shift work – and decisions on whether to use it, and if so which system to use – are complex. Whether shift work will be feasible and profitable in a particular case depends on, for example, the relative costs of labour and capital; the effects on labour costs and productivity; the state of the market; the nature of the production process; the kind of product; the rate of depreciation; or the means and costs of maintenance and repair.

However, under certain circumstances, shift work has such great economic potential that it has spread not only to more industries and undertakings but also to a wider range of activities. A prominent example is electronic data processing, where the high cost of computer systems and the fast rate of depreciation imposed by the speed of technological advances has made it imperative to use the equipment as intensively as possible. One result has been the extension of shift work to categories of office and non-manual workers to whom it was virtually unknown before. The introduction of automated equipment in industry has also stimulated the wider use of shift work for both economic and technical reasons.

In industrialised countries, shift work has often been introduced in response to such factors as the high level of investment imposed by rapid technological progress and the resultant need for more intensive capacity utilisation and faster amortisation.

In developing countries, the key factors have been the high cost of capital as compared with that of labour and the consequent need to ensure the fullest practicable use of equipment. Shift work may also be used to increase the output of products for which the country has a competitive advantage and, most important, to expand employment. Yet there are often considerable obstacles to the extension of shift work, since it poses complex technical and managerial problems which many developing countries have, so far, a limited capacity to handle.

Some of the questions to be considered by managers and others concerned with the introduction or extension of shift work, particularly in developing countries, are set out below.

– What are the relative levels of labour and capital costs?

– What is the expected rate of depreciation and economic obsolescence of the machinery and equipment?
– How will shift work affect overhead and other costs?
– How do the wages of shiftworkers compare with those of other workers?
– How does the productivity of shiftworkers compare with that of other workers?
– Is the production process largely controlled by machines or by the workers?
– Can supervisory functions be split up?
– Is there an adequate and reliable supply of power?
– Is there an adequate and regular flow of raw materials?
– Can maintenance and repair be easily carried out?
– Are there sufficient spare parts?
– Are sufficient numbers of workers, particularly skilled workers, available?
– Will the workers and their unions accept shift work?
– Will the workforce readily adapt to it?
– What are the implications for training?
– Are there any legal obstacles in the composition of the workforce (e.g. restrictions on the night work of women and young persons)?
– Can adequate transport, food services and other facilities be provided?
– What is the present and projected demand for the product?
– Are there adequate storage and distribution facilities to prevent serious bottlenecks?

This (by no means exhaustive) list of questions shows the complexity of evaluating the feasibility and potential profitability of shift work in a specific situation – even before the possible social disadvantages are considered.

Effects of shift work on workers

The disadvantages are of two types: the effects on the workers' health; and the effects on family life, social relationships, trade union work and community activities.

There is substantial evidence that shift work – particularly night shifts and, more particularly, night shifts under rotating shift systems – can have adverse health effects, although the extent and gravity of such effects remain a matter of controversy. The specific physiological problems associated with shift work derive from the disturbance of normal biological rhythms (i.e. the changes in the body's various functions and reactions linked with the alternation of day and night). Manifestations may include digestive disorders, fatigue, irritability, nervous disorders and, most important, sleep disturbances. Evidence on other problems is inconclusive, but there is no doubt that shiftworkers have difficulty in getting adequate, restful sleep. Daytime sleep is affected by noise (especially in urban housing) and other disturbances; it is of shorter duration; and adjustment to changes in daytime and night-time living patterns may interfere with rest. Such difficulties appear more serious in rotating shift work than in systems with permanent night shifts. Many of the other health problems, especially nervous disorders and fatigue, may derive largely from sleep disturbances.

Two further points about the health effects of shift work should be emphasised. First, the effects are not the same for all workers: some workers are seriously affected, some mildly and some slightly. Age, general state of health, capacity to adjust and other individual characteristics influence both the physiological reactions and the attitudes of workers. Second, shift work is only one of a complex series of factors that may be interacting to cause or aggravate health problems. These may be internal to the job (e.g. working hours, physical working conditions, work content and organisation, stress, relationships with other workers and supervisors) or external to it (e.g. housing conditions, transport, family life, place in the community). Any evaluation of the effects of shift work in a given situation should not ignore the possible inter-relationships of such factors. In many developing countries, the effects may be aggravated by general health and environmental problems, although this has been inadequately investigated: severe climate, poor housing and hygiene, polluted water supply, inadequate nutrition, debilitating diseases.

Although the health effects of shift work may be open to debate, the possible adverse effects on family and social life are widely recognised. Shiftworkers often have more difficulty organising household life (arranging meals, performing domestic chores, carrying out errands) and maintaining normal relationships with their spouse, parents and children. Night work and especially weekend and holiday work can make family activities, recreation and simply being together a practical problem. Changes in the shifts worked under rotating systems add to the difficulties. When both husband and wife work, the disruption may be severe.

The interference with social and community life may also be serious. Contacts with friends are harder to maintain; participation in clubs, in associations, in sporting, cultural, educational or recreational groups and in religious, civic and trade union activities is at best irregular; public entertainment, such as concerts, theatres, cinemas and popular television shows, must often be forgone.

All these factors affect the quality of life. Like health problems, they will vary in nature and importance with the individual and the circumstances. The family situation, for example, is usually a crucial element in determining a worker's attitude both towards shift work in general and towards particular schemes. Workers' social habits and recreational preferences will influence their reaction: those who want to participate in clubs, team sports or other group activities or who like watching television and other forms of entertainment at fixed times will be more disturbed than those who pursue individual hobbies or unstructured activities.

The outside environment may also play a significant role. Apart from the material conditions of life and the availability of public services and facilities, the social and cultural patterns of the community must be considered. In many countries (especially in the developing world), where family ties are strong, religious and other traditional observances are important and community obligations are strict, shift-work practices (or indeed other employment practices) that conflict with local customs will have damaging effects on the undertaking or on the workers, or on both.

Finally, the effects of shift work are not limited to the undertakings and workers directly concerned: the implications for the community and the public authorities must also be considered. The need for transport, utilities and other public services may be affected; when these must be provided beyond the usual working hours, the workers running them will also have to do shift work.

Improving the conditions of shiftworkers

Given the variety and complexity of the problems that shift work may create, what measures can be taken to alleviate its ill effects and improve the conditions of shiftworkers? Action in two areas is needed: first, improving the specific shift-work arrangements themselves; second, improving the other conditions of work and life of the workers concerned.

Improving shift-work arrangements

Each of the three basic shift systems and its variations has advantages and drawbacks. Much of the research shows that the shift-work schemes used in specific undertakings are often not the result of careful evaluation of the factors and consideration of alternatives but derive mainly from habit, from insufficient knowledge of the possibilities and from inability to appreciate all the relevant factors. On the workers' side, preferences are often expressed for the system that is most familiar: change and the necessity of adapting to it are anticipated with misgiving. Yet the development and application of shift-work schemes with the most favourable combination of advantages and disadvantages in a particular context offers considerable scope both for increasing operating efficiency and for reducing the adverse effects on workers.

It is not possible here to examine the various practices in detail. To generalise, however, the two-shift system is the simplest to operate, allows the greatest flexibility to management and entails the least inconvenience for workers; the continuous system permits the highest output and the fullest utilisation of capacity (which may or may not mean highest profitability and optimal utilisation), provides the most jobs but is the most difficult to manage and the most disadvantageous for workers; the semi-continuous system falls somewhere in between.

Under the two-shift system, the basic choice is between fixed shifts and alternating shifts. Fixed shifts virtually eliminate the problem of adjustment, but they do mean that one crew of workers will always be on the afternoon or the night shift. Some workers may prefer that shift. Giving workers the choice of shift, wherever possible, would practically solve the problem. Since this is not always feasible, the next best solution is to give workers with a certain number of years of service the option to change their shift. To the extent that fixed shifts are acceptable to the workers in terms of their family and social needs, they seem to have real advantages from the point of view of health.

Where rotating shifts are used, particularly in the semi-continuous and continuous systems, a key problem is the frequency of shift changes. Is it better to have, at one extreme, a rapid rotation with only two or three days spent on each shift or, at the other extreme, very infrequent rotation with periods of two weeks or longer on each shift? No clear answer can be given because the question is a source of great disagreement. Traditionally, long periods are claimed to facilitate physiological adjustment and permit greater regularity and order in family and social life. However, opinion now seems to favour rapid rotation because it reduces the period spent on night shifts and may reduce the need for adjustment. The most widely used practice lies between the two approaches: a weekly rotation. Being the most common, however, does not mean that it is the best: it is probably retained because it is familiar and because no conclusive evidence has been adduced in favour of either alternative. This is an area in which there is a clear need for more intensive research.

The direction of rotation (morning shift – afternoon shift – night shift – morning shift again, or afternoon – morning – night – afternoon again) is another variable. It is of little significance under the semi-continuous system, but many workers find a change from night shift to afternoon shift easier than that from night shift to morning shift. Under the continuous system, the direction of rotation may affect the length of rest periods and the extent to which they coincide with weekends. The question should be carefully examined in connection with the specific rotation schedule envisaged.

The number of crews is also an important question in the continuous system. When this system first came into use, only three crews were used – one for each daily shift. Since the plant operated the full 168 hours a week, the average working week for each crew worker was 56 hours, and continuous shift work was often a major exception to the legal limit on normal weekly hours of 48. Crews rotated shifts once a week. The only way to provide a rest period of 24 consecutive hours for two of the crews at the change-over would be for the third crew to work a double shift – 16 consecutive hours. This system has become manifestly out of date but is still occasionally found in some developing countries. High priority should be given to eliminating continuous shift work with only three crews where it does exist. It clearly imposes excessive weekly and daily hours of work which are prejudicial to health and safety and of doubtful value in terms of productivity.

The use of a fourth crew, now widespread, has permitted the maintenance of more reasonable average weekly hours of work, the granting of longer rest periods, the abolition of double shifts and the development of many different rotation schedules. In some industrialised countries, proposals have been made or implemented for a fifth crew: this would permit a sharp reduction in average weekly hours. However, since the four-crew system is prevalent, a few examples of rotation schedules for continuous shift work with four crews are given in tables 1 to 9.

It will be noticed here that the three consecutive rest days that occur once in the cycle for each crew coincide with a weekend; there are two free Sundays every eight weeks.

A variation on the system just described is one that rotates every two days, which gives an eight-week cycle. The rest periods are all of 48 hours and also include two Sundays during the cycle. It should be noted, however, that if a backward shift rotation is introduced (each crew moves from the afternoon shift to the morning shift, and so on), a rest period of 56 hours is obtained when the crews change over from the night shift to the earlier shift.

Table 1. Four-crew continuous shift-work system with a four-week cycle of rotation

Assignment	First week MTWTFSS	Second week MTWTFSS	Third week MTWTFSS	Fourth week MTWTFSS
Morning shift	aaaaaaa	bbbbbbb	ccccccc	ddddddd
Afternoon shift	ccddddd	ddaaaaa	aabbbbb	bbccccc
Night shift	bbbbccc	ccccddd	ddddaaa	aaaabbb
Day off	ddccbbb	aaddccc	bbaaddd	ccbbaaa

Crews *a*, *b*, *c* and *d* change over every seven days in the four-week cycle, two rotations being followed by rest periods of 48 hours each, while the third rotation is followed by a rest period of 72 hours, including Sunday. The average length of the working week is then 42 hours.

Source. M. Maurice: *Shift work* (Geneva, ILO, 1975).

Table 2. Four-crew continuous shift-work system with a 20-week cycle of rotation

Assignment	First week MTWTFSS	Second week MTWTFSS	Third week MTWTFSS	Fourth week MTWTFSS
Morning shift	aaaaabb	bbbcccc	cddddda	aaaabbb
Afternoon shift	ccddddd	aaaaabb	bbbcccc	cddddda
Night shift	bbbcccc	cddddda	aaaabbb	bbccccc
Day off	ddcbbaa	dccbbad	dccbaad	dcbbaad

Here again, the average length of the working week is 42 hours. Crews *a, b, c* and *d* change over every five days. The rest periods following the changes are of 48 or 72 hours. Each crew has five free Sundays over the total cycle of rotation.

Source. M. Maurice: *Shift work* (Geneva, ILO, 1975).

Table 3. Four-crew continuous shift-work system with irregular frequency of rotation

Assignment	First week MTWTFSS	Second week MTWTFSS	Third week MTWTFSS	Fourth week MTWTFSS
Morning shift	aaddccb	bbaaddc	ccbbaad	ddccbba
Afternoon shift	bbaaddc	ccbbaad	ddccbba	aaddccb
Night shift	ccbbaad	ddccbba	aaddccb	bbaaddc
Day off	ddccbba	aaddccb	bbaaddc	ccbbaad

The systems described so far have constant frequencies of rotation, every seven, six, five or four days, but others have varying frequencies, as in the case of what is sometimes called the "Continental" system. Under that system, crews *a, b, c* and *d* change over sometimes after two days and sometimes after three, the rest periods being also of either two or three days.

Source. M. Maurice: *Shift work* (Geneva, ILO, 1975).

Table 4. Mixed continuous shift-work systems

Assignment	First week MTWTFSS	Second week MTWTFSS	Third week MTWTFSS	Fourth week MTWTFSS
Morning shift	aaaaaaa	ggggggg	fffffff	ddddddd
	dddbbbb	bbbeeee	eeecccc	cccaaaa
Afternoon shift	ecccccc	caaaaaa	aggggggg	gffffff
	ffffddd	ddddbbb	bbbbeee	eeeeccc
Night shift	bbeeeee	eeccccc	ccaaaaa	aaggggg
	ggggff	fffffdd	dddddbb	bbbbbee
Day off	cebdfgg	acebdff	bacebdd	fgacebb
				etc.

Among the possible variations on the systems described above, the system with seven half-crews is of particular interest. This system is so called because each shift (morning, afternoon and night) is worked by two half-crews, each of which has its own rotation schedule. As a result there are six half-crews working in every period of 24 hours, while one half-crew is resting; hence the reference to seven half-crews. The cycle is seven weeks in length, divided up into 42 working days and seven rest days.

In the above example, the seven half-crews are designated by the letters *a, b, c, d, e, f* and *g*.

Source. M. Maurice: *Shift work* (Geneva, ILO, 1975).

Table 5. Four-crew continuous shift-work system with rotation every two days (average working week: 42 hours)

Day of the week	Rest	4 a.m.-12 noon	12 noon-8 p.m.	8 p.m.-4 a.m.	Rest	4 a.m.-12 noon	12 noon-8 p.m.	8 p.m.-4 a.m.
Monday	c	a	b	d	b	d	c	a
Tuesday	c	a	b	d	b	d	c	a
Wednesday	d	c	a	b	a	b	d	c
Thursday	d	c	a	b	a	b	d	c
Friday	b	d	c	a	c	a	b	d
Saturday	b	d	c	a	c	a	b	d
Sunday	a	b	d	c	d	c	a	b
Monday	a	b	d	c	d	c	a	b
Tuesday	c	a	b	d	b	d	c	a
Wednesday	c	a	b	d	b	d	c	a
Thursday	d	c	a	b	a	b	d	c
Friday	d	c	a	b	a	b	d	c
Saturday	b	d	c	a	c	a	b	d
Sunday	b	d	c	a	c	a	b	d
Monday	a	b	d	c	d	c	a	b
Tuesday	a	b	d	c	d	c	a	b
Wednesday	c	a	b	d	b	d	c	a
Thursday	c	a	b	d	b	d	c	a
Friday	d	c	a	b	a	b	d	c
Saturday	d	c	a	b	a	b	d	c
Sunday	b	d	c	a	c	a	b	d
Monday	b	d	c	a	c	a	b	d
Tuesday	a	b	d	c	d	c	a	b
Wednesday	a	b	d	c	d	c	a	b
Thursday	c	a	b	d	b	d	c	a
Friday	c	a	b	d	b	d	c	a
Saturday	d	c	a	b	a	b	d	c
Sunday	d	c	a	b	a	b	d	c

The length of the complete cycle is eight weeks; it takes eight days for each crew (*a*, *b*, *c* or *d*) to come back to the same shift, and eight weeks to come back to the same shift on the same day of the week. The order of the shifts and the frequency of change-over are as follows: two morning shifts, two afternoon shifts, two night shifts and two rest days. This routine gives two free Sundays every eight weeks, and these Sundays are accompanied, in turn, by a Saturday and a Monday.

Source. M. Maurice: *Shift work* (Geneva, ILO, 1975).

Table 6. Four-crew continuous shift-work system with rotation every three days and then every two days (average working week: 42 hours)

Day of the week	Rest	4 a.m.-12 noon	12 noon-8 p.m.	8 p.m.-4 a.m.
Monday	b	a	d	c
Tuesday	b	a	d	c
Wednesday	c	b	a	d
Thursday	c	b	a	d
Friday	d	c	b	a
Saturday	d	c	b	a
Sunday	d	c	b	a
Monday	a	d	c	b
Tuesday	a	d	c	b
Wednesday	b	a	d	c
Thursday	b	a	d	c
Friday	c	b	a	d
Saturday	c	b	a	d
Sunday	c	b	a	d
Monday	d	c	b	a
Tuesday	d	c	b	a
Wednesday	a	d	c	b
Thursday	a	d	c	b
Friday	b	a	d	c
Saturday	b	a	d	c
Sunday	b	a	d	c
Monday	c	b	a	d
Tuesday	c	b	a	d
Wednesday	d	c	b	a
Thursday	d	c	b	a
Friday	a	d	c	b
Saturday	a	d	c	b
Sunday	a	d	c	b

The length of the complete cycle is 28 days; it gives each crew (*a*, *b*, *c* and *d*) one free Sunday every four weeks. Shifts are rotated after two days, again after another two days, and then after three days. Then three days of rest (72 hours) complete the cycle.

Source. M. Maurice: *Shift work* (Geneva, ILO, 1975).

Table 7. Four-crew continuous shift-work system with rotation every four days (average working week: 47 hours)

Day of the week	Rest	4.30 a.m.-12.30 p.m.	12.30 p.m.-8.30 p.m.	8.30 p.m.-4.30 a.m.	Rest	4.30 a.m.-12.30 p.m.	12.30 p.m.-8.30 p.m.	8.30 p.m.-4.30 p.m.
Monday	d	a	c	b	b	a	d	c
Tuesday	c	a	d	b	a	b	d	c
Wednesday	b	a	d	c	d	b	a	c
Thursday	b	a	d	c	c	b	a	d
Friday	a	b	d	c	c	b	a	d
Saturday	d	b	a	c	b	c	a	d
Sunday	c	b	a	d	a	c	b	d
Monday	c	b	a	d	d	c	b	a
Tuesday	b	c	a	d	d	c	b	a
Wednesday	a	c	b	d	c	d	b	a
Thursday	d	c	b	a	b	d	c	a
Friday	d	c	b	a	a	d	c	b
Saturday	c	d	b	a	a	d	c	b
Sunday	b	d	c	a	d	a	c	b
Monday	a	d	c	b	c	a	d	b
Tuesday	a	d	c	b	b	a	d	c
Wednesday	d	a	c	b	b	a	d	c
Thursday	c	a	d	b	a	b	d	c
Friday	b	a	d	c	d	b	a	c
Saturday	b	a	d	c	c	b	a	d
Sunday	a	b	d	c	c	b	a	d

Monday	d	b	a	c	b	c	a	d
Tuesday	c	b	a	d	a	c	b	d
Wednesday	c	b	a	d	d	c	b	a
Thursday	b	c	a	d	d	c	b	a
Friday	a	c	b	d	c	d	b	a
Saturday	d	c	b	a	b	d	c	a
Sunday	d	c	b	a	a	d	c	b
Monday	c	d	b	a	a	d	c	b
Tuesday	b	d	c	a	d	a	c	b
Wednesday	a	d	c	b	c	a	d	b
Thursday	a	d	c	b	b	a	d	c
Friday	d	a	c	b	b	a	d	c
Saturday	c	a	d	b	a	b	d	c
Sunday	b	a	d	c	d	b	a	c

The length of the complete cycle is 16 weeks; it gives each crew (*a*, *b*, *c* and *d*) four free Sundays per cycle, the last being preceded by a free Saturday. The order of the shifts is as follows: four morning shifts, one rest day, four afternoon shifts, one rest day, four night shifts, two rest days.

Source. M. Maurice: *Shift work* (Geneva, ILO, 1975).

Table 8. Four-crew continuous shift-work system (average working week: 40 hours)

Order of weeks	Days of the week							Number of shifts worked
	M	T	W	T	F	S	S	
1	–	M	M	A	A	N	N	6
2	–	–	–	M	M	A	A	4
3	N	N	–	–	–	M	M	4
4	A	A	N	N	–	–	–	4
5	M	M	A	A	N	N	–	6
6	–	–	M	M	A	A	N	5
7	N	–	–	–	M	M	A	4
8	A	N	N	–	–	–	M	4
9	M	A	A	N	N	–	–	5
10	–	M	M	A	A	N	N	6
11	N	–	–	M	M	A	A	5
12	A	N	N	–	–	M	M	5
13	M	A	A	N	N	–	–	5
14	–	M	M	A	A	N	N	6
15	N	–	–	M	M	A	A	5
16	A	N	N	–	–	M	M	5
17	M	A	A	N	N	–	–	5
18	–	M	M	A	A	N	N	6
19	N	–	–	M	M	A	A	5
20	A	N	N	–	–	M	M	5
21	M	A	A	N	N	–	–	
								Total 105
								Average 5

This system, which has the advantage of giving a 40-hour week for each crew, has a cycle of 21 weeks. The cycle given in the table is for worker no. 1 starting from week no. 1.

M = morning shift (6 a.m.-2 p.m.)
A = afternoon shift (2 p.m.-10 p.m.)
N = night shift (10 p.m.-6 a.m.)

Source. M. Maurice: *Shift work* (Geneva, ILO, 1975).

Table 9. Mixed four-crew continuous shift-work system (three rotating shifts and one fixed shift)

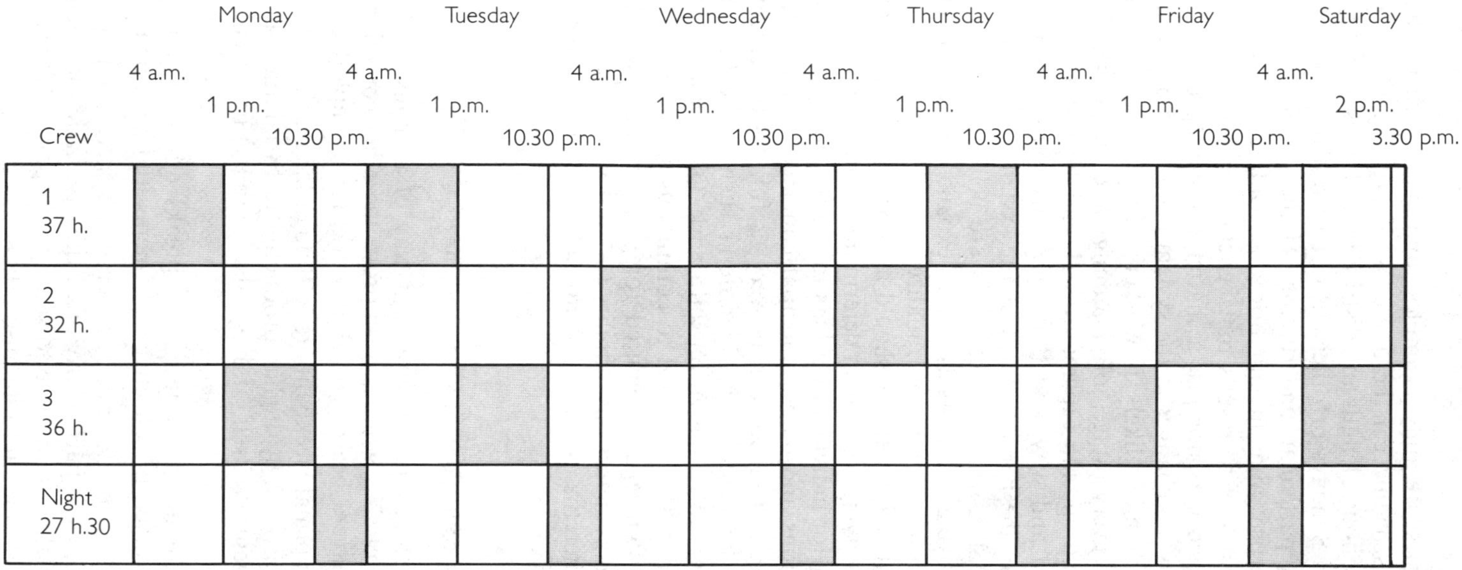

Crew	Monday			Tuesday			Wednesday			Thursday			Friday			Saturday	
	4 a.m.	1 p.m.	10.30 p.m.	4 a.m.	1 p.m.	10.30 p.m.	4 a.m.	1 p.m.	10.30 p.m.	4 a.m.	1 p.m.	10.30 p.m.	4 a.m.	1 p.m.	10.30 p.m.	4 a.m.	2 p.m. – 3.30 p.m.
1 37 h.	■			■				■			■						
2 32 h.							■			■				■			■
3 36 h.		■			■								■			■	
Night 27 h.30			■			■			■			■			■		

Rotation period: three weeks

Weekly working time for rotating crews: 32-37 hours, i.e., an average of 35 hours per week for each period of three weeks: night crews work 27.5 hours.

This system was adopted in the Annecy plant of Gillette-France, allowing a preliminary study by the medical and personnel staff with worker involvement. Its main interest is that workers are able to choose their shift (only volunteers work the fixed night shifts), and that working time for night-workers is reduced.

These tables are only examples, and many other schedules are possible. The schedule can be a means of arriving at the pattern of average weekly hours, length and distribution of weekly rest periods and number of free Sundays or other customary rest days considered most suitable for a particular undertaking. Higher levels of management training, expertise and imagination – leading to well-considered decisions – can lead to results that benefit everybody.

But shift work is not just a matter for management. When it is introduced or when changes in the system being used are contemplated, success is much more likely if workers and their unions are closely associated at every stage of planning and implementation. Full information and specific explanations should be provided on what is being proposed and why, and thorough consultations should be held. Where possible, the main features of the system should be the subject of agreement. Shift-work arrangements are being dealt with more and more in collective agreements as unions expand their efforts to improve working conditions and the quality of working life. Individual preferences and characteristics should also be taken into account as far as possible. Involving workers in determining shift schedules and leaving some flexibility for individual workers to change shifts or crews can help to relieve tensions and defuse potential conflicts. Trial periods when a change is being introduced can be very useful. To the extent feasible, efforts should be made to apply criteria such as health and age in selecting workers to be placed on shift work. Older workers and young workers should not be assigned to continuous shift work (a minimum age of 18 is often legally required for shift or night work). Medical examinations may detect unsuitability for shift work and form the basis for advice on diet, the use of stimulants or tranquillisers and other health matters.

Improving conditions of work and life

A second and equally important way of minimising the possible adverse effects of shift work is to improve the other conditions of work and life that have a bearing on shift work, for example by:

- reducing normal weekly hours: although far-reaching, this measure may be justified by higher productivity. In addition to being an improvement in itself, it may also permit further shift-schedule refinements;
- fixing adequate meals and other breaks during a shift: these are widely specified in collective agreements and are often paid;
- organising adequate transport services: where public services are inadequate, as is particularly often the case in developing countries, the undertaking may have to supply transport so as also to reduce commuting time and minimise fatigue;

- ensuring medical attention: although medical or nursing services should preferably be open throughout working hours, the requirement is adequate first-aid equipment and emergency services with trained personnel;
- providing canteens or other facilities for hot meals, snacks and hot drinks;
- providing other welfare facilities at the undertaking: these may include places to rest and relax during breaks, recreational equipment, employment of social workers, and so on;
- giving information and advice to workers: advising workers on eating habits, rest requirements, conditions most conducive to sleep, prevention of common medical disorders, and so on, may help to reduce shift-work problems;
- improving housing conditions: better housing, including sound-proof rooms, is normally a matter for the public authorities rather than the undertaking and forms part of a wider problem;
- scheduling leisure activities to meet the needs of shiftworkers: this is again desirable but difficult in practice; and
- paying shift allowances and bonuses: although a common practice, this is not a substitute for real improvements in shift arrangements and working conditions.

Should shift work be encouraged or discouraged?

In many industrialised countries, shift work has been cited as an example of the way in which work can be ill adapted to human needs, and proposals have been made to circumscribe its use and prevent its extension for reasons other than technical necessity. On the other hand, it has been claimed that shift work can bring clear economic benefits and, in turn, social benefits – not least by expanding employment opportunities. In many developing countries, shift work is looked upon more and more as a possible means of raising output and productivity, making fuller use of scarce capital resources and creating more jobs.

There can be no single answer to the policy question of whether shift work should be promoted or restricted, and perhaps the question itself is not valid. Situations do arise in which conflicting aims are both or all desirable and a choice has to be made. The decision must depend on specific conditions and the priorities they indicate. Opinions about those priorities may differ, especially when material interests diverge. Few would deny the drawbacks of shift work: the dispute is over their gravity and their relative importance as compared with the benefits. Efforts have been made to find optimal solutions representing the best blend of economic and social elements, and they are worth pursuing, difficult though the task may be. Further efforts should also be made to improve the conditions of shiftworkers through the means mentioned earlier. In many developing countries, the extension and efficient use of shift work may depend on the resolution of managerial, technical and social problems.

Night work

Many shift-work problems derive specifically from night work which, although unavoidable in some activities and even liked by some workers, is, as a general rule, to the workers' disadvantage, as explained above. Whether it is intrinsically harmful to health is a controversial question; however, some authorities contend that the potential health risk is serious enough for non-essential night work to be discouraged and even prohibited. Others maintain that the drawbacks relate more to convenience than to health and are not severe enough to warrant measures that would prejudice not only the potential economic benefits to the undertaking but also employment opportunities.

The controversy has immediate relevance because of one feature of the legislation of many countries: the prohibition of night work for women in industry. Enshrined in international labour Conventions [5] and long embedded in national legislation, this prohibition has recently been attacked as an example of an ostensibly protective measure that, in practice, has a discriminatory effect. It limits women's employment opportunities and excludes women from industries where shift work is common and from certain better-paid jobs.

Two questions are asked by the opponents of this measure. First, why is night work banned for women but not for men? If it is intrinsically harmful, it should be prohibited for everyone; if not, women no less than men should be free to accept or reject it. There is no evidence that women are biologically more vulnerable. Second, why is the night work of women prohibited in industry and not in other sectors? Many forms of industrial work are no different in character from certain non-industrial activities which impose worse conditions. Again, if night work is intrinsically harmful, it should be generally prohibited; if not, industry should not be singled out.

Two main counter-arguments are advanced by those in favour of retaining the prohibition. First, they maintain that night work *is* harmful and should be restricted to unavoidable cases. Even if a wider prohibition or limitation is not immediately feasible, that is no reason to lift existing restrictions and give less protection instead of more. Second, removing the prohibition would lead to serious abuses. Undertakings which now work ordinary hours or use double day shifts because they depend on women for a large part of their workforce might introduce night shifts, and even continuous shift work, leading to a deterioration of working conditions for all. Many women at present on day work would be coerced or even obliged to accept night work. They would have no real choice, since the alternative might be unemployment.

In spite of this, the trend seems to be towards removing the prohibition, and certain countries have rescinded their laws on night work for women, thus rejecting the international labour Conventions they had previously ratified. For these countries, the overriding consideration was to eliminate discrimination against women in employment and occupation.

At the international level, there is no consensus whether the ILO standards should be modified and, if so, in what way. It has frequently been suggested that night work in industry should no longer be prohibited for women: instead, night work in all sectors should be more strictly regulated for both sexes. Agreement on this still seems remote, but the question is becoming more and more urgent.

Holidays and leave

Annual holidays with pay

Paid annual holidays for workers in general did not become widespread until after the Second World War. Once the privilege of limited categories of workers – such as managerial staff, public servants and office employees – they are now accepted as an essential feature of conditions of work. In the great majority of countries, entitlement to annual holidays with pay is provided for by laws or regulations that are often supplemented by collective agreements, etc. However, large groups of workers (e.g. in agriculture and in the informal sector of industry, commerce and services), especially in developing countries, do not benefit from such provisions, either because they are formally excluded from coverage or because enforcement is inadequate. Extending annual holiday entitlement to these workers is a major policy issue for the future.

Virtually all workers in the more organised sectors receive annual holidays with pay, and over the past three decades substantial improvements have been made in many countries. The coverage of legislative and other provisions has been widened, the length of the holiday has been increased and the conditions governing entitlements have been liberalised, which has been reflected in, and also stimulated by, advances in international labour Conventions and Recommendations.

International labour standards on holidays and leave

The earliest standards were laid down in the Holidays with Pay Convention, 1936 (No. 52), which provided for an annual holiday with pay of at least six working days (usually constituting one working week) after one year of continuous service; a holiday of at least 12 working days after one year of continuous service for workers under 16 years of age; and increases with length of service under conditions to be prescribed by national laws or regulations. The Holidays with Pay Recommendation, 1954 (No. 98), laid down somewhat higher standards, including a minimum holiday of two working weeks.

Convention No. 52 was revised in 1970 and the new instrument – the Holidays with Pay Convention (Revised), 1970 (No. 132) – prescribes a holiday of at least three working weeks for one year of service and provides for improvements in various other respects. The provisions of Convention No. 132 indicate the main principles for establishing and administering holiday entitlements (see Panel 30).

Length of annual holidays

In any consideration of the length of holidays, a distinction should be made between the basic holiday entitlement and any entitlement to longer holidays under specified conditions.

The basic holiday entitlement is the amount of holiday granted to workers in general after one year of service. This may be the minimum prescribed by national legislation or it may be a higher amount provided for by collective agreements or other means; however, it is still the amount granted after one year of service. It should not be confused with longer holidays granted, for example, to workers below a certain age or who have served a certain number of years in the undertaking. A basic holiday of less than two weeks is now the rule in only a few countries. Many countries have prescribed three weeks – the minimum standard laid down in Convention No. 132 – and, in recent years, some have moved to four or even five weeks. Often, the legal minimum has first been exceeded under collective agreements in particular industries or branches; the new level has then spread to other industries and, ultimately, it has been consolidated and generalised by legislation. Non-manual workers sometimes enjoy, by tradition or convention, longer holidays than manual workers, but the tendency is towards equalisation.

Beyond the basic entitlement, it is a common practice to grant longer holidays to workers fulfilling specified conditions, such as:

- workers below a certain age or ages (ranging from 15 to 21 but usually 18);
- workers above a certain age or ages;
- workers doing arduous or hazardous work;
- workers working under severe climatic conditions;
- workers doing unusually long hours of work or shift work;
- workers with a high output; and
- workers with a significant length of service.

Length of service is probably the most widespread and the most important criterion and is found both in legislation and, very often, in collective agreements. The pattern of increases with length of service

PANEL 30

Holidays with Pay Convention (Revised), 1970 (No. 132)

Article 2

1. This Convention applies to all employed persons, with the exception of seafarers.

. . . .

Article 3

1. Every person to whom this Convention applies shall be entitled to an annual paid holiday of a specified minimum length.

. . . .

3. The holiday shall in no case be less than three working weeks for one year of service.

. . . .

Article 4

1. A person whose length of service in any year is less than that required for the full entitlement prescribed in the preceding Article shall be entitled in respect of that year to a holiday with pay proportionate to his length of service during that year.

. . . .

Article 7

1. Every person taking the holiday envisaged in this Convention shall receive in respect of the full period of that holiday at least his normal or average remuneration (including the cash equivalent of any part of that remuneration which is paid in kind and which is not a permanent benefit continuing whether or not the person concerned is on holiday), calculated in a manner to be determined by the competent authority or through the appropriate machinery in each country.

2. The amounts due in pursuance of paragraph 1 of this Article shall be paid to the person concerned in advance of the holiday, unless otherwise provided in an agreement applicable to him and the employer.

. . . .

varies. At one extreme, the increases may start fairly early – after two or three years – and may continue progressively over ten, 20 or even 30 years until a stated maximum is reached. At the other extreme, a few supplementary days of holiday may be given after 20 or more years of service. The former seems more appropriate where the basic holiday is relatively low. In some countries, workers with a basic holiday of two weeks or less can eventually gain entitlement to four weeks or more.

The need for longer holidays under conditions such as those indicated above is perhaps less where the basic entitlement is relatively high. It is a matter of judgement on the part of the public authorities and of employers, workers and their organisations. This is an area of considerable potential for negotiations and imaginative solutions.

Holiday pay

The next most important question after length of holidays is the determination of holiday pay. Disputes may easily arise over the remuneration that workers should receive for the duration of the holiday, especially in work where earnings fluctuate or where the basic wage forms only part of the remuneration. To what extent should earnings resulting from overtime be reflected in holiday pay? How should piece-rate or other incentive earnings be counted? Which bonuses or allowances should be included? What about payments in kind?

Convention No. 132 (Article 7, paragraph 1) lays down the principle that a worker should receive "at least his normal or average remuneration (including the cash equivalent of any part of that remuneration which is paid in kind and which is not a permanent benefit continuing whether or not the person concerned is on holiday), calculated in a manner to be determined by the competent authority or through the appropriate machinery in each country". While this principle is clear, its practical application may be a source of uncertainty. To minimise grounds for dispute, national laws, regulations, collective agreements or other texts should specify how remuneration is to be determined for the purposes of holiday pay. Overtime, for instance, might be treated differently, depending on whether it is occasional and represents an insignificant part of earnings or whether it is frequent and represents a regular and substantial element in the pay packet. A cost-of-living allowance which forms a major part of total remuneration might be counted towards holiday pay; various minor allowances might not. Several countries take an average of actual earnings over a specified period preceding the holiday – 13 weeks, six months, even a year. But in this case account might have to be taken of major fluctuations in earnings or interruptions of work, perhaps by excluding from the calculation of the average any time not worked or any periods of short-time working and the like.

Obviously, no single formula can meet all circumstances. What is important is that the relevant provisions should be as clear and as explicit as possible and that they should respect the principle of relating holiday pay as closely as possible to normal remuneration.

Other problems of definition and regulation

Whether annual holidays are governed by laws and regulations or by collective agreements or by other methods, a number of additional difficulties may arise in administering them. Some of the most important are outlined below. The less ambiguous the provisions, the less acute these difficulties are likely to be.

1. The terms in which the entitlement is defined should be precise. Simply stating the number of "days" or "weeks" may lead to confusion. Does "day" mean "working day" or "calendar day"? Is a "week" the same regardless of whether it has five, five-and-a-half or six days? Probably the term that is most precise and easiest to handle (for instance, in calculating partial entitlements or compensation) is "working day". However, equality should be ensured among groups of workers with working weeks of different lengths.

2. For reasons both of fairness and of clarity, it is also essential to specify how public holidays are to be treated in relation to annual holidays. Can a public holiday falling outside the annual holiday be set against the latter entitlement? What if a public holiday occurs while the worker is on annual holiday? Should a country which has 15 public holidays a year act differently from one with seven or eight? It must be remembered that annual holidays and public holidays do not have the same purpose or the same effect on the worker. However, it should also be recognised that both represent time off from work, and that beyond a certain point there may be a need for correlating, one way or another, paid public holidays and paid annual holidays. Where that point lies is a matter for judgement, but Convention No. 132 lays down a minimum standard: public holidays, whether or not they fall during the annual holiday, must not be counted as part of the minimum annual holiday prescribed in the Convention, namely three working weeks.

3. Similarly, periods of absence from work owing to illness or injury – whether they occur during or outside the holiday – should not, in principle, be counted against the annual holiday. Yet the application of this precept may be subject to conditions determined by national laws or regulations or other methods. This is again a point on which clarity is essential.

4. An important issue in determining holiday entitlement is the qualifying period of service required before any entitlement arises. Traditionally, one full year of continuous service was required before the worker became entitled to a holiday at all, but there has been a strong trend towards easing this requirement. Many countries now provide for a shorter qualifying period or for the accrual of holiday entitlement on a proportional basis. This does not mean that the holiday can actually be taken without any delay: it means that the entitlement starts accumulating and that the employer must grant the appropriate amount

of holiday at a suitable date or, in the case of termination of employment, pay appropriate compensation. Convention No. 132 reflects this trend by stipulating that the qualifying period must not exceed six months.

The calculation of the qualifying period may itself pose problems: it is important to specify how length of service should be determined for this purpose and, in particular, under what conditions absences for such reasons as illness, injury or maternity are to be counted towards the period of service.

5. Entitlement to an annual holiday with pay is meaningless without measures to ensure that, under normal conditions, the holiday is actually granted and taken.

In determining the specific dates of the holiday, account should be taken both of the needs of the undertaking and of those of the worker. The final decision normally rests with the employer, but it should be taken only after consultation with the workers or their representatives. Some countries have regulations under which at least part of the holiday must be granted during a specified period of the year, that is, during the usual holiday months in the country concerned. Often, the whole undertaking closes down during the holiday period. However, there is growing realisation that such annual close-downs may be detrimental to the undertaking and to the economy in general, and may not always meet the interests or preferences of the workers. In countries where annual holidays are traditionally concentrated over one or two months, attempts have been made to encourage staggering of the holidays with the dual purpose of avoiding unnecessary interruptions in production and of relieving strain on transport and holiday facilities. These have not, in general, proved very successful because of the understandable desire of people to benefit from good weather and to go on holiday with their family (in particular their children) and friends. A programme of staggering holidays must take these factors into account. A few countries also have provisions in legislation or collective agreements that are designed to give workers incentives, usually in the form of extra days of holiday, to take part of their holiday in the off-season.

Apart from the question of dates, various other problems can arise in connection with the actual granting and taking of the holiday. It is important for laws, regulations, collective agreements or other texts to be clear on this matter.

The purpose of the holiday is to give the worker a reasonable amount of time away from work to permit rest, recreation and recuperation from accumulated physical and mental strain. An annual holiday is not just a pleasant facility but an important measure in protecting the worker's health and well-being. Its object would be defeated if the holiday were broken up into scattered days off or if it were postponed from year to year or, worst of all, if it were given up altogether in exchange for payment. Provisions designed to guard against such practices – even if they occur with the consent, or the apparent consent, of the worker – form an essential part of holiday regulations. The provisions can, and should, be flexible – on condition that a reasonable degree of protection is ensured.

There is nothing wrong, for example, with taking a year's holiday entitlement in two or more instalments: to do so when the entitlement is three to four weeks or longer may even be in the interests of all concerned. However, at least one of the instalments should be substantial enough to represent a significant, usable period of leisure; and Convention No. 132 calls for at least two uninterrupted working weeks.

Similarly, when the entitlement is relatively long, it may well suit both the worker and the undertaking to postpone part of the holiday to the following year and perhaps even beyond. But at least part of the holiday – and in particular the major, uninterrupted instalment – should be granted and taken every year, and reasonable limits should be placed on how much can be postponed and until what date.

The most serious risk is that the worker may, under pressure or through temptation, give up his holiday entitlement in exchange for cash compensation. If this were to happen, the whole purpose of the holiday would be negated. Measures to prevent such occurrences are indispensable. Article 12 of Convention No. 132 contains a strong provision: "Agreements to relinquish the right to the minimum annual holiday with pay prescribed in Article 3, paragraph 3, of this Convention [i.e. three working weeks] or to forgo such a holiday, for compensation or otherwise, shall, as appropriate to national conditions, be null and void or be prohibited".

Compensation in lieu of a holiday is, however, appropriate and indeed necessary in the event that the worker's employment is terminated before he has used his entitlement. That entitlement has been earned over his period of service. If his employment is terminated, he should be given an opportunity to use it or should be compensated. In some countries, measures have been taken to permit the transfer of accrued holiday "credit" from one employer to another: for instance, national or industry holiday funds to which employers contribute and which are then responsible for administering holiday pay. Such schemes are especially useful in industries (such as construction) in which workers frequently move from job to job.

The scope for further improvements in holiday entitlements through legislation and collective agreements, in both industrialised and developing countries, seems considerable. The extent and form of such improvements will depend on the undertaking, the economy as a whole and the priorities of the employers, workers and governments concerned. An increase in holidays with pay is a reduction in working time: it must be evaluated as such. Other improvements in holiday entitlements will also have economic and operational implications but, where a reduction in working time is an issue in negotiations, it may be easier to agree on improved holiday entitlements – either for their own sake or as a step in the progressive reduction of hours of work – than on an immediate reduction in normal hours of work. For the worker, the benefit may be more tangible or more attractive in that a longer holiday means more usable free time than, for example, a relatively small reduction in the working day. For the employer, improved holiday schemes may create fewer operational difficulties than shorter hours and may offer personnel management advantages. Increases with length of service, provision for extended vacation or sabbatical leave, linkage with retirement plans . . . the possibilities are many, and an imaginative, flexible approach could be beneficial to all concerned.

Public holidays

All countries observe certain days as public, statutory or legal holidays. These usually have some religious, historical or cultural significance. The specific dates are fixed by law or regulation, often supplemented by collective agreements. The number of public holidays on which workers must be paid is laid down in a similar manner and may vary widely from country to country, ranging from about five to about 18.

An increase in the number of paid public holidays, though a relatively minor issue in itself, is a possibility to be kept in mind in negotiations on the reduction of working time. An agreement may, for example, add to the number of days already prescribed or give an extra day off before or after certain of the holidays so as to allow a longer break. Many variations can be devised.

Other points for regulation or negotiation may include:

- overtime rates for those who must work on a paid public holiday;
- days off in lieu of public holidays falling on Sundays or other weekly rest days:
- treatment of unauthorised absences on days preceding or following public holidays (i.e. when absenteeism may rise); and
- articulation with annual holidays with pay (see the subsection "Other problems of definition and regulation", above).

Other forms of leave

Although annual and public holidays are the main forms of leave, many others may be encountered, such as casual leave, compassionate leave, and so forth. Such practices are often related to the culture and way of life in the country concerned. Where, for example, there are strong religious traditions, where family responsibilities weigh heavily or where workers have economic or social obligations in their home villages, approved time off from work may be extremely important to the workers and may be the only alternative to uncontrolled absenteeism or excessive labour turnover.

Paid educational leave

A recent but potentially significant development is paid leave for education or training. This is not to be confused with annual holidays intended for rest and recreation, for paid educational leave allows workers to improve their knowledge and skills or to acquire new ones, whether directly related to jobs or connected with civic or trade union activities. Paid leave for training may also help to promote adjustment to structural change and to combat unemployment.

Public servants, teachers, and similar occupations have long been allowed paid leave for further education or training, but only recently have efforts been made to extend this opportunity systematically to workers in general. These efforts were encouraged by the adoption of the Paid Educational Leave Convention, 1974 (No. 140) (see Panel 31), and Recommendation, 1974 (No. 148). It should be noted that, in both instruments, paid educational leave is defined as "leave granted to a worker for educational purposes for a specified period during working hours, with adequate financial entitlements". Three main purposes of educational leave are indicated:

- training at any level;
- general, social and civic education; and
- trade union education.

The Convention is intended to encourage member States of the ILO to formulate and apply a policy on paid educational leave. Such a policy may be developed by stages and through methods that are appropriate to national conditions and practice. The Convention stresses the importance of the involvement of public authorities, employers' and workers' organisations, and education and training institutions, in the formulation and application of policies to promote paid educational leave.

Depending on the country, provision for granting paid educational leave may form part of national laws and regulations, collective agreements, arbitration awards or informal arrangements.

Paid educational leave can take many forms and include practices variously described as educational leave, study leave, examination leave, day-release, block-release, extended leave, sabbatical leave, and so on. What is important is that the leave should be for educational purposes, attract payment fully or in part and take place for a specified period during working hours. Education and training under such arrangements might take place in company, group or government training centres, perhaps located near worksites; in educational institutions, training establishments or conference centres remote from places of work; or in

PANEL 31

Paid Educational Leave Convention, 1974 (No. 140)

Article 1

In this Convention, the term "paid educational leave" means leave granted to a worker for educational purposes for a specified period during working hours, with adequate financial entitlements.

Article 2

Each Member shall formulate and apply a policy designed to promote, by methods appropriate to national conditions and practice and by stages as necessary, the granting of paid educational leave for the purpose of —

(*a*) training at any level;
(*b*) general, social and civic education;
(*c*) trade union education.

Article 3

That policy shall be designed to contribute, on differing terms as necessary —

(*a*) to the acquisition, improvement and adaptation of occupational and functional skills, and the promotion of employment and job security in conditions of scientific and technological development and economic and structural change;
(*b*) to the competent and active participation of workers and their representatives in the life of the undertaking and of the community;
(*c*) to the human, social and cultural advancement of workers; and
(*d*) generally, to the promotion of appropriate continuing education and training, helping workers to adjust to contemporary requirements.

. . . .

Article 9

As necessary, special provisions concerning paid educational leave shall be established —

(*a*) where particular categories of workers, such as workers in small undertakings, rural or other workers residing in isolated areas, shiftworkers or workers with family responsibilities, find it difficult to fit into general arrangements;
(*b*) where particular categories of undertakings, such as small or seasonal undertakings, find it difficult to fit into general arrangements, it being understood that workers in these undertakings would not be excluded from the benefit of paid educational leave.

. . . . ▷

Article 11

A period of paid educational leave shall be assimilated to a period of effective service for the purpose of establishing claims to social benefits and other rights deriving from the employment relation, as provided for by national laws or regulations, collective agreements, arbitration awards or such other means as may be consistent with national practice.

. . . .

temporary centres, perhaps with the use of mobile training equipment.

The Convention further specifies special assistance for workers – such as those in small undertakings or in rural areas, or shiftworkers – who have difficulty in taking advantage of general arrangements for paid educational leave.

Paid educational leave policy is closely related to employment, education, training, hours of work and other policies and may contribute significantly to a policy of recurrent or lifelong education.

The continuing development of paid educational leave should assist in improving the quality of working life, in enhancing the worker's contribution to the undertaking and the community, and in increasing efficiency and productivity as a whole.

Reduction of working time

Should working time be reduced?

That is the most basic – and the most contentious – question in this whole area. Whatever the existing level of working time – whether normal hours are 48 or 40, whether annual holidays are two weeks or four weeks – it is a question that sooner or later will be asked. Trade unions may give it high priority or low priority but, in one form or another, most of them will raise it some day.

When it is raised, the objective may be social but much of the debate will inevitably be conducted in economic terms. Both advocates and opponents will be obliged to respond to questions such as the following.

What will be the effects of a reduction of working time on the undertaking or the industry concerned? Will costs be raised, the competitive position weakened and viability threatened? Will there be an increase in productivity and a more efficient utilisation of resources? What will be the effect on the economy as a whole? Will there be an inflationary push, damaging to the country's economic health? Will a reduction of working time be an effective tool in fighting unemployment?

In a period of economic crisis marked by persistent high levels of unemployment, this last question has come to the forefront of social and economic debate in a number of industrialised countries. Trade unions, especially in Western Europe, have been vigorously arguing for a reduction of working time as a means of creating jobs. Employers have just as vigorously opposed this idea as embodying a fallacy. Some governments have responded positively; others quite sceptically; still others with cautious reserve. The issue is controversial since great interests are at stake and sharp differences of opinion exist. Although these questions have long been debated, nothing resembling a consensus has emerged. Not even the methodology for analysing the economic effects of a reduction of working time is beyond discussion.

There is no clear-cut answer, nor even an easy approach, to the problem, and in this section we can merely outline some of the points for consideration and some of the questions that should be asked. A more in-depth picture can be obtained from a recent ILO study which discusses the possible economic effects of a reduction of working time in industrialised countries.[6]

What is meant by reduction of working time?

The key factors here are whether the reduction entails a corresponding, or partly corresponding, reduction in wages or whether it is made without loss of pay. Although the two concepts are completely different, they are commonly the subject of confusion, and this leads to confusion of the whole debate.

Usually, "reduction of working time" or "reduction of hours of work" refers to a reduction without a corresponding reduction in wages. This is the sense in which the provisions of the Reduction of Hours of Work Recommendation, 1962 (No. 116), are framed and the sense in which the term is normally used. To avoid any misunderstanding, a compulsory reduction in working time with a reduction in pay should be described as "short-time working" or "partial unemployment", and voluntary working of hours substantially below normal hours with correspondingly lower pay should be termed "part-time work".

"Short-time working" may be contemplated as a means of preventing lay-offs, that is, partial unemployment for all the workers in an enterprise instead of total unemployment for some. Arrangements of this sort are often termed "work-sharing" (as is also reduction without loss of pay aimed at redistributing a fixed amount of work). They may be complemented by social security payments so that the actual loss of income is minimal. Such arrangements have given mixed results and can perhaps best be regarded as temporary expedients useful in riding out short-term difficulties.

The actual reduction of working time without a reduction in wages may be achieved in different ways, the most common of which are the reduction of normal hours, the reduction or stricter limitation of overtime and an increase in holidays with pay. To these may be added the lowering of the retirement age and possibly the raising of the school-leaving age or minimum age for admission to employment. Reductions for special purposes may also be considered, for example, paid educational leave.

Effects of reducing working time

The proponents of the reduction of working time as a means of fighting unemployment do not contend, as a common caricature of their views would suggest, that the work hours cut would be directly reconstituted into new jobs which would be filled by unemployed workers. Rather, they argue that certain forms of reductions in working time will, in a period of crisis, hold down job losses and stimulate growth, thus leading also to job creation. Long hours, and in particular extensive overtime, enable employers to reduce the workforce more drastically than they could otherwise. They thus save on certain non-wage costs (e.g. social charges, fringe benefits) and establish lower manning levels at the price – for the workers – of heavier workloads and greater intensity of work. The effective limitation of overtime should be given special emphasis to counter such practices. Where overtime is systematically used and hence where overtime earnings are virtually built into the usual wage, a revision of the wage structure may be necessary. Another measure strongly urged is the introduction of an additional crew in shift work (normally a fifth crew in continuous shift work): this would not only have a direct effect on employment but could also improve productivity. In recognition of the importance of productivity, it is further argued that greater concern for employment would lessen the fear of technological change; this would facilitate the introduction of advanced technology and more efficient methods, and result in higher productivity. A key point is that the reduction in working time is intended as part of a policy of economic growth and employment expansion, not as a substitute for such a policy.

The opponents of the idea stress that a reduction without loss of pay will increase unit costs, reduce competitiveness and profitability, and hence threaten rather than promote employment. Increased productivity, profitability and investment are necessary to increase employment; shorter hours, like higher pay, must be paid for by higher productivity. Moreover, the hours cut cannot simply be made into new jobs and the workers themselves are not interchangeable: manning patterns are determined by job content, workforce structure and geographical mobility and, especially, the availability of skilled workers. There may still be shortages of skilled workers during periods of high unemployment: a reduction in hours would aggravate the effects of these shortages and reduce the work available to semi-skilled or unskilled workers "downstream"; the result would then be more, not less, unemployment. If the reduction is in normal hours, the practical consequence might merely be increased overtime paid at premium rates – in other words, a disguised wage increase meaning higher production costs. An effective reduction in hours might mean reduced productive capacity leading to further job loss.

These arguments have been presented in over-simplified form, but two general observations can still be made. First, when considering specific industries or undertakings, a detailed case-by-case analysis must be made. Circumstances differ too much and there are too many variables to permit general postulates on the effects at the level of the industry or the undertaking. Second, when considering the effects on the economy as a whole, attention should be paid to possible wider consequences and to economic and social policy measures to integrate such a reduction into a coherent national policy.

In assessing the possible effects at industry or enterprise level, it may be necessary to ask:

- What form will the reduction take and what will be the operational consequences?
- What is the current competitive position domestically and internationally?
- How great is the dependence on exports or the competition with imports?
- What share of total production costs is accounted for by labour?
- How much latitude is there in setting prices?
- How important are competitive factors other than prices?
- Will existing regulations on overtime, shift work and night work be changed if working time is reduced?
- Will current patterns of overtime, shift work and night work be changed?
- What is the structure of the labour force and how much skilled labour is available?
- Will productive capacity be used more fully, at the same level or less fully?
- Will there be any gains in productivity that may be attributed to the reduction as such?
- Will a reduction in working time accelerate rationalisation, mechanisation or automation and result in higher productivity with the same number of workers, more workers or fewer workers?

When all these factors have been considered, there is still another variable: different enterprises may react differently to the challenge and may find different solutions depending on the quality and style of the management.

The effects on the economy as a whole are difficult to predict, and even the forecasting methodology is insufficiently developed. Numerous factors are at play; many are hard to forecast, and the inter-relationships are complex and sometimes hidden. Account must be taken of other policies relating to employment, entry into work, training, retraining, retirement and social security and of other aspects of individual and social behaviour, such as the utilisation of leisure and the effects of increased leisure on consumption, savings and demand for new goods and services. The area for study is vast and exploration has only just begun. Amidst all this complexity, three general points should be kept in mind when a possible reduction in working time is examined.

First, the problem is essentially one of interests and attitudes. Positions may be supported by more or less solid arguments, but to think that an objective, scientific solution will be found is illusory. That is why collective bargaining and other comparable methods, including discussions between the government and the social partners, are particularly appropriate for dealing with this question.

Second, one of the main themes of Recommendation No. 116 remains fundamental: reductions in working time are best carried out progressively. There are many ways of applying this idea but, as a rule, a progressive approach is more prudent, allows more time for evaluation and gives more room for adjustment than abrupt action.

Third, the reduction of working time is a social objective that is valid in itself. Whether or not one regards it as an appropriate tool against unemployment, it must also be debated on its own merits as a means of improving the well-being of workers and, indirectly, of their families. The degree to which it is needed, the extent to which it is practicable, the priority it should be given and the methods by which it should be applied will all vary and will all be open to argument.

New patterns in working time

The pattern of work and leisure discussed so far has for many years been common to practically all wage earners and salaried employees outside agriculture. It comprises full-time work for some eight hours a day and five or six days a week; set times for starting and finishing work and for meal breaks; and the placing of all workers or groups of workers on the same schedule, as fixed by the management. Working hours are determined in terms of the workforce, not the individual worker; for the individual, work schedules are a daily constraint.

However, notable departures from this pattern have recently taken place, especially in certain industrialised countries, with efforts being made to develop new patterns that are better adapted to modern conditions and that are more responsive to individual needs.[7] The reasons are varied: greater interest in the qualitative aspects of working life; technological change affecting the content and organisation of jobs; a better-educated workforce; practical problems of traffic, public transport and other services in urban or suburban areas; and the need to attract workers, in particular women who have left the workforce and for whom the traditional pattern would be impracticable. The practices that have resulted are also varied: the compressed working week, staggered hours, *à la carte* hours, flexible hours, part-time work, and so on. Some of these are already widely applied, whereas others remain largely theoretical. The more important ones are discussed below.

Compressed working week

The compressed working week is a system in which the same weekly number of hours are compressed into a smaller number of days. It does not entail a reduction in weekly hours (though it may be introduced together with such a reduction). The most common example is a 40-hour week compressed into four days instead of the conventional five, with a longer working day of ten hours instead of eight. Other typical working weeks are 36 hours over four days of nine hours; 36 hours over three days of 12 hours; 37.5 hours over three days of 12.5 hours; and so forth. Where such schedules are in operation, they often form part of shift-work arrangements.

The ostensible advantages for the workers include a weekend of three days or more; fewer trips between home and work, and hence more usable leisure (not least, wider possibilities for education and training); and, for those on shift work, longer periods of adjustment after night shifts. The undertaking may gain from reduced heating and lighting costs, less time lost in starting up and finishing, easier maintenance and repairs, simpler organisation of shift work and (though this is very uncertain) less absenteeism.

Yet the problems entailed by the compressed week probably often outweigh the advantages. Trade unions have tended to oppose it since a regular working day of ten hours or more can be detrimental to the worker's health and well-being. The extent will depend on the type of work; but there will be a real likelihood of fatigue and strain which, it has been found, may lead to diminished quality, a greater risk of accidents and reduced output; trends towards a reduction in absenteeism tend subsequently to be reversed.[8] In addition to the increasing health and safety risk, the long working day is often psychologically and socially oppressing and damages the workplace atmosphere. Moreover, the long weekend entails less free time and greater fatigue on working days, and this means less time for family and friends — a very high price to pay.

In view of these disadvantages, the compressed week has not spread very widely and is unlikely to do so as long as the normal working week remains close to 40 hours. However, should normal hours of 35 or less become common, the question will undoubtedly be re-examined.

Flexible hours

Flexible hours are perhaps one of the most radical and important of recent innovations in working conditions. They are a means of partly adapting work schedules to meet individual needs and preferences and of making them less constraining. Such schemes (which are also known as flexitime, variable hours, discretionary hours, and so forth) were first introduced in the Federal Republic of Germany and Switzerland in the late 1960s and early 1970s; they have since become widespread there and have been adopted, to varying degrees, in other industrialised countries.

The following is suggested as an adequate all-purpose definition: flexible hours schemes are those which give workers considerable latitude, within stated limits, in fixing their own work schedules from day to day.

These last four words are important because there are other approaches worthy of consideration under which workers have a measure of discretion but not on a day-to-day basis – for instance, staggered hours and *à la carte* hours.

With staggered hours, different units or departments within an undertaking may have different starting and finishing times which, if they are set with the agreement of the workers concerned, do take into account collective preferences, although they allow no individual discretion.

With *à la carte* hours, each worker chooses a schedule from several alternatives proposed by the employer, or chooses starting and finishing times within stated limits, and must then keep to that schedule or to those times. In a variant of this approach, the workers may change schedules at intervals of, say, one month. This system does provide for some individual discretion but does not have the flexibility of allowing the worker to vary his hours from day to day. It is in effect a compromise between a fixed schedule and true flexibility.

Where flexible hours do not seem feasible, such solutions may well be a means of easing the rigidity and impersonality of traditional schedules.

True flexible hours schemes take many forms. Not all can be described here; however, they usually have the following type of structure in common: a flexible period in the morning, during which workers may arrive at any time; a "core" period in the morning, when all workers must be present; a minimum and maximum period for the midday break; another "core" period in the afternoon; and another flexible period in the evening, when workers may leave at any time. Workers must maintain, over a specified period (a week or a month), an average number of hours making up the normal working week. Under some schemes, the hours worked are considered to be in "credit" or in "debit" against the normal hours, and the "credit" or "debit" up to a specified limit may be carried over to the next period. Normally, "credit" hours would be compensated by "debit" hours, but some schemes allow them to be taken in the form of days or half-days of leave.

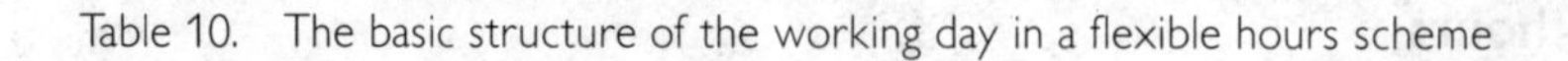

Table 10. The basic structure of the working day in a flexible hours scheme

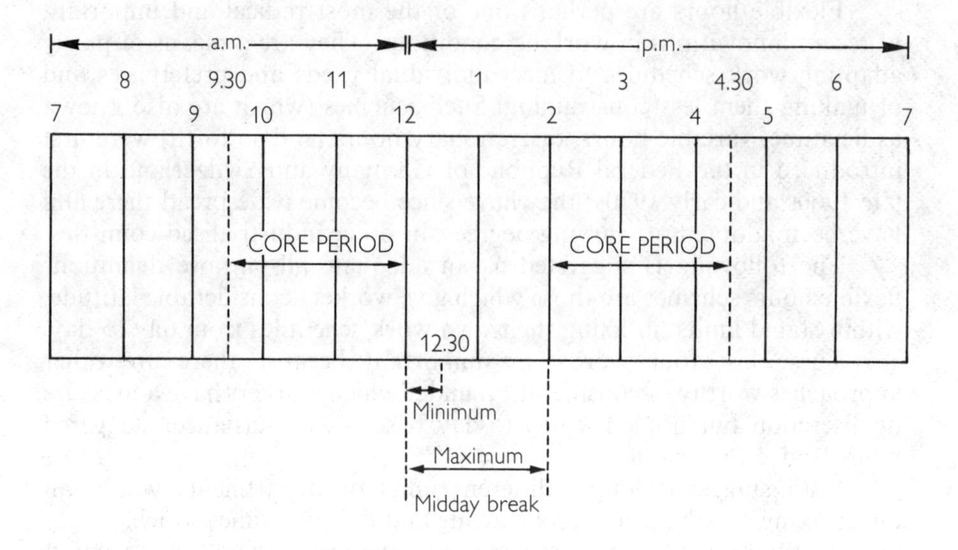

The basic rules of a typical scheme are shown in table 10.

Workers may arrive any time between 7 a.m. and 9.30 a.m. and must be present between 9.30 a.m. and 12 noon; at 12 noon they may take from 30 to 120 minutes as a break; they must again be present between 2 p.m. and 4.30 p.m.; and they may leave at any time between 4.30 p.m. and 7 p.m. Thus, on any one day they could theoretically work as little as five hours or as much as 11.5 hours. Where the normal working week of the undertaking is 40 hours, workers must arrange their hours in such a way that they are never more than ten hours in debt (i.e. have worked fewer than 30) or more than ten in credit (i.e. more than 50). At the end of each four-week period, they may carry over up to ten hours of credit or debit or, if they have four or eight hours of credit, may take a half-day or a day off.

This is, of course, just one system; numerous variations are possible.

Flexible hours give workers a considerable amount of freedom in organising their own lives – in taking care of family responsibilities, in meeting personal needs, or simply in following individual preferences. They relieve traffic congestion, strains on transport facilities and overloads on welfare services, such as canteens. Undertakings have generally found them popular with workers and useful in attracting new workers. In some cases, they may make it possible to maintain certain operations or to provide services over longer periods in the day. Flexible hours also seem to lead to reductions in overtime and – subject to the caution previously expressed about such conclusions – to lower absenteeism and staff turnover.

While experience with flexible hours has almost always proved successful, a number of possible problems should be kept in mind. Flexible hours are not readily applicable to all types of work – for example, where the workers are highly interdependent (as on assembly lines), where strict continuity of manning is imperative or where services must be assured during fixed periods. If some workers in an undertaking enjoy flexible hours and others do not, a potential source of friction may exist. The introduction of flexible hours may cause resentment among lower- and middle-level managers or supervisors, who find their authority and prerogatives (for example, in controlling attendance, giving permission to leave early and the like) threatened. More serious management problems may arise in distributing and supervising work and checking output and quality when the daily spreadover is as long as 12 hours.

A potential problem for workers is the possible length of the working day, as was observed above in the discussion on normal hours. However, a more immediate problem is distinguishing between flexible hours and overtime. If extra hours are worked, when do they count as flexible hours and when do they count as overtime? Can flexible hours be a way for undertakings to deal with fluctuations in workload without having to pay overtime rates? The gain in freedom may be offset by the loss of little periods of lateness or short absences which previously were overlooked or tolerated. Are flexible hours a means of exercising greater control over timekeeping and thus increasing productivity without paying any more?

Timekeeping can itself be a problem, and the introduction of time-clocks where they did not previously exist will undoubtedly be resisted. Other methods, such as attendance sheets, or individual meters which keep a cumulative tally of hours worked but do not record actual arrival or departure times, may be more acceptable.

For trade unions, a vital requirement is that flexible hours should not impair the exercise of trade union rights. When, for example, workers' representatives are allowed time off for union duties, should this be taken in core periods or in flexible periods? To what extent should time spent on such activities in flexible periods be counted as time worked? Should union meetings be held in core periods or flexible periods? How should the time spent on them be counted? Clear understandings on such questions are indispensable.

In spite of these difficulties, flexible hours are an important advance which can make working life easier and more pleasant. To be successful, they should be introduced only with the agreement of the workers and their representatives and after the fullest possible consultation on details. The scheme should be carefully prepared, management at all levels should be properly briefed and the workers should be thoroughly informed. Gradual introduction, including a trial period, is often helpful. Above all, expectations should not be placed too high: flexible hours are usually popular but they are not, and will not be accepted as, a substitute for a reduction in normal hours or other quantitative improvements.

Staggered hours and staggered holidays

If the rearrangement of working time within an undertaking, through staggered hours or flexible hours or other means, represents one way of easing the problems of traffic congestion and over-burdened public transport at certain peak hours, another approach is to promote the staggering of hours among undertakings. Instead of having all workers arrive and leave at practically the same time, different firms or even different branches of activity could adopt different starting and finishing times. Traffic would thereby flow more smoothly, public transport would be utilised more rationally and the physical and nervous strain on all concerned would be lessened. Although efforts in this direction have been made over many years in a number of cities, they have not had very much success because of the difficulties of co-ordination, of reconciling different interests and preferences, of changing habits and attitudes, and of meeting operational and commercial requirements. Within fairly narrow limits, however, many firms in large cities have been able to alter the usual starting and finishing times with positive results.

In many developing countries faced with the problems of urban sprawl, traffic jams and grossly inadequate transport, the co-ordinated staggering of hours may well offer a measure of relief. The difficulties should not be underestimated; but, given the magnitude of the problem, the idea is worth exploring.

The staggering of annual holidays has often been suggested in countries where these are usually concentrated in a period of one or two months (in Europe, the months of July and August) to benefit from good weather and so that parents' holidays may coincide with their children's school holidays. This concentration has two consequences: first, a significant drop in production during the holiday period; second, extreme congestion of roads, public transport and holiday facilities. Some attempts have been made to promote staggering by offering incentives, by seeking to change attitudes and by trying to stagger school holidays; however, in the countries where the tendency towards concentration is most pronounced, these have had little success.

Part-time work

A different type of departure from traditional patterns that has become more and more popular in some countries is part-time work. Under this system, certain workers choose, for their own reasons, to work fewer than the normal number of hours. Part-time work may be defined as work undertaken (on a voluntary and regular basis) for a number of hours that is substantially lower than the stipulated normal hours of work in the enterprise. It should not be confused with short-time working or with temporary work. Part-time work can take various forms: so many hours a day, so many days a week, every other week, and so forth.

For the undertaking, part-time work offers the advantages of flexibility in personnel management, a reduced need for overtime at premium rates and, possibly, lower rates of absenteeism and turnover. For the workers concerned, it offers the possibility of having more free time; of earning an income while exercising family responsibilities, continuing studies or pursuing other activities; and of meeting the special needs of, for example, disabled or older workers. For society, it permits the insertion or reinsertion into economic activity of workers who are unavailable for full-time employment and hence the maintenance, development and utilisation of their skills.

Part-time work also raises a number of problems, however.

From the standpoint of the undertaking, the main difficulties are technical (changes in work organisation, job content, assignment of tasks) and financial (proportionally higher cost of recruitment, training, supervision and social charges).

For workers and trade unions, the main disadvantages are the lack of opportunity for advancement, the possible disturbance of the employment market, the potential threat to the position of full-time workers and the danger of undermining trade union action. Part-time work may also lead to a certain marginalisation of workers who are assigned to low-level tasks requiring few skills and who may be constrained to accept lower salaries and less favourable working conditions.

In any event, given the millions of people engaged in part-time work and the likelihood of its expansion, there is a need to examine its implications carefully and to take measures aimed at regulating working conditions, ensuring social protection (for example, in respect of illness, accident and retirement) and preventing discrimination.

Towards a more integrated approach

As this chapter has shown, working time has many dimensions. Some are essentially quantitative, whereas others are qualitative. In the past, the different aspects of working time were generally treated as separate issues. Little attention was paid to the inter-relationships between the various quantitative elements or between them and the qualitative dimensions. Relatively little attention, in fact, was paid to the qualitative side at all.

This is beginning to change, and in recent years there has been a growing tendency towards dealing with working time in more integrated terms. Negotiations between employers and trade unions and policy discussions between governments and the social partners tend more and more to encompass the whole range of issues. If the reduction of working time is in question, various measures – such as the reduction of normal hours, the limitation of overtime, the extension of paid educational leave, the lengthening of the basic annual holiday, the improvement of holiday entitlements based on seniority – may be considered simultaneously as options from which the appropriate combination may be selected. Proposals have been made to regulate or to negotiate the total number of working hours in a year; however, this concept of "annual hours of work" remains controversial because of concern that it might lead to a weakening of protection as regards the length of the working day or the working week.

In some cases, working time is seen in a still longer perspective: that of a working life. The adjustment of the minimum age for admission to employment – primarily a means of protecting children and young persons – and of the age of retirement – primarily a question of social security – can also be seen as a way of adjusting overall working time and controlling movements into and out of the employment market.

Such problems relating to the level, or quantity, of working time are often discussed in conjunction with others relating to the organisation of working time. Employers may be more willing, for example, to concede a reduction in normal hours or a limitation of the amount of overtime if they are given greater freedom in using and organising overtime. The easing of restrictions on part-time employment may facilitate adjustment. Trade unions may wish to give priority to improving the conditions of shiftworkers by first reducing their hours, by introducing an extra crew or by changing rotation schedules. The possibilities are many. What is significant is the growing awareness that the reduction and the rearrangement of working time are closely linked, and that improvements in working conditions may be achieved by acting on one or the other, or both.

Greater attention is also being paid to the relationships between working time and living conditions. Long distances between home and the workplace, which mean at least two long journeys a day on top of working hours; the strain, discomfort and fatigue caused by congested roads or over-crowded public transport; school starting times that coincide with the usual starting times of workplaces and thus add to the burden on public transport at peak hours; opening hours of shops, banks and public services that coincide with those of factories and offices and make it difficult for working people to take care of their shopping or other business – problems such as these have a clear connection with working time. Although flexible hours and other measures have helped a little, it cannot be said that a great deal of success has been achieved in resolving them. The problems are complex, entrenched patterns are hard to change and the interests of the people concerned often conflict.

But in the future it will be more and more important to keep the relationships in mind and to approach the different issues in an integrated way.

Notes

[1] Reduction of Hours of Work Recommendation, 1962 (No. 116), Paragraph 11.

[2] Convention No. 1, Article 6 (2); Convention No. 30, Article 7 (4); and Recommendation No. 116, Paragraph 19 (2).

[3] ILO: *Hours of work: A world survey of national law and practice*, Extract from the Report of the 37th (1967) Session of the Committee of Experts on the Application of Conventions and Recommendations (Geneva, 1967), para. 238.

[4] Hours of Work and Rest Periods (Road Transport) Convention, 1979 (No. 153) and Recommendation, 1979 (No. 161).

[5] The Night Work (Women) Convention, 1919 (No. 4) which has been revised twice, on the last occasion by the Night Work (Women) Convention (Revised), 1948 (No. 89). International labour Conventions and most national legislation also have a less controversial prohibition on the night work of young persons.

[6] Rolande Cuvillier: *The reduction of working time: Scope and implications in industrialised market economies* (Geneva, ILO, 1984).

[7] For more information, see ILO: *New forms of work organisation* (Geneva, 1979), Vols. 1 and 2; and George Kanawaty (ed.): *Managing and developing new forms of work organisation* (Geneva, ILO, 2nd ed., 1981).

[8] Innovations, by their very nature, often lead to immediate improvements in morale, turnover and productivity and to reductions in lateness and absenteeism. However, after a certain time, the situation returns to what it was before. It is therefore necessary to be cautious in drawing major conclusions from short-term measurements.

WAGES

4

"Everyone who works has the right to just and favourable remuneration ensuring for himself and his family an existence worthy of human dignity, and supplemented, if necessary, by other means of social protection."

Universal Declaration of Human Rights, Article 23 (3)

Pay is a matter of fundamental importance for wage earners and their families, no matter what the country's level of development may be. Wages influence the general conditions of life of workers and their families; consequently, an adequate wage is a basic need. However, like employment, wages also contribute to the maintenance of economic equilibrium, social peace and progress. Because they are closely related to all aspects of the work and life of workers, undertakings, trade organisations and communities, they involve many complex and variable human and social parameters.

Wages are one of the main factors having an influence on working conditions, and their level has a definite effect on work motivation and job satisfaction: moreover, wages affect and are affected by other working conditions.

Since the above interactions are examined more extensively elsewhere in this book, this chapter will deal solely with the relationship between wages and the creation of good working conditions, and the worker's acceptance – through necessity or self-interest – of bad working conditions. It will concentrate on the main problems of remuneration; however, this general survey should be supplemented by a careful reading of the other chapters in the book, especially Chapter 8, "How to improve working conditions and environment". Reference should also be made to other relevant publications, such as the recent ILO books on wages and minimum wage fixing.[1]

Wage-fixing machinery

Of all the issues to which the question of pay gives rise, undoubtedly the most important is the fixing of wage levels since, to a large extent, these determine workers' living standards. Accordingly, one of the first questions that must be asked is: How, and through what machinery, can an adequate wage be fixed for each worker?

Individual contracts

Wages may be defined as payment by an employer for work done or services rendered by a worker under a contract or an agreement.[2] The simplest and, traditionally, the most common procedure has therefore been the establishment of an individual contract between the employer and the worker. The employer offers a wage based on the going rate for the job in question, and the worker, by accepting the conditions offered, agrees to make the physical and mental effort required to do it. In theory, workers are free to negotiate wages with their employers and to refuse conditions they consider unacceptable. In practice, however, they can rarely bargain on an equal footing with an employer, since the latter knows that, provided that the wages he offers reflect the relevant going rate, other candidates can be found on the local employment market.

Where there is a labour surplus – as is now the case in most developing countries – an individual unskilled worker cannot hope to negotiate wage rates with an employer. Such workers often accept a low wage because they have no other choice; to refuse a job may mean privation for both the worker and his or her family.

The history of workers' movements abounds with examples of employers who paid their workers only a meagre pittance. To cope with such social injustices, workers set up trade unions where possible, to bargain with employers or their associations for higher wages and better working conditions. Yet in vast sectors of the economy in both the industrialised and the developing countries, the individual contract is still the only method of wage fixing (except for legal protection). There may be no trade union because of – for example – the small size of the undertaking or the geographical dispersion of the worksites; or because, in some countries, freedom of association has been suppressed. The absence of strong, well-structured unions may mean that, in addition to wages and other working conditions being precarious, the protection of workers' legal rights, such as that of a minimum legal wage, is weak; this is particularly so in developing countries where labour inspection is often inadequate.

Collective agreements

Through collective bargaining and collective agreements, an organised group of workers can protect and improve its members' working conditions, including wages. Thus one of the ILO's basic tasks is to promote effective recognition of the right to collective bargaining.[3]

The protection afforded by collective agreements, and their content and scope, vary enormously from case to case and from country to country. At one end of the scale, collective agreements may be concluded at the national level; at the other, they may apply solely to a specific category of workers in a given plant. The contents may also vary considerably: in some, bargaining procedures and working conditions are set out in detail; in others, only basic procedures are laid down.

Broadly speaking, collective agreements may be drawn up at one of three levels: at the national level; by branch of activity; or at the level of the plant or undertaking. The scope of collective agreements in any given country will depend on the structure of the occupational organisations, tradition and the legal framework.

National collective agreements may sometimes be used to fix minimum wage levels applicable to all workers, but only rarely are they used in an attempt to fix wage levels for specific occupational categories, such as highly skilled workers. However, a collective agreement for a specific branch may fix wage levels for each category of worker. In certain countries, these occupational rates may be only a minimum, as for example in France, where individual undertakings will often offer workers higher rates. In other countries, collective agreement rates are closer to the going rates actually paid. Finally, agreements at the plant or undertaking level usually fix a rate for each category of worker. Each level of negotiation has its advantages, either by laying down national minimum rates or by fixing actual wage levels by collective bargaining.

Collective agreements at the national or branch level usually apply to the majority of workers in question; however, agreements at the level of the undertaking seldom extend to workers in small and medium-sized undertakings which are not unionised.

No matter what the level of the collective agreement, wage fixing will be effective only if bargaining is resumed periodically to adapt wage levels so as to take account of inflation and changing economic circumstances. Collective bargaining has the advantage of being a dynamic process that enables workers, through their representatives, better to defend their interests before their employer.

Voluntary or compulsory arbitration

In many countries, the right to collective bargaining in certain sectors has been limited. Where the breakdown of negotiations in the public services sector could lead to a stoppage of work with particularly undesirable effects for the national economy or the population in general, legislation may stipulate an arbitration procedure by which the two parties, i.e. the employer and the workers' organisation, request a third party to arbitrate in the matter.

In some countries – Australia and New Zealand, for example – arbitration is the traditional procedure for fixing wages and other conditions of work in all sectors. Each year, labour arbitration tribunals deliberate on cases submitted to them by one of the parties and issue awards that fix wages and working conditions for the sectors in question.

Minimum wage

Minimum wage fixing is a procedure for direct government intervention to reduce, if not eliminate, poverty and social injustice. Originally, minimum wages were intended for workers who did not benefit from the protection of collective agreements and who had very low wages. The first international labour standard on the subject, the Minimum Wage-Fixing Convention, 1928 (No. 26), stipulates the need "to create or maintain machinery whereby minimum rates of wages can be fixed for workers employed in certain of the trades or parts of trades (and in particular in home working trades) in which no arrangements exist for the effective regulation of wages by collective agreement or otherwise and wages are exceptionally low". The Minimum Wage Fixing Convention, 1970 (No. 131), specifies, inter alia, the factors to be considered in determining the level of minimum wages (see Panel 32).

Many countries have set up tripartite bodies to fix minimum wages or propose them to governments. In the United Kingdom and certain countries which have been influenced by British practice, wages councils cover narrowly defined trades such as those of tailor or laundry worker. In other countries, especially in Latin America, a wages council may often cover a whole industry or, as is the case in Mexico, a whole region.

Recently, minimum wages have assumed a greater role in many countries and have become a major component of economic policy. By enacting or decreeing a national minimum wage, a government can stipulate the lowest wage below which no employer may go and can thus provide those workers who have no trade union in their undertaking or branch of activity with a "safety net" protection for their wages. This type of national minimum wage now exists both in many highly industrialised countries (e.g. Canada, France, the Netherlands, the United States) and in certain developing countries (most French-speaking African countries, Brazil, Colombia, Costa Rica, Honduras, Jamaica, Kenya, the Philippines, the United Republic of Tanzania, Thailand, etc.).

Given its broad impact, the fixing of a national minimum wage generally has repercussions on the overall level of wages and on other macro-economic variables such as employment. In developing countries, minimum wages often directly determine the wages for unskilled workers on the employment market and, in this way, decisively influence wage levels and structures as a whole. Consequently, minimum wage fixing tends to be a major factor in a country's economic and social policy.

PANEL 32

Minimum Wage Fixing Convention, 1970 (No. 131)

Article 1

1. Each Member of the International Labour Organisation which ratifies this Convention undertakes to establish a system of minimum wages which covers all groups of wage earners whose terms of employment are such that coverage would be appropriate.

2. The competent authority in each country shall, in agreement or after full consultation with the representative organisations of employers and workers concerned, where such exist, determine the groups of wage earners to be covered.

Article 2

1. Minimum wages shall have the force of law and shall not be subject to abatement, and failure to apply them shall make the person or persons concerned liable to appropriate penal or other sanctions.

2. Subject to the provisions of paragraph 1 of this Article, the freedom of collective bargaining shall be fully respected.

Article 3

The elements to be taken into consideration in determining the level of minimum wages shall, so far as possible and appropriate in relation to national practice and conditions, include —

(*a*) the needs of workers and their families, taking into account the general level of wages in the country, the cost of living, social security benefits, and the relative living standards of other social groups;

(*b*) economic factors, including the requirements of economic development, levels of productivity and the desirability of attaining and maintaining a high level of employment.

Article 4

1. Each Member which ratifies this Convention shall create and/or maintain machinery adapted to national conditions and requirements whereby minimum wages for groups of wage earners covered in pursuance of Article 1 thereof can be fixed and adjusted from time to time.

Wage fixing in the public sector

The fixing of wages in the public sector has features that distinguish it from the procedure followed in the private sector. In most countries, the public sector is regulated not by collective agreements but by regulations issued by Parliament or the public authorities. Since the State is an employer in this case, a wage increase may entail additional public expenditure which falls within the domain of Parliament.

However, certain countries have recently recognised *de facto* or *de jure* the possibility of collective bargaining similar to that in the private sector. In Canada and Sweden, salaries have been fixed in collective agreements. In other countries, consultation and bargaining between public servants and the State as an employer have been institutionalised in a system of employer-employee councils. Parliament retains the right to fix wage and working conditions for public sector employees, usually following consultation with the recognised public sector unions.

Municipal or local authority workers in most countries are subject to independent local authority management. However, some or all of these workers in developing countries such as India or Jamaica fall under a national unified scheme.

The degree of autonomy that public undertakings exercise in wage setting varies considerably from one undertaking to another and from one country to another. The railways offer a characteristic example: in some countries (the Federal Republic of Germany, Sweden, Switzerland, etc.), railway workers are classed as public servants whilst, in others, they are subject to official regulations (Poland, USSR). Except where special systems for settling conflicts exist (as in Canada and the United States), these workers have virtually the same collective bargaining rights as private sector employees. Finally, in countries such as France and Belgium, railway workers' wages and working conditions are negotiated unofficially between the management and the trade unions.

Although a wide range of schemes usually operates in the public sector, how can wage rates be adjusted where no collective bargaining exists? In a number of countries, the wage adjustment machinery incorporates a unit which regularly compares wage levels in the private and public sectors, and recommends any wage adjustments that may be considered necessary. Sometimes, however, because of economic and political influences, the machinery works only spasmodically and ineffectually.

Wage-fixing parameters or criteria

Although there is world-wide recognition of the principle that workers should receive "an adequate living wage" (to quote from the Preamble to the Constitution of the ILO), fixing such a wage in practice presents numerous difficulties. Wage fixing normally involves such factors as the cost of living, the undertaking's ability to pay, the employment market situation and the worker's individual characteristics (skills, individual ability, age, seniority, etc.). The wage-fixing machinery described above is merely a way of achieving a balance between these different criteria at a given point in time.

Although wage-fixing factors or criteria can be grouped in different ways, they have been set out below under three major headings: a living wage (sufficient to meet the worker's basic needs); an acceptable wage (by the undertaking or the economy in relation to ability to pay); and an equitable wage (in relation to other job categories in the same undertaking).

A living wage

How does one fix a living wage, that is, one which is sufficient to cover the worker's material, moral and cultural needs? No categorical answer is possible, since the concept of a living wage – however defined – is a relative concept depending, in particular, on the country's level of development, the branch of activity and the worker's personal characteristics. For example, a living wage in a developing country would certainly be unacceptable in a country that enjoys high wage levels. Nevertheless, there are a number of factors or criteria that are often taken into account when deciding what may be considered an adequate living wage. First and foremost, the wage should at least meet the basic needs of workers and their families, give protection against cost-of-living increases, and be reasonably in line with other wages and incomes.

Although it is widely agreed that a wage should meet the worker's basic needs and should normally be set at the minimum acceptable threshold, difficulties are encountered in actually determining this level. This question has often been the centre of controversy in minimum wage fixing.[4] Some governments have tried to quantify this threshold by defining a "basket" of goods and services essential to meet the worker's basic needs. Although this approach looks attractive and logical, experience has shown that it is quite difficult to apply. For example, the basket of goods and services must contain sufficient food for the worker's caloric expenditure – but this expenditure varies according to the type of physical effort required by the work, the worker's age and weight, and the working environment (heat). Even if an acceptable estimate of food requirement is obtained using, for example, the FAO/WHO study which indicates that a person weighing 65 kg has an energy expenditure of 2,700 calories for light work and 4,000 calories for heavy work, it is still necessary to know what kind of food is eaten by the workers in question. It cannot be realistically assumed that workers eat bread, rice or manioc every day, and this inevitably upsets calculations based only on the calorie content of basic foodstuffs. In addition, what about protein and vitamin intake?

It can therefore be seen that there is a large margin of discretion in defining the "basket", and also matter for endless debates. Moreover, in some cases, the calculated theoretical minimum consumption has proved to be far above the minimum wage the country could afford. In India, for example, a national committee analysed consumption and recommended a minimum wage of 196 rupees per month for a federal civil servant; however, other considerations led it to recommend no more than 100 rupees for other sectors.

These difficulties have led many governments to abandon fixing minimum wage rates based on a basic-needs "basket" or a "poverty-line budget" and, instead, to determine periodic changes in minimum wages on the basis of changes in the general consumer price index or a special cost-of-living index for low-income groups. It is thus evident that "basic needs" is still a relative concept that is difficult to use directly and in isolation for fixing minimum wages.

Cost of living

The cost of living is another criterion widely used in determining wages. Since wages are usually the worker's only source of income, they must be adequate to guarantee a certain standard of living even when prices are rising rapidly.

Some of the terms used in this context may cause confusion. For example, "standard of living" evokes a life-style that is closely linked to the wage and disposable income available to the worker. However, if the prices of the products he regularly buys are rising rapidly – i.e. the cost of living is rising – and if wages remain unchanged, there will be a corresponding fall in his standard of living. Thus living standards are related directly both to wage and income levels and to the cost of living.

The cost of living reflects the prices of purchases; for example, a 10 or 20 per cent rise in the price of staple foods such as rice, manioc or bread entails a corresponding increase in household expenditure, that is, the worker's cost of living will increase. For statistical purposes, a consumer price index is used to measure changes over time in the price of goods and services bought by an average family. This index does not indicate how much it would cost to live in a given place at a given time but how much more or less it costs to buy the same goods and services from month to month. The consumer price index is an essential component of national statistics in all countries.

The difference between nominal and real wages also needs to be explained. Nominal wages are what the worker is actually paid; however, what is more important to workers is the amount of goods and services they can buy with their wages. With a rapidly rising cost of living and an unchanged nominal wage, workers can buy fewer goods and services: in other words, their real wages decrease. Calculations of real wages eliminate the influence of price increases.

It is real wages that are of greatest importance to workers. In wage negotiations, they usually put forward cost of living as the argument for an increase in (nominal) wages to maintain and raise their standard of living. In many cases, employers recognise the need to compensate cost-of-living rises, provided that the consequent increase in wage costs does not impede the undertaking's operation.

Compensation for price increases is usually the minimum basis for fixing a satisfactory wage since if pay no more than keeps abreast with the cost of living, real wages will stay exactly the same. Workers usually hope that, through sharing in the economic growth of their undertaking or the national economy, their real wages will increase.

Wage comparisons with employment market rates

Workers also expect a living wage to be at least as good as the going rate for the same kind of work on the employment market. If a worker's wage is clearly below that earned by workers in other undertakings, it is probable that that wage is neither appropriate nor satisfactory. A worker who stays on will be frustrated and show little motivation or interest in his work. The employer should, therefore, normally ensure that the wages paid are not too far below the going rate on the employment market.

Yet the question is somewhat more complex; going rates usually only indicate an order of magnitude. Only rarely do two workers with the same qualifications in two different undertakings do exactly the same work, with the same responsibility and the same output. In other words, there is always a risk that comparisons will be made between items which are not strictly comparable. A comparison of wages with employment market rates is usually only indicative, and the interpretation will be different depending on whether it is done by the worker or the employer. None the less, if rates for certain categories of workers in one undertaking are clearly below those paid for similar jobs in a comparable undertaking, the workers will show their dissatisfaction in some way or another. A living wage must not be lower than that paid to similar workers in similar undertakings.

An acceptable wage (for the undertaking and the economy)

There is a limit to the size of any undertaking's wage bill. If wage costs rise more rapidly than productivity, the undertaking will lose its competitive edge and find it hard to survive. Neither the State nor the unions can continually seek to raise wages above the undertaking's ability to pay.

Ability to pay is a very difficult criterion to assess since it depends on the undertaking's finances and economic position and on the kind of goods it produces. A high productivity undertaking with a good profits record will probably not be able to use its ability to pay as an argument for putting a brake on wage rises. However, a declining industry with a shrinking market will scarcely have any margin in its ability to pay.

It can therefore be seen that ability to pay varies from undertaking to undertaking, even in the same sector. Finally, a certain proportion of wages cannot be compressed; and, for example, where a collective agreement sets minimum wages for various trades, and so on, the undertaking must pay at least the legal minimum rates, no matter what its ability to pay.

Ability to pay also depends on the undertaking's field of activity. For example, a public transport firm may have a chronic deficit; however, this does not limit its ability to raise wages, since fares can be increased to absorb the larger wage bill. An undertaking with a virtual monopoly position can nearly always raise prices to pass on the cost of wage increases, but an undertaking in a highly competitive field cannot raise wages more than its competitors if it does not want to lose its share of the market.

Ability to pay is a criterion that varies considerably from undertaking to undertaking and that is frequently put forward by employers to counter wage claims. Probably the best way of determining ability to pay is to confront opposing opinions, that is, through collective bargaining.

The national economy's ability to pay involves arguments of a different order. In a developing economy, it is necessary to promote productive investment for the sake of future economic development; wage increases, on the other hand, increase consumption. Some governments claim that a substantial wage increase would imperil international competitivity and that increases which outpace productivity lead to inflation and unemployment. Such statements are difficult to confirm or refute, but it cannot be denied that a substantial wage increase that is not balanced by a rise in productivity may impede economic growth. There is a limit beyond which economic growth is hard to sustain and there is a limit to an acceptable wage – even if it is difficult to measure accurately – both for the undertaking and for the national economy.

An equitable wage

Another wage-fixing criterion is that a wage must be equitable, and equal pay for work of equal value is a widely accepted principle in the prevention of wage discrimination based on sex, in particular, but also on nationality, religion, social background, and so on. This criterion was the basis of the Equal Remuneration Convention, 1951 (No. 100) (see Panel 33).

This criterion is also important when the workers compare their wages with those of others, since workers doing the same job for lower pay will never accept that their wages are equitable. The wage hierarchy within the undertaking must be based on objective or justifiable differences related to job skills and ability: if a worker is doing a difficult

PANEL 33

Equal Remuneration Convention, 1951 (No. 100)

Article 1

For the purpose of this Convention —

(*a*) the term "remuneration" includes the ordinary, basic or minimum wage or salary and any additional emoluments whatsoever payable directly or indirectly, whether in cash or in kind, by the employer to the worker and arising out of the worker's employment;

(*b*) the term "equal remuneration for men and women workers for work of equal value" refers to rates of remuneration established without discrimination based on sex.

Article 2

1. Each Member shall, by means appropriate to the methods in operation for determining rates of remuneration, promote and, in so far as is consistent with such methods, ensure the application to all workers of the principle of equal remuneration for men and women workers for work of equal value.

2. This principle may be applied by means of —

(*a*) national laws or regulations;

(*b*) legally established or recognised machinery for wage determination;

(*c*) collective agreements between employers and workers; or

(*d*) a combination of these various means.

Article 3

1. Where such action will assist in giving effect to the provisions of this Convention measures shall be taken to promote objective appraisal of jobs on the basis of the work to be performed.

2. The methods to be followed in this appraisal may be decided upon by the authorities responsible for the determination of rates of remuneration, or, where such rates are determined by collective agreements, by the parties thereto.

3. Differential rates between workers which correspond, without regard to sex, to differences, as determined by such objective appraisal, in the work to be performed shall not be considered as being contrary to the principle of equal remuneration for men and women workers for work of equal value.

. . . .

job which is mentally and physically demanding but receives a lower wage than another worker doing a routine job, there will be a feeling of wage inequity.

Internal wage hierarchy is a difficult matter and involves such factors as tradition and labour-management relations. One trade in a plant may be paid more because it has long had privileges over another trade; or an electrician may be paid significantly more than a mechanic because the electricians have negotiated a national minimum salary. In such situations the soundness of the wage hierarchy may be contested.

One systematic, if not objective, method of establishing a wage hierarchy is through job or post evaluation which analyses the content of a job and not the person doing it, and – by determining the relative value of various jobs – attempts to ensure equal pay for work of equal value. First, the characteristics of the job are identified (job analysis), and then criteria common to all the posts are selected so as to determine the importance of one job in relation to others. These criteria vary from one skill level to another but usually include: knowledge, experience, responsibility, difficulty, physical and mental effort, working conditions, and so on.

There are basically two approaches to job evaluation: the analytical method and the global method. The analytical method breaks the job down into components which are each awarded points using a grading system, and the total number of points awarded determines the level of the job in the hierarchy. The global method compares the significance of different jobs and the final ranking determines the hierarchy; it is usually less sophisticated and easier to implement.

Both methods of job evaluation classify jobs systematically, and the analytical method even has the advantage of having predetermined criteria to explain why one job is classified higher than another. Wage differences are therefore based on clear and specific criteria and, in principle, inequities and anomalies in the wage structure are eliminated. Unfortunately, the technique is not objective since any choice of criteria and ratings is necessarily subjective, although it can claim to be more rational and systematic than traditional methods. Job evaluation entails considerable investment in time and expertise, and, although commonly employed in North America, it is uncommon in many other industrialised and in developing countries.

Wage systems

Discussion of wage systems usually centres on wage regularity, payment by results and productivity bonuses. There are two basic wage systems: time rate, and payment by results. The various systems of payment by results may involve health hazards; they are dealt with in greater detail below.

Time rate system

Under this system, wages are paid according to the number of hours, days, weeks or months that the worker is at the employer's disposal. It takes no account of variations in workers' output and ensures stable and secure earnings, particularly when on a monthly basis. The time rate system is used widely for work that is not quantifiable (e.g. office work and supervisory functions). In some countries, time rate systems are tending to replace payment by results systems, which are difficult to apply and which conflict with workers' aspirations for stable wages. The growing trend towards the monthly payment of wages in many countries implies a monthly wage rate for wage calculations. Monthly payments guarantee the worker greater income stability and security than do hourly and daily rate systems, and tend to harmonise and smooth out the working conditions of both wage earners and salaried employees, especially as regards the length and payment of annual and public holidays.

Payment by results [5]

Payment by results systems fix wages by the measured output of a worker or group of workers. There are several such systems: uniform or differential piece work, productivity bonuses, and merit or personal ratings. Their aim is usually to improve productivity, increase production or, less frequently, quality, whilst at the same time keeping wage costs down.

The various arguments for and against payment by results systems are difficult to assess; they are but one element of the socio-technical system represented by the undertaking and are closely related to work organisation structures. Consequently it is hard to attribute specific effects.

At first sight, payment by results systems may seem just and effective: the higher the output, the higher the income. However, depending on the systems and rates selected, they may be seen as a means of getting workers to maximise their output or of adapting wages.

Measuring the aspects of performance (physical effort, judgement, dexterity or precision) which are required by the work and rewarded or encouraged by these systems is not always easy. Will one aspect of performance be encouraged to the detriment of others and, eventually, to the detriment of quality?

Where workers are organised into more or less autonomous groups, often with the intention of making the work more interesting, improving working conditions and giving greater opportunity for flexibility and innovation, individual performance is difficult to assess since the very principle of group involvement and responsibility is in conflict with remuneration on an individual basis; group bonus schemes are necessary here. The trend is also towards paying workers on the basis not only of output but also of experience and of the contribution that their multiple skills or organisational or adaptive talents make to the team.

It is normal to offer incentives for the achievement of production targets and compensation for extra effort, but numerous safeguards are necessary. The system should not demand undue effort, pace or hours of work to earn a living wage; it should not incite workers to neglect safety requirements and, consequently, wage rates should allow for the observance of safety rules and safety devices; it should guarantee an adequate minimum wage when production is impeded by circumstances beyond the workers' control, such as breakdowns, supply hitches, time lost through other workers, or a machine-dependent slower work pace; it should not create an imbalance between workers paid by results and workers with similar skills who are paid under other systems; in some people's view, it should guarantee a constant minimum wage for a given level of effort in jobs where the effort needed for a given result increases with age; in all cases, it should guarantee a minimum living wage for workers who, for reasons of age, constitution or health, cannot achieve the same output as young, healthy, well-trained workers whose output is used as a base for rate calculations; and finally, it should be easy to understand, so that workers can see the relationship between work and wage, it being realised that the effect of an incentive may vary in line with a worker's needs during his working life.

Earnings under payments by results systems seldom make up the whole of the wage package; rather, they are used as a productivity bonus. The size of the output-related wage component should be limited in order to guarantee a degree of wage stability. Where a change-over is being made from payment by results to a time rate system, it is necessary to guarantee a wage at least equal to the average output-based wage that the workers had received over a given period prior to the change-over. At the same time, other incentives will have to be sought in the work itself and in the workers' physical and social environment.

Straight piece-rate systems

These are the most simple system of payment by results and give individual or group earnings related directly to production, with each unit of output being paid at a constant rate. They assume that output is measurable (by units of production), and that the production process and the product are sufficiently uniform and standardised that the effort demanded of the worker does not vary significantly from one product to another. When production hitches keep workers waiting and therefore not earning, a guaranteed minimum daily or weekly wage should be paid, independent of output. A piece rate that is too low will mean working unduly hard; a rate too high will give a worker an advantage over fellow-workers being paid under, for example, a time rate system. In both cases, there will be a temptation to neglect safety requirements or to bypass safety devices which have not been taken into account in the calculation of time requirements.

Rates should be fixed carefully for each production item so that all workers, including the older and less adroit, can earn a fair wage without undue effort; they should be fixed after an adequate trial period, say of a few weeks, to allow discussion and to ensure that, once accepted, they will not subsequently be questioned and, above all, not revised downwards.

Those engaged in jobbing (e.g. in logging) may fall into this wage category, as may also homeworkers, who are often paid by results, or sales personnel paid according to predetermined rates of commission.

Differential piece-rate systems

Under many systems rates for each product vary according to output. Product rates that rise with output favour the strongest or most adroit and provide a strong incentive for, perhaps, undue effort. However, they may create inequalities and tensions among workers who often consider them unfair and frequently do not understand the calculation of their earnings. Moreover, it is difficult to achieve the degree of accuracy in standard-setting that is required for the effective functioning of these systems.

Premium and task bonus systems

These systems, which are frequently used for groups of workers who have a fixed daily, weekly or monthly output target, try to attenuate the defects of the systems described above by reducing the value of the production incentives, once a certain output has been achieved. They are based, first, on an hourly wage rate and a standard job time – which may be fair either for all workers or for only the best. Workers who finish the job in the set time or who finish early receive a bonus which is added to the hourly pay rate for effective working time; thus, workers are paid for a somewhat longer working time than that actually spent on the job. Workers who do not finish in time are not penalised but are merely paid for their actual working hours.

Merit rating

This has been adopted by a number of undertakings and administrations, and involves the linking of part of a worker's wages to some systematic appraisal of his behaviour or demonstrated ability. The result of the appraisal determines the wage increase rate or progress up the scale. The technique is used mainly for those paid on a time basis, particularly administrative and technical workers.

These schemes are not related to the job evaluation methods described earlier, and they apply to the worker and not the job. They can be even more effective when backed by in-service training and promotion programmes.

Profit-sharing and co-partnership schemes

Although these are not directly related to output and are more a form of social benefit, they are still intended as an incentive for employees (especially those in positions of responsibility) by giving everyone a better general understanding of the undertaking's problems; co-partnership also encourages savings. Profits may be shared by allocating a pre-set percentage of turnover or profits, or by cash bonuses payable after a given length of service – usually at least one year; the sums may increase with increasing length of service, and vary depending on the job or wage level. At first reluctant to accept such innovations, trade unions no longer oppose this type of payment, provided they are genuinely involved at the planning stage and are able to check that payments of this kind are not a substitute for regular wage increases.

Share distributions under co-partnership schemes are related to profits, and wholly or partly replace corresponding cash payments.

Protection of wages

For most wage earners, the only means of subsistence for themselves and their family is their pay. Consequently, once fair wage levels have been fixed, measures must also be taken to guarantee regular wage payment and to protect the worker from the employer, the worker's or the employer's creditors and the worker himself.

Protection from the employer

The Wage Protection Convention, 1949 (No. 95), states that "employers shall be prohibited from limiting in any manner the freedom of the worker to dispose of his wages" (see Panel 34). It also stipulates that "deductions from wages shall be permitted only under conditions and to the extent prescribed by national laws or regulations or fixed by collective agreement or arbitration award" and that workers "shall be informed . . . of the conditions under which and the extent to which such deductions may be made".

In general, deductions may be made only for compulsory contributions (social security contributions, income tax where applicable, trade union dues), the repayment of legal benefits (housing, food), the repayment of advances or pre-payment received by the worker (the amount of which is often limited), and stoppages (sometimes with an upper legal limit).

PANEL 34

Protection of Wages Convention, 1949 (No. 95)

Article 3

1. Wages payable in money shall be paid only in legal tender, and payment in the form of promissory notes, vouchers or coupons, or in any other form alleged to represent legal tender, shall be prohibited.

2. The competent authority may permit or prescribe the payment of wages by bank cheque or postal cheque or money order in cases in which payment in this manner is customary or is necessary because of special circumstances, or where a collective agreement or arbitration award so provides, or, where not so provided, with the consent of the worker concerned.

Article 4

1. National laws or regulations, collective agreements or arbitration awards may authorise the partial payment of wages in the form of allowances in kind in industries or occupations in which payment in the form of such allowances is customary or desirable because of the nature of the industry or occupation concerned; the payment of wages in the form of liquor of high alcoholic content or of noxious drugs shall not be permitted in any circumstances.

2. In cases in which partial payment of wages in the form of allowances in kind is authorised, appropriate measures shall be taken to ensure that —

(*a*) such allowances are appropriate for the personal use and benefit of the worker and his family; and

(*b*) the value attributed to such allowances is fair and reasonable.

Article 5

Wages shall be paid directly to the worker concerned except as may be otherwise provided by national laws or regulations, collective agreement or arbitration award or where the worker concerned has agreed to the contrary.

Article 6

Employers shall be prohibited from limiting in any manner the freedom of the worker to dispose of his wages.

Article 7

1. Where works stores for the sale of commodities to the workers are established or services are operated in connection with an undertaking, the workers concerned shall be free from any coercion to make use of such stores or services. ▷

2. Where access to other stores or services is not possible, the competent authority shall take appropriate measures with the object of ensuring that goods are sold and services provided at fair and reasonable prices, or that stores established and services operated by the employer are not operated for the purpose of securing a profit but for the benefit of the workers concerned.

Article 8

1. Deductions from wages shall be permitted only under conditions and to the extent prescribed by national laws or regulations or fixed by collective agreement or arbitration award.

2. Workers shall be informed, in the manner deemed most appropriate by the competent authority, of the conditions under which and the extent to which such deductions may be made.

Article 9

Any deduction from wages with a view to ensuring a direct or indirect payment for the purpose of obtaining or retaining employment, made by a worker to an employer or his representative or to any intermediary (such as a labour contractor or recruiter), shall be prohibited.

Article 10

1. Wages may be attached or assigned only in a manner and within limits prescribed by national laws or regulations.

2. Wages shall be protected against attachment or assignment to the extent deemed necessary for the maintenance of the worker and his family.

Article 11

1. In the event of the bankruptcy or judicial liquidation of an undertaking, the workers employed therein shall be treated as privileged creditors . . .

2. Wages constituting a privileged debt shall be paid in full before ordinary creditors may establish any claim to a share of the assets.

3. The relative priority of wages constituting a privileged debt and other privileged debts shall be determined by national laws or regulations.

Article 12

1. Wages shall be paid regularly. . . .

. . . .

Article 13

1. The payment of wages where made in cash shall be made on working days only and at or near the workplace, except as may be otherwise provided by national laws or regulations, collective

▷

agreement or arbitration award, or where other arrangements known to the workers concerned are considered more appropriate.

2. Payment of wages in taverns or other similar establishments and, where necessary to prevent abuse, in shops or stores for the retail sale of merchandise and in places of amusement shall be prohibited except in the case of persons employed therein.

Article 14

Where necessary, effective measures shall be taken to ensure that workers are informed, in an appropriate and easily understandable manner —

(a) before they enter employment and when any changes take place, of the conditions in respect of wages under which they are employed; and

(b) at the time of each payment of wages, of the particulars of their wages for the pay period concerned, in so far as such particulars may be subject to change.

Although employers, for example in agricultural or developing regions, may commonly provide housing and subsistence which would otherwise not be available, efforts should be made to bring about economic or social changes to allow everyone to receive full payment of his or her wages in cash, and to have freedom to buy from any supplier he or she chooses.

Labour legislation usually limits the deductions that employers can make from a worker's wages for debts resulting from, for example, cash advances, supplies of materials or food, or compensation for damage caused by the worker. Labour codes also usually prohibit fines or deductions for production defects, indiscipline, professional error, and so on. Where fines or deductions are permitted, they are limited to a very small part of the wage (e.g. 5 per cent). To ensure regular payments, national laws prescribe payment intervals – which will usually not exceed one month – to ensure that workers are not kept short of money for long periods, thus impoverishing their whole family.

The amount reimbursed to the worker's creditors from his or her wages should not be such that the remaining pay is no longer sufficient for subsistence, and Convention No. 95 states that "wages may be attached and assigned only in a manner and within limits prescribed by national laws or regulations" and that "wages shall be protected against attachment or assignment to the extent deemed necessary for the maintenance of the worker and his family".

Wage protection in the event of bankruptcy of the undertaking

Convention No. 95 states that "in the event of the bankruptcy or judicial liquidation of an undertaking, the workers employed therein shall be treated as privileged creditors", that is, the wage earner must be paid before any other creditors. National legislation usually specifies the priority rating amongst privileged creditors should the undertaking's assets be liquidated. Certain European countries have recently taken steps to insure workers' credits through special guarantee institutions in the event of an employer's becoming insolvent. In 1980 the Council of the European Communities adopted a directive requesting member States to strengthen this aspect of wage protection.[6]

Protecting the worker against himself

Measures may be taken to prevent workers from being treated unjustly as a result of their own ignorance or weaknesses; these measures concern obligations relating to the payment of wages (place, date, intervals, etc.) and the prohibition of payment in places where alcoholic beverages are sold or in stores in which the worker may be tempted to spend part of his wages. Reference has already been made to limitations on commitments made by workers and their families such as, for example, credit, hire purchase or loans. Finally, there are legal requirements concerning the payment of a part of wages (usually as the result of a court order) to the wife, following separation or divorce or when the husband fails to support his family.

Wages and other conditions of work and life

Wage level and subsistence

Since wages are the worker's prime means of subsistence, they are a major factor in adequate and healthy nutrition; the healthy and normal development of workers and their families; and the ability of workers to do a satisfactory amount of work over a sufficient period of time at an adequate level of intensity, in the interests of both themselves and the community.

In general, wages directly influence living conditions and hygiene, and a living wage permits good accommodation ensuring the worker better rest and better conditions of family and social life which, in turn, make working life easier.

Wages, skills and employment

Lack of skills limits the wage that workers can hope to aspire to, as well as their chances of progress through promotion and increased earnings. Unskilled workers do not usually have the possibility or the opportunity to learn or progress; nor will unskilled workers usually have sufficient basic education to pass through an apprenticeship.

Lack of skills also affects access to job opportunities. A newly industrialising country will usually offer a large number of low-skilled jobs, and the division of labour or the reduction of skill requirements is aimed specifically at the rapid employment of low-skilled workers whilst at the same time minimising wage costs. Consequently, job opportunities and wage levels are the outcome of choices in work organisation and in job and employment definitions. However, in the event of recession or economic crisis, or the introduction of mechanisation or automation, the least-skilled workers are those whose jobs are at greatest risk and also those who will have the greatest difficulty in taking up new jobs. Lack of vocational training – and often education in general – places many women in the high-risk groups, especially in industries that are in difficulties.

Levels of trade skills – and therefore wage rates and living standards – are influenced by the general and vocational education and training that workers have received. Conversely, development potential and the technical options open to a country depend on the skills of the country's workforce.

Wages and the improvement of working conditions

Poor employment prospects and wage levels often oblige workers to accept arduous, unhealthy or dangerous working conditions and unduly long hours of work. Moreover, wage structures and bonuses (for dirty or dangerous work and for improved output) often impede the implementation of technically feasible improvements in working conditions.

It is not normal for workers to exhaust their strength prematurely, working at poorly designed workplaces or on repetitive and arduous tasks. In many cases, calculations would show that it is more profitable to change the workplace or machine design than to go on forever paying danger money or allowances for arduous work. When the workplace and its dangers cannot be modified, it is sometimes possible to rotate jobs or introduce rest breaks (for recuperation from exposure to noise, fatigue, etc.) which reduce exposure to the harmful agents. The solution of paying compensation for the risk should be used only as a last resort; moreover, compensation for risk should be distinct from payment for job skills since it is of no added vocational value. Finally, the only genuine compensation for physiologically harmful agents is that which reduces the effects of exposure, e.g. an increase in rest times or isolation from a noise or pollution source.

Compensation or bonuses for overtime, work during weekends or public holidays, shift work and, in particular, night work certainly offer financial advantages, but also make the worker accustomed to a higher income which rapidly becomes indispensable. For example, a worker may tend to prolong night shift work irrespective of the resulting fatigue or prejudice to health, since the likelihood of working on another shift without loss of pay is limited. Moreover, loss of income owing to health problems or work incapacity is not always fully covered by social security schemes. Financial advantages may compensate the worker and his family for the undesirable features of longer or unusual working hours or night work, but compensation for deleterious physiological effects should be sought through a reduction in the length of the working day or an increase in rest periods or holidays.

The social aspects of wages and the national social security structure often influence the worker's safety and health. For example, lack of sick pay in the event of non-occupational accidents and diseases may incite workers to neglect their health (and that of their colleagues in the case of infectious diseases), rest or safety in order to protect earnings in the event of sickness. It can be worrying for workers if their medical expenses and those of their families are not reimbursed, especially since health care costs are rising more rapidly than wages in many countries. In some cases, a worker with family responsibilities will refrain from claiming reimbursement (if this is at the employer's expense) for fear of being given notice (illegally) or being subject to discrimination when applying for a new job. In the absence of a pension scheme, workers may have to work beyond their physical capacities to the detriment of their safety and health. When family allowances are paid by the employer, workers may refrain from declaring their full number of children. Finally, where there are no unemployment payments, workers may accept abnormal or illegal working conditions, wages and hours of work, for fear of losing their jobs.

Job and wage classification systems and payment by results schemes may be an obstacle to flexible work organisation based on flexible, self-organised teams. For example, it may be difficult to move from one category of work to another or to evaluate the contribution of each member of a team that organises its own work.

Notes

[1] ILO: *Wages: A workers' education manual* (Geneva, 3rd ed., 1982); G. Starr: *Minimum wage fixing: An international review of practice and problems* (Geneva, ILO, 1982).

[2] Article 1 of the Protection of Wages Convention, 1949 (No. 95), states: "the term 'wages' means remuneration or earnings, however designated or calculated, capable of being expressed in terms of money and fixed by mutual agreement or by national laws or regulations, which are payable in virtue of a written or unwritten contract of employment by an employer to an employed person for work done or to be done or for services rendered or to be rendered".

[3] The Declaration of Philadelphia recognises "the solemn obligation of the International Labour Organisation to further among the nations of the world programmes which will achieve . . . the effective recognition of the right of collective bargaining". See also the Freedom of Association and Protection of the Right to Organise Convention, 1948 (No. 87); the Right to Organise and Collective Bargaining Convention, 1949 (No. 98); and the Collective Bargaining Convention, 1981 (No. 154).

[4] Starr, op. cit., pp. 91-118.

[5] For a more detailed treatment of the subject, see ILO: *Payment by results* (Geneva, 1984).

[6] ILO: *Social and Labour Bulletin* (Geneva), Mar. 1981, pp. 46-47.

ORGANISATION OF WORK AND JOB CONTENT

5

When hired by a modern undertaking, a worker normally finds that his tasks and procedures have been carefully specified in advance. For some jobs, time and motion study may have identified each appropriate movement and the corresponding time available. For others, a more general set of duties and responsibilities may be indicated. In both cases, procedures will specify who supervises the worker, what should be done when a problem occurs, with whom the worker will co-operate, and so forth. Ideally, after appropriate introductory training, the worker will know how to carry out the assigned tasks and what to do in the event of questions or difficulties.

These initial tasks and procedures are critical to the worker in many ways. The skills implied by the job specifications help to determine base pay; output standards may be connected to an incentive system; the job may be monotonously simple or excessively demanding; the worker may have opportunities to co-operate and communicate with co-workers or may be isolated; supervision may be close or relaxed.

Until recently, it was assumed that all these job characteristics were economic or technical necessities. However, attempts are now being made in many workplaces to combine the demands of efficient production with methods of work organisation which provide better jobs. The implementation of these improvements is the subject of this chapter.

However, before we consider the improvement of work organisation and job content, it is necessary to understand the principles of efficient work design and explore the characteristics of a "desirable" job. This chapter has, therefore, been divided into three sections: the efficiency principles of work organisation; the social principles of work organisation; and methods of improving work organisation and job content.

Efficiency principles of work organisation

Division of labour

There are numerous reasons of cost-effectiveness for bringing workers together in one place: to share tools and machines, exchange knowledge, co-operate on a task beyond one person's capacity, or use common services. However, the key to major increases in efficiency is specialisation. This principle is so fundamental to economic analysis that Adam Smith's *The wealth of nations* begins as follows: "The greatest improvement in the productive powers of labour, and the greater part of the skill, dexterity, and judgement with which it is anywhere directed or applied, seem to have been the effects of the division of labour." [1] Smith estimated that the division of pin manufacture into 18 distinct operations by specialised workers raised production by between 240 and 4,800 times; he attributed this gain to three factors: "the increase in dexterity in every particular workman", "the saving of time which is commonly lost in passing from one species of work to another", and the greater ease of technological improvement (it was easier, for example, to devise a specialised tool for a single, well-defined task). In analysing the division of labour, Smith was mainly concerned with separating a work process into a series of tasks which could be executed independently, that is, what is now known as the technical division of labour.

Although there may seem to be no limit to the simplification of work and the work process by the technical division of labour, there are nevertheless a number of constraints. First, some kinds of work cannot easily be subdivided into a series of separate tasks which can be executed independently. For example, a series of related construction work tasks are often done by a single worker, since to bring different workers to a specific place at the worksite in order to execute each individual task would be inefficient. A second limitation to the subdivision of work is the need for two or more workers to co-operate on a task, as is commonly the case in construction work, mining or the operation of large, complex machinery. A third limitation results from mechanisation and automation, since, with automated machines which require only intermittent intervention by workers, a worker may supervise several machines simultaneously; to subdivide this work would be inefficient. A final limit stems from human limitations: people cannot execute some tasks as rapidly and accurately as machines. In short, although the technical division of labour has the potential for productivity increases, there are limits to productive efficiency that subdivision can achieve.

The social division of labour, on the other hand, aims not so much at subdividing tasks as at assigning specific tasks and responsibilities to an individual who can then develop the necessary job skills and knowledge. In specialising, people usually concentrate on a coherent set of related skills which can be developed gradually over a period of time. By specialising in a specific craft, trade, occupation or profession, a worker increases his personal competence and his value to the undertaking. Moreover, the investment made in a worker can be transferred to another undertaking doing similar work.

Since the social division of labour entails investing in an individual and teaching him certain skills, it is sometimes opposed to the technical division of labour, which subdivides tasks so that they can be done by semi-skilled or unskilled workers. Much of the history of industrial relations relates to the opposition of these two forms of the division of labour which, although necessary to efficient production, are easily brought into opposition by technological progress.

The division of labour can also increase efficiency by reducing costs rather than by increasing production. The person hired to do work which is not subdivided must be sufficiently skilled to do the most complex operation involved. When the work is subdivided, the simpler tasks can be assigned to less skilled and lower-paid workers, so that each task is paid at the lowest possible wage rate and the total wage bill is thereby minimised. Where skill-dependent wage differences are marked, wage minimisation can have quite dramatic effects. Charles Babbage, who first put forward this principle in the early nineteenth century, described a pin-manufacturing plant in which, by employing ten persons at wage rates ranging from 4-5 pence to 6 shillings a day, total wage payments had been reduced to just over a quarter of the previous level when all the workers were highly skilled and paid commensurately. Babbage points out that a worker who carried out all the pin-making operations would have to be sufficiently skilled to do the most difficult and highest-paid task. Separating out the less difficult tasks and assigning them to unskilled, lower-paid workers considerably reduced the total labour costs per unit of production.

Standardisation was another early and basic organisational approach in raising efficiency, since it was found that, by using standardised interchangeable parts, individual handcraft could be replaced by assembly work. Henry Ford's establishment of the Model-T assembly line is usually seen as a demonstration of the overwhelming advantages of this organisational method – standardisation had reinforced the benefits of specialisation, and opened the way to the large undertaking.

Nowadays, it is difficult to demonstrate clearly the production and efficiency gains that can be achieved by the division of labour, since no modern undertaking fails to apply this approach. It is difficult to imagine how long a worker would take to carry out singly all the tasks involved in the production of a large, complex product such as a motor car: without the division of labour, modern industrial wealth would not exist.

Figure 13. Steep and flat hierarchies in the pyramidal chain of command

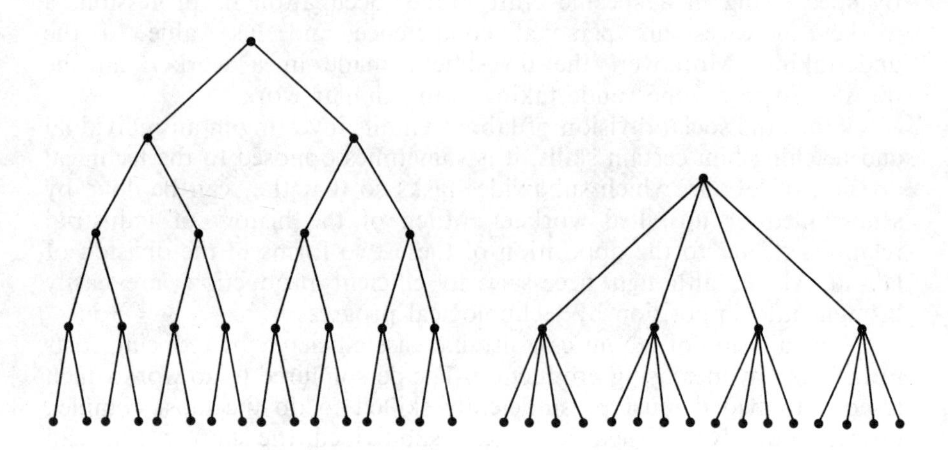

Hierarchies

How are all the individual tasks in a modern undertaking co-ordinated and controlled? Most organisations use the hierarchical principle in which each worker reports to a single supervisor, who, in turn, reports to another supervisor in a pyramidal chain of command, the length or "steepness" of which may vary (see figure 13).

Like the division of labour, the hierarchical principle is so efficient that it is found, in some form, in almost every organisation; however, hierarchies – like the division of labour – have their limits. Moreover, a number of organisational choices may be made in structuring hierarchies, and, in some circumstances, non-hierarchical alternatives may be adopted. Hierarchies which are too steep or too rigid may introduce inefficiencies:

(a) steep hierarchies increase the ratio of supervisory staff to workers and raise operating costs;

(b) a proliferation of supervisory layers impedes communication between groups of workers, increasing lost time and introducing inaccuracies when messages have to be passed through many intermediaries;

(c) a steep hierarchy can mean that a supervisor does not have responsibility for all the workers who are in the same work area or who need to co-operate with each other – the steeper the hierarchy, the narrower the "span of control" of each supervisor.

It has been suggested that the efficiency gains attributable to the division of labour make specialisation inevitable. However, to be effective, the resultant complex work systems must be correctly managed.

Since a prime goal of management is the optimal use of labour, hierarchies have been developed to co-ordinate these complex production processes. Nevertheless, there are still limits to the efficiencies that can be achieved by increased division of labour and hierarchical organisation.

Scientific management

Although the basic principles of work organisation cannot be ignored without impairing efficiency, certain extremes may be criticised on both human and efficiency grounds. For example, much criticism has been directed at the principles of "scientific management", developed over a century ago by F. W. Taylor and his followers, the most important of which are:

- the workers should be completely separated from the design of work procedures and the setting of production standards – managers should do the thinking and workers should only execute orders;
- each job should be simplified to the maximum through division of labour and specialisation, in order to minimise training time, and ensure the easy replacement of workers and that wages correspond to unskilled work;
- the worker's every action should be specified in detail by what came to be known as "time and motion study" (now more commonly referred to as "work study");
- payment should be by a progressive piece-rate system.

These principles were based on the concept that workers are merely a type of machine designed to apply muscular force and that money is their only motivation.

The principles of scientific management gradually came to dominate thinking about shop-floor workers and many lower-level clerical jobs. Their application was further stimulated as organisations grew in size and hierarchies became steeper. Ultimately, the proponents of scientific management exceeded the degree of division of labour that could be justified on grounds of efficiency, and were confronted with technical problems:

- the rigid specification of job content makes adaptation to changing conditions very difficult: for example, since workers receive no generalised training, they are unable or unwilling to deal with disruptions caused by variations in the supply of raw materials, machinery and equipment problems, changes in product specifications or the absence of fellow-workers;
- the detailed specification of job content is costly and often technically difficult, leads to complaints when the above-mentioned variations occur, entails delicate and difficult balancing in the allocation of time to different jobs, and requires revision following any change in product, processes, machinery or procedures;

- classifying workers as replaceable components of the production process takes no account of inter-individual variations; and, in fact, Taylor demanded the "best man" for each job, even though his concepts were ultimately supposed to apply to the entire workforce;
- no utilisation is made of the worker's knowledge, experience, adaptive abilities, inventiveness or other capacities.

In view of this, a trend away from scientific management might have been expected. However, the engineer designing the work at shop-floor level often failed to realise the problems inherent in treating workers as machines. Nevertheless, many undertakings were operating at less than optimal efficiency, and the potential was present for gains in efficiency through organisational innovation.

Social principles of work organisation

Given the extremes to which scientific management was sometimes carried, it is not surprising that the importance of social factors was soon "discovered". Studies on women telephone assemblers at the Western Electric Company in the 1920s clearly demonstrated the importance of human factors. It was found that, when working conditions such as lighting and the length of rest pauses and the working day were improved, there was a simultaneous and significant increase in productivity; however, when the women were returned to their original working conditions, productivity fell back to previous levels. It thus became clear that, under certain circumstances, social factors could influence productivity to a greater extent than physical and economic conditions. Subsequently, other studies were to demonstrate, in particular, that interpersonal relations in small groups of workers had a major influence on work norms and productivity.

Studies on coalminers by researchers at the Tavistock Institute in the early 1950s confirmed these findings. The introduction of the new long-wall mining technique broke up the existing small, close-knit and relatively autonomous groups which had developed a particular cohesion in the face of the dangers confronting them in their work and as a result of the supportive community that the miners had formed outside the workplace. Consequently, efforts were made to re-create these autonomous groups whilst using the new technology; the achievement of this objective was greatly appreciated by the miners and contributed substantially to productivity. Other studies followed on the potential for group work or alternative organisational arrangements under various technological and economic conditions.

By the 1960s, there was growing realisation of the advantages of offering work which was attractive to the worker and which, for example, had characteristics similar to the following: [2]

- variation and meaning in the job;
- continuous learning on the job;
- participation in decision-making;
- mutual help and support from fellow-workers;
- meaningful relations between the job and social life outside; and
- a desirable future in the job – not only through promotion.

Although there is widespread agreement that work dimensions need to be extended, it should not be forgotten that workers have very different backgrounds, skills and preferences and that their objective needs, personal strengths, individual weaknesses and attitudes towards work are multifactorial. For example, one worker may emphasise pay, another companionship, and yet another responsibility, and the same individual may evaluate different aspects of his job in very different ways. No single job or organisational improvement will be appropriate for all workers, especially since needs and preferences change with time.

The improvements that would result from the implementation of the above-listed characteristics would make jobs more challenging and demanding. There is always the danger of overloading a job and that excessive work intensity will result in occupational stress. Research has shown that fast-paced but boring jobs are more stressful than mentally challenging ones, and that manual workers can show greater symptoms of stress than managers or office workers.

Although different workers will evaluate the above job characteristics differently and although desirable levels of variety may change from one worker to another, questionnaire surveys and behavioural studies of productivity, product quality, absenteeism or labour turnover have shown that job improvements in line with these characteristics can significantly modify the worker's subjective reactions.

Methods of improving work organisation and job content

After the above review of the economic and social reasons for organising work and designing jobs so as to avoid the extremes of scientific management and the dangers of rigidly hierarchical and centralised control, some examples of how to achieve the desired objectives will now be given, ranging from the simplest to the most complex – that is, from marginal improvements in individual jobs to major organisational changes.

Decoupling people from machines

Work in which the machine or assembly line sets the worker's pace is particularly stressful and demotivating. Not only is the work repetitive and monotonous but, moreover, the worker is unable to adjust the pace to deal with temporary fatigue or circadian rhythms. Often, a break is possible only when a relief worker takes a scheduled turn at the job.

The most direct means of decoupling workers from machine pacing is to provide an upstream stock of unfinished products that the worker can draw on when wanting to work faster, and another stock, downstream, of finished goods to act as a buffer when the worker takes a break or is working at a slower pace, and thus to release him from the minute-by-minute demands of machine-paced activity. A more advanced solution is to install automatic machinery to carry out the machine-paced tasks so that the worker can devote time to tasks which are not machine paced, such as machine loading and machine maintenance and adjustment.

A convenient way of determining the extent of the worker's freedom from the rigours of machine pacing is by measuring "time of autonomy" or "temporal span of control", that is, the time a worker can leave his workplace without perturbing production. This is dependent on two factors: the total duration of work breaks during the working day; and the level of the downstream buffer stocks the worker can build up.

The use of incoming and outgoing buffers in assembly work is shown in figure 14. Figure 15 illustrates a number of other possible buffer techniques.

Optimisation of cycle time

Many jobs (e.g. in assembly lines) comprise a series of tasks to be done in a specified fashion. The operation of most machines (whether lathes or cash registers) entails the repetition of a series of tasks. Many office jobs, especially those handling large numbers of files, as in insurance or social security administrations, consist in repeatedly applying the same procedure. The period between the start of two consecutive series of operations is the "cycle time". Very short cycle times are associated with low skill levels, monotony, boredom and occupational stress. Where a simple task such as tightening a bolt is repeated several times within each cycle, the cycle is said to have "internal repetition". Jobs which have both short cycle times and considerable internal repetition are particularly tedious.

If cycle times are continually shortened, there will clearly come a point at which the job is too repetitious for most workers; on the other hand, jobs with very long cycle times may be excessively difficult to learn and complicated to execute, especially when each task must be executed swiftly.

The optimal cycle time for assembly work is probably between 90 seconds (the minimum specified in a collective agreement in the Federal Republic of Germany) and 15 to 20 minutes (which has proved to be a practical upper limit in a number of situations – although longer cycles

Figure 14. Incoming and outgoing buffers on the motor-car assembly line of the Volvo plant in Kalmar, Sweden

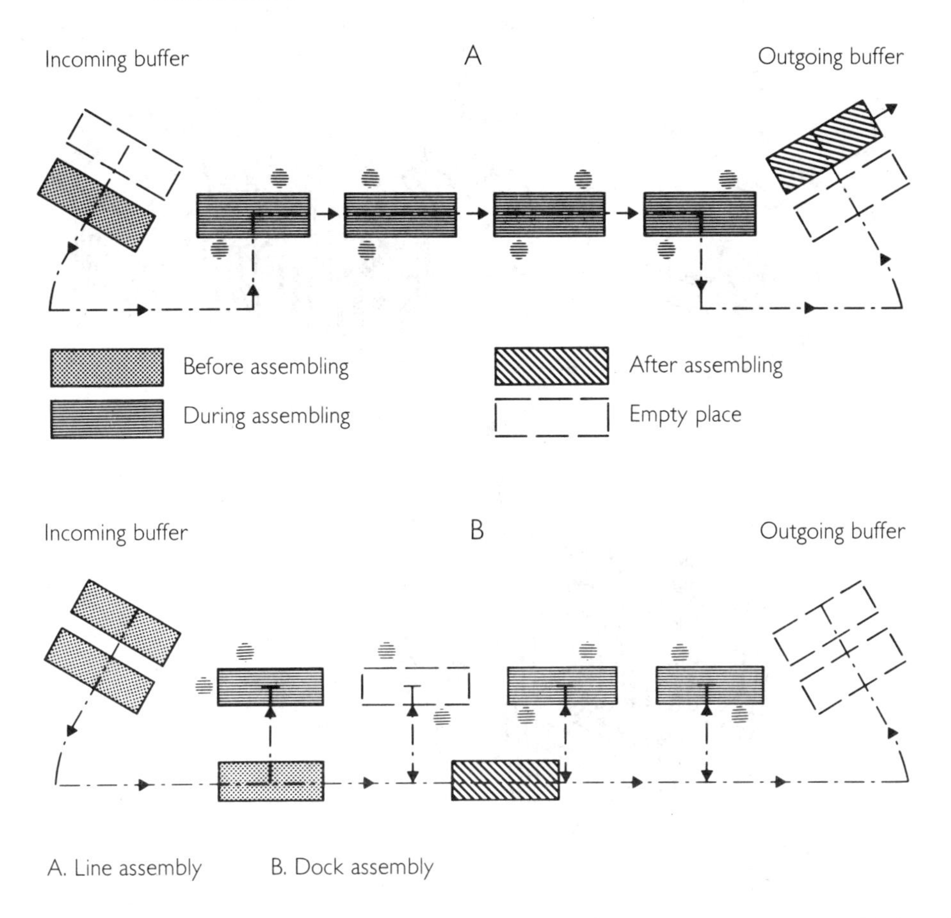

A. Line assembly B. Dock assembly

Source. P. Dundelach and N. Mortensen: "Denmark, Norway and Sweden", in ILO: *New forms of work organisation*, Vol. 1 (Geneva, 1979), p. 29.

have been successful). A major factor in determining the optimal cycle time is whether the worker is able to complete an entire product or a significant sub-assembly.

One way of lengthening the cycle time on a production line, in which the work is subdivided many times, is to create shorter parallel lines, as shown in figure 16.

Providing workers with the opportunity of completing an entire product or sub-assembly may require redesigning the product so that it is built up of a small number of major modules, each of which can be assembled by a worker, as is the case with the domestic oven unit shown in figure 17.

Figure 15. Examples of typical buffer stock techniques in manufacturing operations

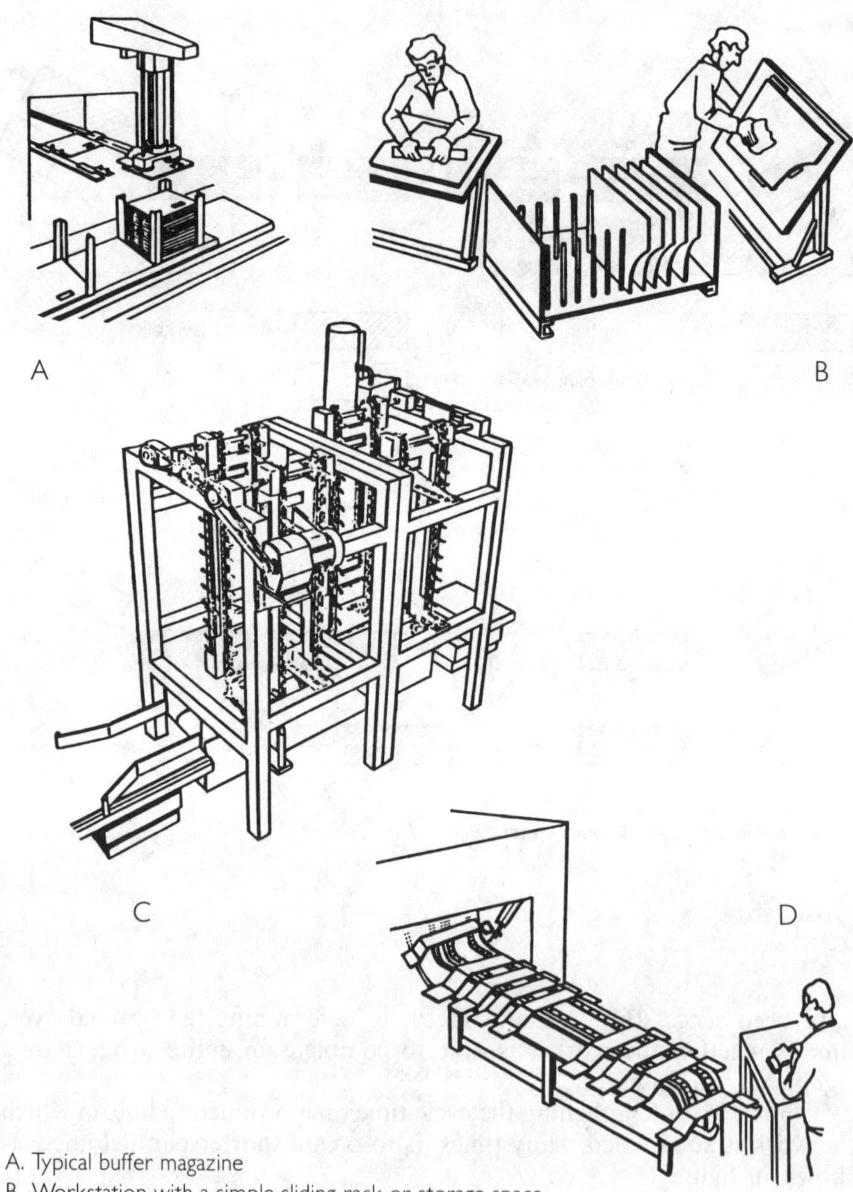

A. Typical buffer magazine
B. Workstation with a simple sliding rack or storage space
C. High-stacking machine used as a buffer
D. Buffering track

Source. ILO: *Introduction to work study* (Geneva, 3rd (revised) ed., 1979), p. 399.

Figure 16. When the operations on a single, long assembly line are distributed over a number of short, parallel lines, the cycle times can be lengthened

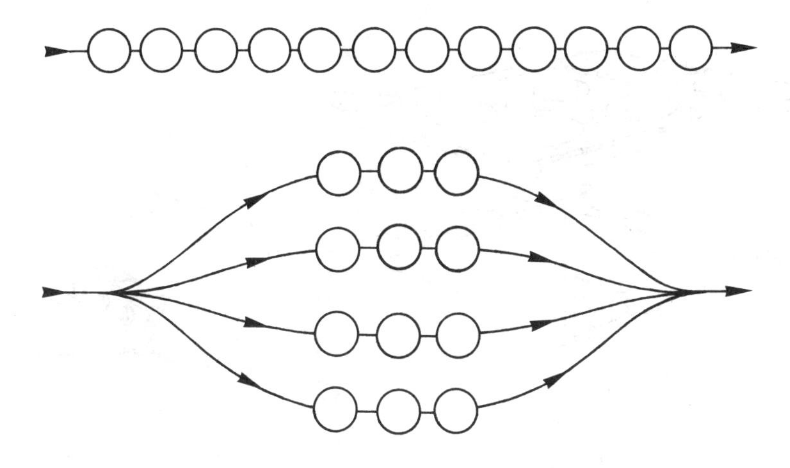

Source. ILO: *Introduction to work study,* op. cit., p. 392.

Job rotation

A simple technique for increasing work variety is to rotate workers among similar jobs in the same production area. Where the tasks are simple, as on an assembly line, training costs will be low and the equipment and work processes will not need modification. Job rotation is often used to reduce the time a worker spends on a job which is particularly fatiguing or stressful, by spreading it among a relatively large number of workers. However, since job rotation can do little more than spread a disagreeable burden or substitute one boring task for another, and since the constant repetition of the same task requires less mental effort, workers often decide that job rotation schemes are not to their advantage and prefer not to participate in them.

Job enrichment

In this technique, tasks and responsibilities are added to a worker's job to increase satisfaction and motivation. A number of job dimensions are listed and defined below; increasing one or more of these dimensions will tend to enrich the job:

- skill variety: the degree to which a job requires a variety of different activities that involve the use of a number of different skills and talents;

Figure 17. By redesigning this domestic oven unit as a number of separate modules, each worker could be given a whole module to produce and the cycle time lengthened accordingly

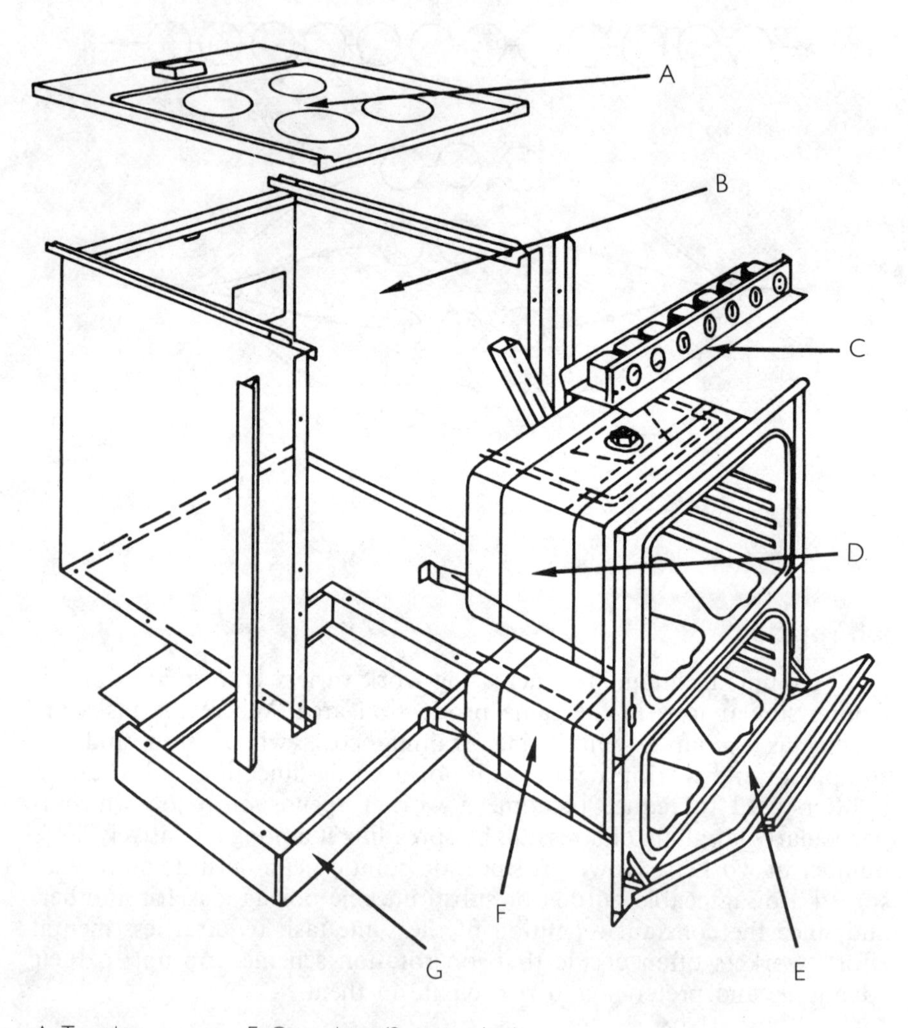

A. Top plate
B. Housing
C. Front panel
D. Upper oven
E. Oven door (2 per cooker)
F. Lower oven
G. Base

Source. R. Lindholm and S. Flykt: "The design of production systems: New thinking and new lines of development", in G. Kanawaty (ed.): *Managing and developing new forms of work organisation* (Geneva, ILO, 2nd (revised) ed., 1981), p. 64.

- task identity: the degree to which the job requires the completion of a whole and identifiable piece of work, that is, doing a job from beginning to end with a visible outcome;
- task significance: the degree to which the job has a substantial impact on the lives or work of other people, whether in the immediate organisation or in the external environment;
- autonomy: the degree to which the job provides substantial freedom, independence and discretion to the individual in scheduling the work and in determining the procedure to be used in carrying it out; and
- feed-back: the degree to which carrying out the work activities required by the job results in the individual obtaining direct and clear information about the effectiveness of his performance.[3]

One example of job enrichment by expanding job dimensions relates to the tools, machinery and equipment involved in the process. Workers can be made responsible for tool set-up before production starts; they can adjust machinery which is not operating correctly; and they can be entrusted with regular maintenance or even, in some cases, repairs. Another concerns the product itself, since the worker can be given responsibility for inspecting his work and repairing any defects. Still another example is to delegate responsibility for work planning and scheduling, for inventories of parts or work in progress, for safety in the immediate work area, and so on. In office work and in many service occupations, where technological constraints may be less significant, workers may be allowed to set their own immediate priorities, plan their own work and choose their own work methods.

Group work

Semi-autonomous groups

The improvements discussed so far are all applicable to individual work stations or jobs, and require no collaborative efforts among workers. They have, however, at least two important limitations. First, they fail to take advantage of people's desire to work together and the motivational potential of co-operation. Second, owing to individual limitations, workers may be unable to carry out the entire set of activities for making a product. In co-operative work, the weaknesses of one individual can be compensated by the strengths of another. Moreover, groups can make internal adjustments to compensate for differing preferences, temporary problems, variations in the nature of the work, absences, and so on. Groups can therefore take on larger and more meaningful responsibilities, including certain supervisory responsibilities.

A widely discussed innovation is the semi-autonomous work group which forms the basis for most experiments in new forms of work organisation in Scandinavia and elsewhere. Semi-autonomous work groups can offer workers jobs which combine all the advantages listed in the section on job enrichment. They have the following characteristics:

- the product of the group is specified (and, as in the case of job enrichment, should be a significant and identifiable output) but the group has considerable scope in deciding on work methods, the scheduling of various activities, the work assignment of each worker and the way of coping with problems that arise in the work;
- the group has considerable autonomy in choosing and changing its internal structure and in dealing with the rest of the organisation: for example, many semi-autonomous groups can choose their own leaders or spokesmen and sometimes their members as well;
- each worker is, in principle, able to carry out all the tasks required to make the group's product – or at least there is a training programme directed at assuring that each worker is developing the necessary range of skills; and
- members of the group have joint responsibility for results, and any incentives relate to the group as a whole.

Relatively major and expensive technical and organisational changes are necessary for the introduction of semi-autonomous work groups. Consequently, it has usually proved desirable for planning for group work to take place when an entire new plant or production facility is being designed. Figure 18 shows an example of semi-autonomous group working in the assembly of engines at a Saab-Scania plant in which small groups are used to assemble an entire engine in approximately 30 minutes. The normal manning level is three operators per group, but up to six workers can be assigned to each loop; the number of loops in operation can be varied and, if necessary, assembly can be carried out by a single worker.

This example relates only to assembly work; however, semi-autonomous work groups have been used successfully in many industries, such as coalmining and the manufacture of fertilisers, paper pulp, petrochemicals, synthetic fibres, aluminium and food products.

Matrix groups

Matrix groups can be used to bring together workers with different and only partially overlapping skills. Figure 19 shows a matrix organisation designed for a ship, in which it is difficult to ensure that all crew members have acquired the full range of skills necessary for the vessel's operation. A matrix group is not, however, the same as a small hierarchical organisation. Each worker is expected to have skills which overlap with those of at least some other workers and to carry out those workers' functions whenever necessary. Matrix groups lack some of the

Figure 18. Three of the seven semi-autonomous work-group engine-assembly loops at a Saab-Scania plant

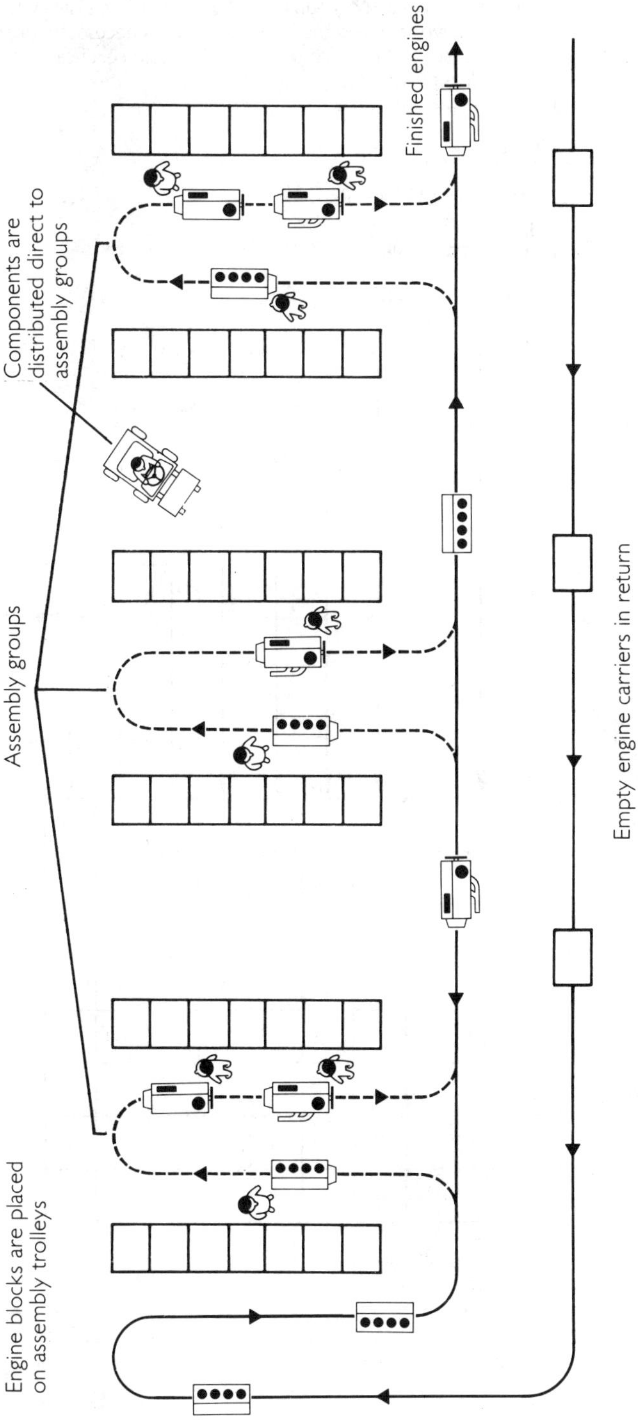

Engine blocks are placed on assembly trolleys

Assembly groups

Components are distributed direct to assembly groups

Finished engines

Empty engine carriers in return

Source. Lindholm and Flykt, op. cit., p. 59.

Figure 19. The organisation of a matrix group for a ship, in which it is difficult to ensure that all crew members have acquired the full range of skills necessary for the ship's operation. In this ship, several members of the crew hold certificates outside their own department (dual junior-officer training)

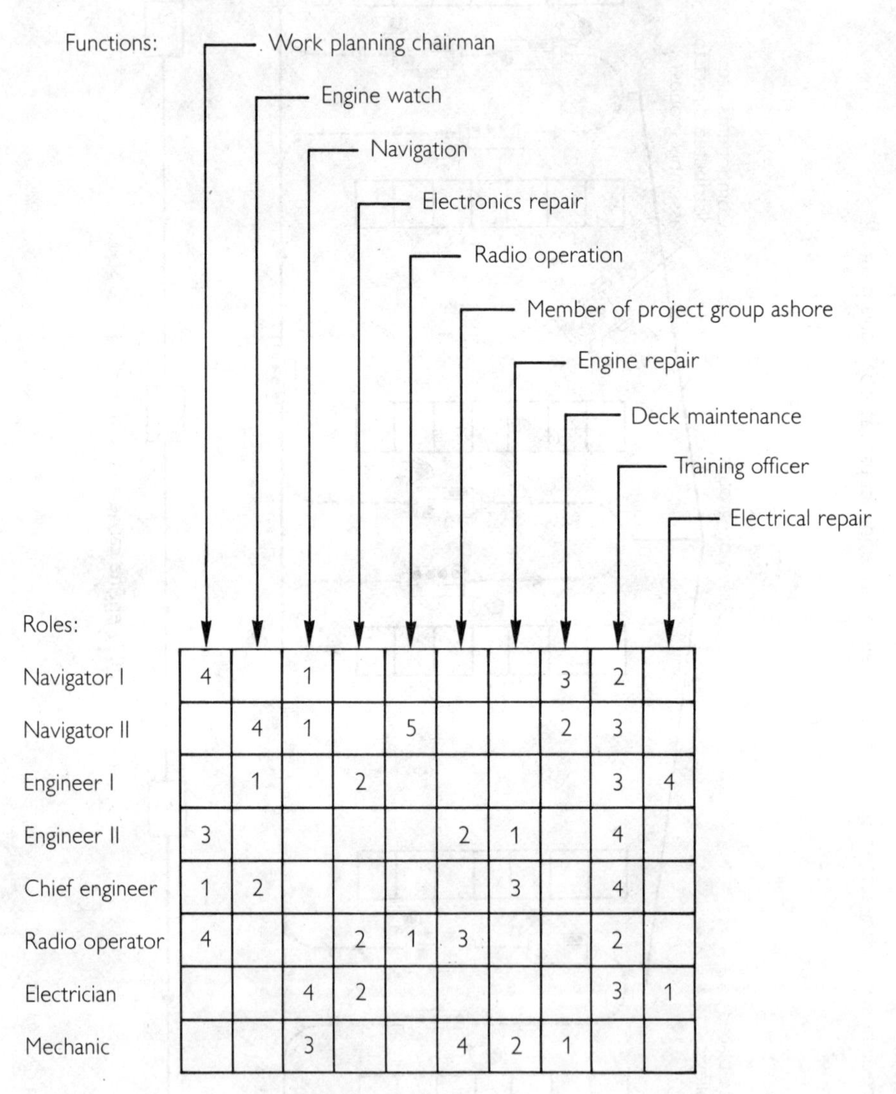

Roles	Work planning chairman	Engine watch	Navigation	Electronics repair	Radio operation	Member of project group ashore	Engine repair	Deck maintenance	Training officer	Electrical repair
Navigator I	4		1					3	2	
Navigator II		4	1		5			2	3	
Engineer I		1		2					3	4
Engineer II	3					2	1		4	
Chief engineer	1	2					3		4	
Radio operator	4			2	1	3			2	
Electrician			4	2					3	1
Mechanic			3			4	2	1		

1 means primary role, 2 secondary, etc.

Source. E. Thorsrud: "The changing structure of work organisation", in Kanawaty (ed.), op. cit., p. 22.

operating flexibility of semi-autonomous groups, but they have several advantages:

- they have no size limit, whereas the maximum size of a semi-autonomous group is usually about 12 persons;
- matrix groups can bring together a wider variety of skills than semi-autonomous groups;
- matrix groups have a basic structure that is determined by the pattern of overlapping skills or competences, but, within this, a variety of structures may be adopted, depending on the task at hand; and
- it is easier for matrix groups to operate on a temporary or intermittent basis whilst remaining embedded in a larger organisational structure.

A new form of work organisation which is found widely in countries with a centrally planned economy is based on "work brigades", and illustrates well the characteristics of matrix groups. These work brigades tend to be relatively large with up to 50, 60 or even, in exceptional cases, 100 members, and are given responsibility for a major project, such as the construction of a building or the operation of a large segment of a factory. They are made up of workers of different occupations and skill levels, although considerable emphasis is placed on multiple-skill development to increase flexibility. Workers elect their team leader or brigadier, but there is often a foreman as well. They negotiate a collective agreement with the management of the undertaking or with the relevant ministry, specifying inputs, production quantity and quality, and wage and other incentives. In some cases, semi-autonomous work groups may be formed within the brigades.

Organisation of industrial work: Flow grouping versus functional grouping

When a manufacturing process routes a product through the hands of several different tradesmen, each of whom is using a different machine, the work has traditionally been organised on the basis of functional grouping in which each department or shop contains job specialists. In flow grouping, by contrast, production equipment is arranged in the same sequence as the operations in the production process, and the organisational unit or department carries out the complete set of production operations. Figure 20 diagrammatically compares these two forms of organisation.

When flow groups are used, each major organisational unit has many different tasks, and jobs are easier to enrich. Moreover, each department has complete control over all the operations in product manufacture and can be given clearer responsibilities, since problems of product quality and quantity cannot be easily blamed on other organisational units. Figure 21 gives another illustration of the

Figure 20. Diagrammatic comparison of functional grouping and flow grouping for the organisation of an industrial process involving a sequence of sheet-metal work, machining, painting, assembly and inspection

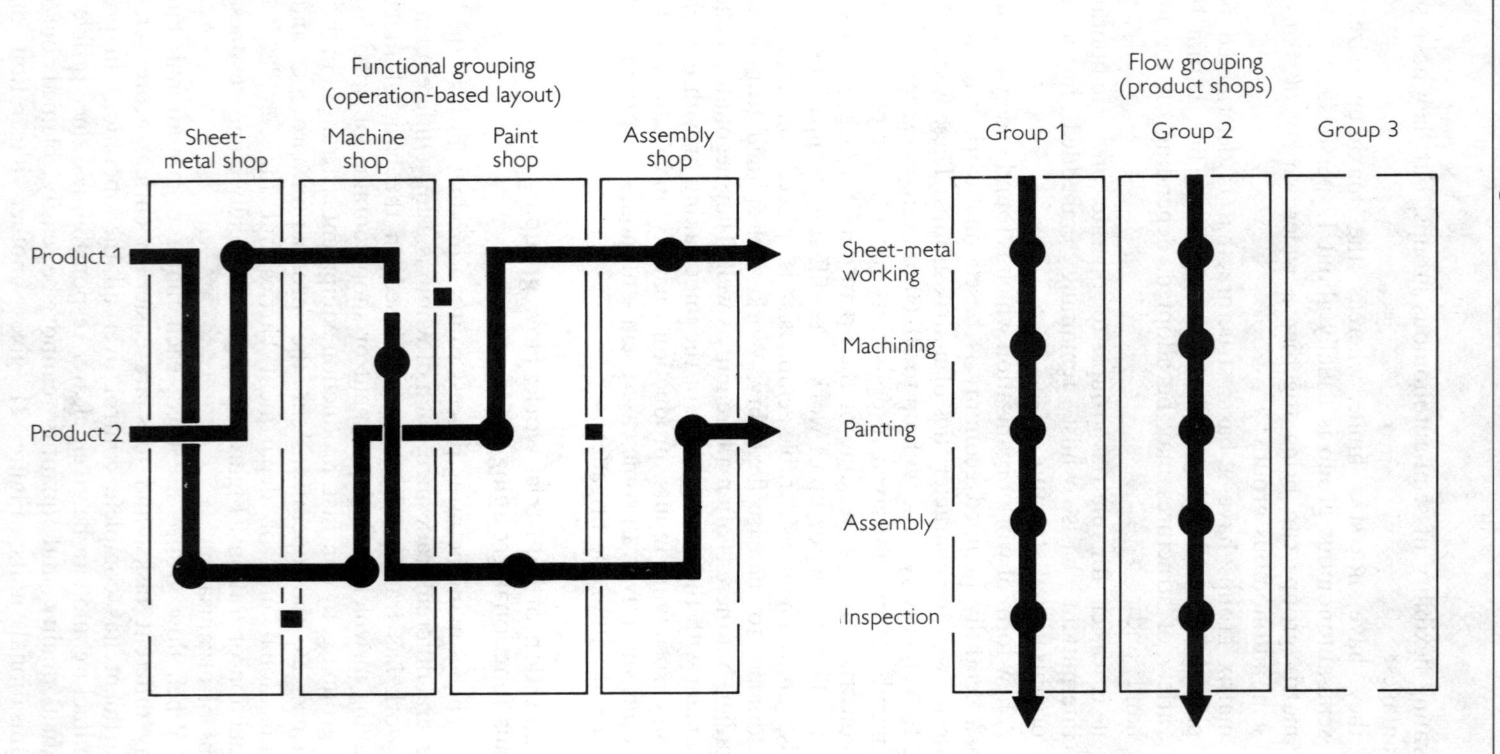

Source. Lindholm and Flykt, op. cit., p. 46.

Figure 21. A comparison of structures for the organisation of production based on function versus flow

Organisation of production activities by function

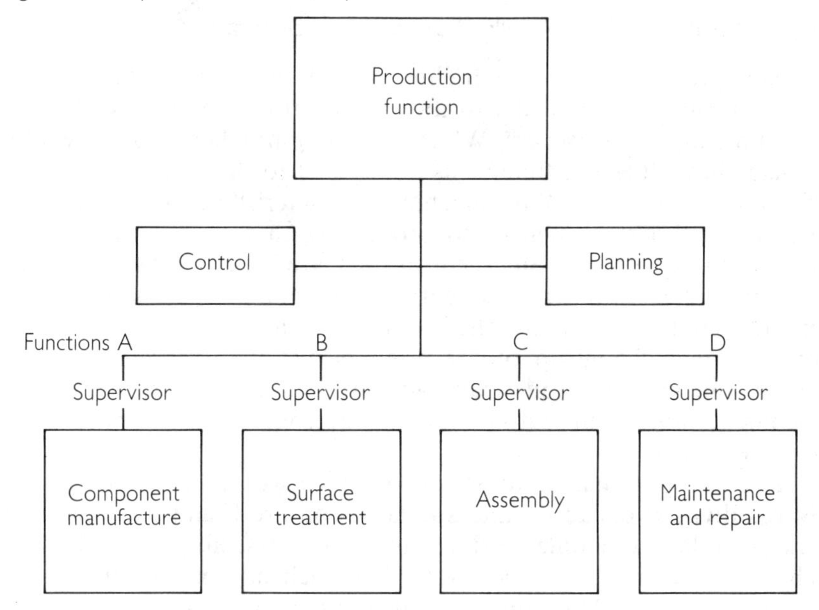

Organisation of production activities by process

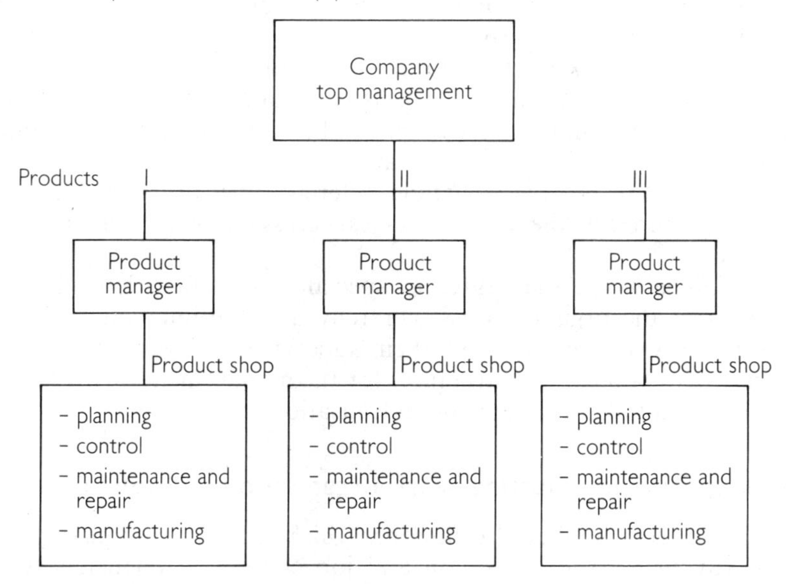

Source. Lindholm and Flykt, op. cit., pp. 75-76.

comparison between organisational structures based on flow versus function and of its implications at various levels of the organisational hierarchy.

Organisation of office work: Forms of decentralisation

The principles of job enrichment and group working described earlier can also be applied to office work, although the technological constraints are not so severe. While a growing number of office workers use machines either full time or as an adjunct to their work, most office jobs are not determined by machine characteristics, and it is usually organisational arrangements and procedures that require modification, rather than equipment. Since constraints are weaker, there is greater variation in the way in which decentralisation and delegation of authority can be handled, and in the forms of intra-group co-operation. Organisational change in offices often occurs in a more ad hoc and gradual manner than in manufacturing, and such terms as "job enrichment" and "semi-autonomous work group" may not have exactly the same meaning.

In the traditional organisation of office work, the same pressures towards division of labour and specialisation are found as in industrial work. This has sometimes resulted in "paper assembly lines" in which each worker carries out a separate task such as opening and routing incoming correspondence, making calculations, preparing replies, typing outgoing correspondence and dispatching. In particular, where office machines are used for calculating or typing, work is commonly organised according to functional groups.

The office work analogy to a "flow group" in manufacturing is a work group responsible for all aspects of a major segment of the office's activities. For example, a group might be responsible for handling all insurance claims from a particular geographical region, and would therefore need to be able to handle different types of claim and to take each claim through the various stages necessary to give a satisfactory response.

In office work, the degree of specialisation often depends on the steepness of the organisational hierarchy and the interactions between workers or work groups and their supervisors. Figure 22 illustrates alternative organisational structures for the management of work groups and the potential for greater decentralisation.

Other techniques affecting work organisation and job content

There are a number of techniques which, although not mainly aimed at improving work organisation and job content, nevertheless tend to produce improvements in at least some cases. For example, the introduction of the so-called Scanlon plans (undertaking-wide incentive

Figure 22. Alternative organisational structures for the management of work groups

1 A traditional hierarchy with work groups in the office

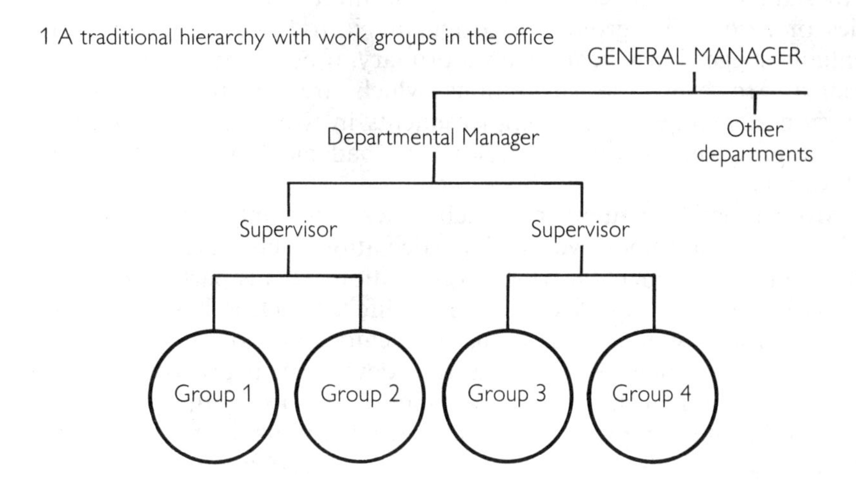

2 A traditional but flatter hierarchy

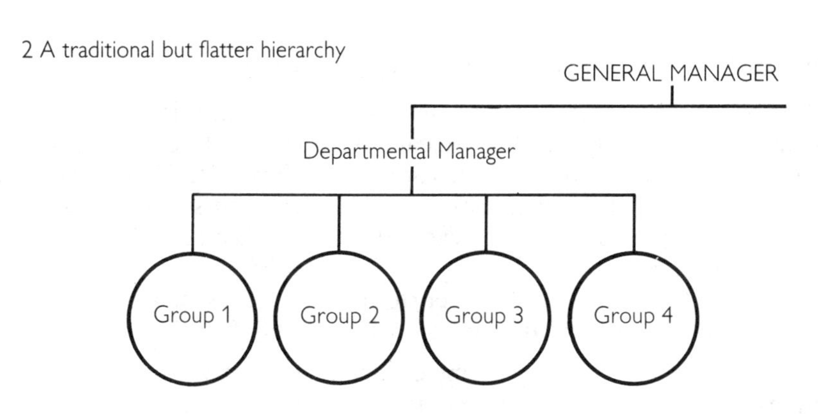

3 Group organisation

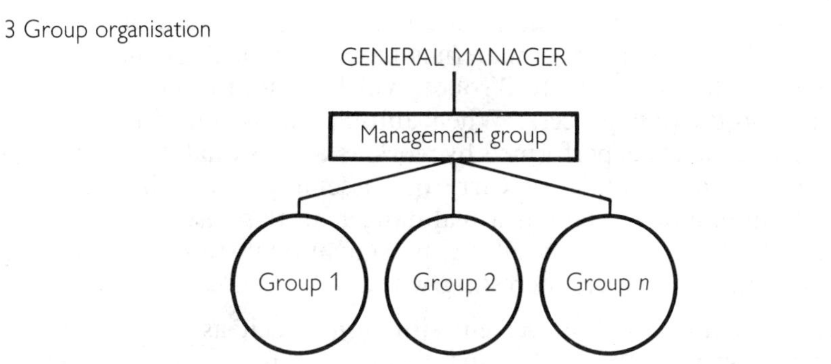

Source. D. Birchall and V. Hammond: *Tomorrow's office today: Managing technological change* (London, Hutchinson Business Books, 1981), p. 61.

schemes with particularly extensive suggestion systems) have sometimes led to significant changes in work organisation. Similarly, the quality circles or zero-defect groups, as originally introduced in Japan, were essentially suggestion schemes with a primary, if not exclusive, economic objective. However, the suggestions which are generated in group discussion sometimes lead to improvements in work organisation, and recent applications of quality circles have had more overall effects on work organisation.

More broadly, a number of techniques commonly grouped under the heading "shop-floor workers' participation" have also eventually focused on improvements in work organisation. This is particularly true of initiatives on the "quality of working life", which is based on both socio-technical design and the industrial democracy movement. Finally, personnel management and organisation development practitioners are more and more realising the importance of many of the techniques described in this chapter. Ultimately, what counts is not the label attached to a particular improvement but the actual efficiency and human benefits.

Automation and its effects on work organisation

A number of the improvements in work organisation described above are closely linked to technological developments. Recent trends – especially the growing use of microprocessor-based technologies – suggest that automation will have growing repercussions on work organisation and job content. However, automation is a general term for complex and differentiated technological changes and, in industrial work, it is important to distinguish between four major types of automation, each of which has a different impact on job quality.[4]

Type 1 automation. The piecemeal introduction of various forms of computer-assisted machines gives rise to what might be called "electronically augmented mechanisation". This includes the use of isolated packaging machines, materials-handling machines, numerically controlled machines or industrial robots which do not form part of a fully integrated production process. When automation is introduced in this way, many of the tasks performed by workers are residual. Workers load the machines, for example, or carry out assembly tasks which are not yet worth automating. There is a real danger, in this case, that the pace of work will be accelerated, resulting in occupational stress and blocking the opportunity for more interesting work.

Type 2 automation. In certain situations, such as arise in power plants or refineries, the underlying technology is predictable and its control is easily programmed. This allows a close approximation of the fully automatic factory in which workers spend most of their time monitoring the process without any need actually to intervene. The problems associated with this type of automation are monotony and boredom, followed by insufficiently rapid reaction when a problem or an emergency occurs.

Type 3 automation. When the underlying technology is less certain but considerable automatic control equipment has been installed, a type of automation occurs which requires frequent human intervention in the form of control tasks. This type of automation is commonly found in paper-mills, printing shops and automated transfer lines with a varied product mix. In type 3 automation, the work is inherently more interesting than in type 1 or type 2. However, the programming and operation of the new automatic control equipment require kinds of knowledge and skill that are completely different from those associated with the traditional crafts or trades in the industries affected. Workers may find that their traditional skills have been obviated, together with the higher rate of pay they commanded. Moreover, these workers may lack the general education and mental flexibility to master the algorithmic procedures involved in control tasks.

Type 4 automation. This combines the technological uncertainties that are characteristic of type 3 automation with the need for manual work. Type 4 automation is characteristic of steel pipe, synthetic fibre and aluminium production, since these industries still require manual skills. They provide opportunities for saving certain traditional skills and for combining manual and intellectual work. However, these tasks may also be assigned to completely different workers – a process which polarises workers into separate categories.

The above classification should make it clear that automation does not inevitably lead to improvements in work organisation and job content. If automation is to result in improvements rather than greater problems, its implementation will require careful planning. Nor should it be assumed that, as more and more of the work process is automated, the problems will progressively disappear. All the different types of automation have their associated problems. Similarly, the automation of office work may result in either improvements or greater problems. For example, the job of specialist word-processor or data-entry operator may combine high work intensity and stress with low career opportunities. Procedural rigidities may be programmed into computer software in ways that give workers little opportunity to handle unusual situations in an appropriate way. Especially in the case of service organisations, this can mean that close worker/client collaboration is replaced by "red tape" or a "faceless bureaucracy". On the other hand, computers may enable workers at lower levels of the hierarchy to have access to a greater range of information and to increase both their knowledge of the total work system and their ability to take on greater responsibilities.

Introducing improvements in work organisation and job content

Not all attempts at improving the organisation of work will be successful. Many pitfalls must be avoided if both efficiency and worker satisfaction are to be increased. A first, basic, point is that even small and seemingly simple organisations have a variety of "subsystems" which help to assure their continuing operation. Among those most relevant to work organisation and the improvement of job content are:

- task-oriented systems: equipment design and layout; standard work methods;
- human resource systems: personnel selection, training and assignment;
- systems of formal rules: contractual obligations; work norms or quotas; work rules and established practices;
- reward and sanctioning systems: payment or allocation of privileges by results; career systems based on performance; rate setting; disciplinary actions; and
- persuasive systems: leadership; emulation; promotional activities.

These subsystems have a critical impact on workers. Changes in work organisation which conflict with these subsystems may generate considerable problems. For example, a system of payment by results for individuals may conflict with the adoption of group production methods; new production methods may require changes in established work rules or even contractual obligations; or the different skill requirements of new jobs may require changes in personnel selection and training.

In many organisations, different people are responsible for the operation of these subsystems. Because of this, attempts to design or redesign work are often undertaken from conflicting points of view. Equipment and plant layout are designed by an engineer with one approach, the workers are then selected and trained by a personnel department with a different approach and, finally, operations are started using work standards and supervisory practices deriving from yet another approach. The potential for problems is obviously considerable. In these circumstances, early planning with the participation of all parties is essential. This participation necessarily includes many basic management functions; for example, the introduction of work groups may require at least some modification in accounting procedures in order to change from individual to group incentive plans.

Worker and trade union participation

Worker and trade union participation is critical to the success of any attempt to improve work organisation and job content, for at least four reasons. First, the workers themselves are best placed to indicate their own preferences as regards job and organisational characteristics. Second, workers have intimate knowledge of production procedures and can give much useful information on the most effective equipment and procedures; however, this entails more than a formal request for help or the introduction of a suggestion scheme. Third, the improvements described in this chapter are directed not only at improving working conditions but also at greater efficiency; greater worker motivation is essential to efficiency improvements, and behavioural research has shown the powerful motivational effect of actual participation in work design and performance-goal setting. Finally, worker participation has been shown to help to overcome workers' apprehension concerning new technology or procedures and therefore to smooth the process of introducing changes.

Until recently, most trade unions tended to leave work organisation to management and to use the defined job characteristics and skill requirements as a basis for negotiating wage rates; work organisation was classed with "soft" issues such as personal satisfaction. Recent experience with rapid technological change has demonstrated that many issues of critical importance to trade unions – such as job numbers, skills and training, job evaluation and pay, and pace of work – are largely influenced by decisions taken early in the process of work and job design. In recognition of this, the trade union attitude is no longer "whether to participate" but rather "under what conditions to participate". Trade unions now often negotiate agreements containing trade union participation, an equitable share for workers in production increases or cost reduction, and the right to withdraw from experiments on new forms of work organisation and to revert to pre-existing conditions.

Sources of pressure for change

These are numerous, but perhaps the most common is the perception by management of an opportunity for greater efficiency. Management may wish to take advantage of developments offering greater manufacturing flexibility; it may perceive the possibility of increased motivation and productivity; or again, it may react to growing dissatisfaction on the part of workers, as expressed by a rise in absenteeism and labour turnover. This last case is an example of pressure for improvements coming indirectly from the workers themselves, but workers and their trade unions may also take the initiative by submitting suggestions or making specific demands in collective bargaining negotiations. Many governments are adopting a more active approach to the promotion of improvements in work organisation, either by funding research, by setting up institutions to formulate and support improvements, or even by legislation. Work organisation standards and models are important means of promoting work organisation improvements in centrally planned economy countries.

Table 11. Comparison of the main features that distinguish old and new designs of work organisation

Old design principles	New design principles
Tasks are broken down into unrelated bits	Tasks vary in complexity and are kept as "complete wholes"
Job training and knowledge are minimised	Training and knowledge are broad and cover future needs
Most workers' jobs are repetitive	Few jobs are repetitive
Sharp job demarcations	Partly overlapping jobs
One person, one job	Individual *or* group work
Every person, one boss	Some report to one, others to more
Groups are only "informal"	Groups share responsibility
Task and responsibility are relatively permanent	People may rotate between roles – horizontally and vertically
Information and control are mostly vertical, top-down	Information and control are vertical or horizontal depending on problem situation
Planning and decision-making are centralised	Decentralisation: planning and decision-making are part of all jobs
Technology is taken as given	Technology is adapted to social and organisational needs
"Tall" organisation chart	"Flat" organisation chart
Few links with outside	Many links with outside
Centralised and little risk-taking	Extensive innovation and risk-taking
Man as "exchangeable part"	Man and organisation are learning to take on new functions
Administrative infrastructure prevents self-regulation	Infrastructure promotes self-regulation and self-management

Source. Thorsrud, op. cit., p. 13.

Conclusion: Old and new design principles

Undertakings no longer operate in a simple, predictable environment; flexibility and adaptability are often as important as cost minimisation. Few undertakings can afford to assume that their workers are merely replaceable machines, and an ever more educated and knowledgeable workforce is aware of the existence of alternative approaches to work organisation and job design. Pressure is building up to abandon hierarchical, over-specified forms of organisation and adopt flexible, decentralised forms requiring multiple skills. Table 11 compares the main features that distinguish old and new types of work design and work organisation.

The new design principles have many advantages for both undertakings and workers. For undertakings, these advantages include greater operational flexibility, improved internal co-ordination and reliability and a better-motivated, better-trained workforce. For workers, improvements such as a reduction in occupational stress, greater opportunities for co-operation, a better use of skills and improved career structures are combined with the benefits of a more productive undertaking, not least of which are improved pay and greater job security.

Notes

[1] A. Smith: *An inquiry into the nature and causes of the wealth of nations* (Oxford, 1776; reprinted many times by various publishers), Book I, Chapter I.

[2] P. Dundelach and N. Mortensen: "Denmark, Norway and Sweden", in ILO: *New forms of work organisation*, Vol. 1 (Geneva, 1979), p. 23.

[3] J. Richard Hackman: "Work design", in J. Richard Hackman and J. Loyd Suttle (eds.): *Improving life at work: Behavioral science approaches to organizational change* (Santa Monica, California, Goodyear Publishing, 1977).

[4] For a full review of the implications of automation for the quality of working life, especially in industrial work, see F. Butera and J. E. Thurman (eds.): *Automation and work design* (Amsterdam, North-Holland Publishing, 1984).

WORKERS' WELFARE FACILITIES 6

The term "welfare facilities" is used in this chapter to describe services, facilities or measures provided or taken by employers, individually or as a group, on their own initiative or to meet a legal requirement, in order to improve worker welfare during working hours or to ensure better conditions of life.

This definition indicates that there are two different types of welfare facilities: those intended to improve worker welfare during working hours (including, for instance, commuting time); and those designed to improve wage earners' conditions of life not so much at the workplace as at or near their homes, and for the benefit of both workers and their families.

Facilities for workers' welfare during working hours

These may be divided into three groups:

- facilities which form part of a workplace health-protection policy, such as adequate and suitably located sanitary facilities (WCs, urinals), drinking water, washing facilities, cloakrooms and cupboards for storing town clothes safely and hygienically during working hours, and for drying working clothes;
- facilities which help to prevent or limit workers' fatigue, such as seats for persons doing work that can largely be carried out seated, rest-rooms where workers can, during working hours, take a momentary break from arduous working conditions (heat, cold, noise, vibration, toxic substances in the air, etc.), and facilities to reduce commuting time and the resultant fatigue;
- facilities which form part of a productivity improvement policy, such as those enabling workers to take meals during working hours, and day-care centres for the young children of women workers.

All the above-mentioned services are important, but some (such as facilities for taking meals at or near the workplace and measures to reduce commuting time and fatigue) are more important than others. The latter will be dealt with in greater detail below.

Catering services

The need for a proper, adequate and balanced diet has already been emphasised (see Chapter 1); however, in the developing countries in particular, workers only too often have a poor and inadequate diet. Cases have been described where "a large proportion of workers came to work without eating and, in the absence of facilities at the worksite, attempted to work a full shift".[1] A meeting of experts on the nutrition of workers, organised jointly by the United Nations Food and Agriculture Organisation, the World Health Organisation and the International Labour Office pointed out that, in many countries, food supplies are limited and vary from season to season; the legal minimum wage is often not enough for workers to buy food for themselves and their families and to cover other essential living costs; manpower is used in intensive labour schemes requiring energy expenditures of up to 4,500 calories per worker for eight hours of heavy work, and this must be met through food intake.[2]

In view of the value of catering services in workers' welfare, health and productivity, the International Labour Conference, ILO Regional Conferences and the majority of ILO industrial and other committees have repeatedly emphasised the need for facilities, at or near the workplace, where workers can take their meals during the working day. The Welfare Facilities Recommendation, 1956 (No. 102), specifies a number of requirements, as shown in Panel 35.

Although plant catering services in the industrialised countries are not yet satisfactory in all cases, they have undergone considerable development since the Second World War. This may be attributed to a number of factors, including:

- the gradual adoption of the continuous working day which, by reducing the midday meal break, has prevented many workers from returning home at midday owing to long commuting distances and increasingly dense traffic, in large urban centres in particular;
- the increase in the number of wives who go out to work and are away from home at midday, which means that the whole family has to eat out;
- the spread of shift work, which upsets family eating patterns and encourages workers to eat at least one meal at the plant wherever possible;
- the importance of the canteen in the undertaking's efforts to attract, recruit and keep workers; and
- the fact that the works canteen, like the school canteen and other forms of group catering, has become an established feature of modern life in many communities.

PANEL 35

Welfare Facilities Recommendation, 1956 (No. 102)

4. Canteens providing appropriate meals should be set up and operated in or near undertakings where this is desirable, having regard to the number of workers employed by the undertaking, the demand for and prospective use of the facilities, the non-availability of other appropriate facilities for obtaining meals and any other relevant conditions and circumstances.

. . . .

10. (1) In undertakings where it is not practicable to set up canteens providing appropriate meals, and in other undertakings where such canteens already exist, buffets or trolleys should be provided, where necessary and practicable, for the sale to the workers of packed meals or snacks and tea, coffee, milk and other beverages. Trolleys should not, however, be introduced into workplaces in which dangerous or harmful processes make it undesirable that workers should partake of food and drink there.

. . . .

11. (1) In undertakings where it is not practicable to set up canteens providing appropriate meals, and, where necessary, in other undertakings where such canteens already exist, messroom facilities should be provided, where practicable and appropriate, for individual workers to prepare or heat and take meals provided by themselves.

. . . .

12. In undertakings in which workers are dispersed over wide work areas, it is desirable, where practicable and necessary, and where other satisfactory facilities are not available, to provide mobile canteens for the sale of appropriate meals to the worker.

13. Special consideration should be given to providing shift workers with facilities for obtaining adequate meals and beverages at appropriate times.

In contrast, workers in the developing countries have little or no possibility of taking their meals at or near the undertaking. Although the description given in Panel 36 is now some ten years old, it is far from outdated.

The lack of adequate catering services for workers in most developing countries demonstrates a poor knowledge of the beneficial effects that improved nutrition can have on labour productivity. For example, an ILO survey showed that average calorie consumption per head is the best qualitative indicator of labour productivity growth. By

PANEL 36

It is a pity that in Dar-es-Salaam, where we pride ourselves on so many luxurious hotels, a section of our population goes through a whole day on an empty stomach. These people go without food not because they don't have money to buy their meals, but because they don't have a place to buy them. They are the factory workers who toil at the machines and in the offices of our factories in the industrial area where there are no canteens to buy cheap food. . . .

What those workers eat, particularly during their so-called lunch-time break, is nowhere near anything that can be called food. It merely consists of dirty, half-cooked, non-nutritious substances. Their "canteens" are what are generally known as *migahawa* or *magenge*. These eating "stalls" are usually patched up under coconut trees. The roof is made of *makuti* or coconut leaves, while the walls are made of broken pieces of cardboard and hard paper. Sometimes, it is even hard to find such "canteens" and some workers are forced to eat in "open-air stalls" which are even worse, for they are dotted wherever there is a tree or some other shrubbery. The menu of a typical *genge* contains *ugali*, beans, *mandazi* (buns) and tea. The more expensive ones serve rice and meat as well. But these foods are not fresh. They merely consist of left-overs which are warmed up every time the workers want their meals. The warming up is repeated several times until the so-called food rots.

A careful investigation has revealed that only three factories in the whole of Pugu industrial area have provided their employees with canteen facilities. For the hundreds of men and women employed by the rest, the "stalls" are the only place they can rush for their lunch. Many walk for more than a mile to the nearest "stall" and some go without lunch because they might return to work late. . . .

These eating-places, many of which operate without a licence, are a constant headache to the city council. Officials sometimes make surprise check-ups, as a result of which many of them are closed down and some are even demolished to the agony of the poor worker who may rush expectantly to his usual eating-place only to be greeted by . . . nothing at all! The eating "stalls" are also a headache to the Ministry of Health and Social Welfare. Most of them are erected at places where there is no running water nearby. Apart from that, the operators of these places are very ignorant about public health. They throw dirt and crumbs all over the place, thus inviting flies and other disease-spreading insects to the stalls. . . .

Judicate Shoo: "The hour that serves Dar workers least", in *The Standard* (Dar-es-Salaam), 23 Feb. 1971.

an analysis of economic growth between 1950 and 1959 in 52 countries divided into six groups on the basis of annual income per head, the survey showed that an increase of 1 per cent in the number of calories per head resulted in a growth of 2.27 per cent in general labour productivity, and that the lower the national income per head, the greater the effects of this parameter.[3] The workers' catering services that do exist in developing countries are usually found only in large undertakings. They are singularly lacking in smaller undertakings, although it is here that they are most needed since the wages that small undertakings pay are usually the legal minimum (sometimes even less) and rarely permit an adequate diet.

In industrialised countries, there are various arrangements for providing workers in small and medium-sized undertakings with access to catering services during working hours. The most common are restaurants set up jointly by a number of employers; works canteens which have no kitchens but are supplied each day with cooked or pre-cooked meals from a central kitchen, by a caterer or restaurant; community restaurants in industrial centres run by non-profit-making associations and open to all, especially workers in undertakings that do not have a canteen; authorisation of workers in an undertaking with no canteen to use a canteen in a nearby undertaking (very common in Eastern European countries); restaurant vouchers issued by the undertaking to pay a part of the meal price in a restaurant of the worker's own choosing (relatively common in Western European countries, especially in France); and, finally, payment by employers of a negotiated subsidy to restaurant owners near the firm who, in return, serve low-price meals to the employers' workers.

In developing countries, with only a few exceptions, small and medium-sized undertakings have made no effort to co-operate in financing and managing catering services for their workers. Moreover, governments have seldom taken the initiative to set up community restaurants or central kitchens for workers in industrial centres who do not have their own works canteens. Finally, trade unions themselves seem to have shown little interest in this problem.

The catering services that do exist in developing countries often suffer from deficiencies in their operation owing to:

- the lack, in most countries, of a public body to advise and guide employers in setting up and operating canteens;
- the lack of stringency or regularity in the inspection of existing catering services, or perhaps no inspection at all;
- the inadequate training of industrial canteen managers and staff;
- poor canteen management (lack of interest by the employer; insufficient involvement of workers' representatives in monitoring the canteen's operation; abuses resulting from inadequate supervision of canteen operators out for quick and high profits).

Ways of promoting the creation and adequate operation of workers' catering services in developing countries were examined by a joint FAO/ILO/WHO meeting which recommended, in particular: [4]

- the promulgation, in these countries, of laws and regulations requiring the establishment of workers' feeding programmes (this requirement now applies in only a small number of these countries);
- the establishment, in each of these countries, of a technical agency to facilitate the administration of these laws and regulations, to act as an information centre on the benefits of workers' feeding programmes and to provide consultant services for solving problems in the organisation and operation of these programmes;
- the assessment, by the competent authorities, of present and future staffing requirements of workers' feeding programmes and the immediate adoption of measures to ensure suitable training for adequate numbers of managers and skilled workers;
- in areas with a high labour concentration, the participation of public authorities in financing group feeding facilities (central kitchens, low-cost restaurants, etc.) to meet the needs of those working in these areas, both those employed in small undertakings and self-employed workers;
- the organisation by governments of food services for workers in government offices and undertakings in order to promote the establishment of workers' feeding services in other sectors of the economy;
- the carrying out by developing countries of studies to evaluate the results of workers' feeding programmes on, inter alia, workers' health and productivity;
- the participation of health services, in particular nutrition and occupational medicine services, in the organisation and operation of workers' feeding programmes to ensure adequate consideration of local sanitary and nutritional conditions and the nutritional needs resulting from special conditions of work, climate and environment;
- the adoption of measures to reduce the price of meals prepared in workers' canteens and to keep as low as possible the prices the workers pay;
- finally, the promotion of the active participation of workers' representatives in the running of workers' feeding programmes.

Commuting facilities

The impact of reduced working hours on workers' leisure and rest has been diminished or even cancelled out by the progressive increase in commuting distances and travelling times.

PANEL 37

Everything that is done to reduce hours of work, both by the ILO and all those persons concerned, is aimed at ensuring that people have sufficient time for rest and relaxation. However, the ratio between work and rest presents problems in big towns. . . . At the beginning of the century the claim for the reduction of hours of work was reflected in the slogan "8 hours work, 8 hours leisure, 8 hours rest". It could not be foreseen at the time that many workers were going to spend 2, 3 or 4 hours per day on travel that would be as long and tiring for them as it would be costly for the community. This is a new problem to be borne in mind, for one cannot turn a deaf ear to the claims of those who think that such a large proportion of time devoted to travel should be taken into consideration when hours of work are being determined.

ILO: *Making work more human: Working conditions and environment.* Report of the Director-General, International Labour Conference, 60th Session, Geneva, 1975, p. 33.

For example, a survey by the Statistics Office of the Commission of the European Communities showed that, in the then nine member States, 21.3 per cent of industrial and service workers took 30-59 minutes to travel to work (i.e. one to two hours per day) and 8.6 per cent of industrial workers and 9.8 per cent of service workers commuted for more than two hours.

These figures give only an imperfect picture of the situation of workers travelling into large towns from suburbs or more distant areas. In Japan, for example, a study on urban transport users by the Ministry of Transport published in 1982 showed that, in the Tokyo region, 20 per cent of season-ticket holders spent more than 90 minutes commuting. There are generally few data on commuting times in developing countries. However, commuting is time-consuming there too, and, for example, many workers in Lagos (Nigeria) and Dakar (Senegal) commute for between three and four hours every day.

Commenting on this problem in his Report to the 60th Session of the International Labour Conference in June 1975, the Director-General of the ILO made the statement shown in Panel 37.

Commuting conditions have a clear effect on conditions of life. For most workers, and in particular those of modest means in large cities, commuting by public transport at rush hours is a major cause of physical and/or nervous fatigue. Finally, the financial cost of daily commuting, especially for many workers employed in large towns in developing countries, represents a sizeable percentage of the wages they earn.

Improving this situation is not an easy task; it requires more rational land-development policy, more decentralisation of industry, quantitative and qualitative improvements in public transport facilities and better co-ordination between transport timetables and working hours, including the rearrangement of working hours (staggered hours, flexible hours, etc.). Workers' organisations are more and more concerned at the slowness and inadequacy of steps taken to deal with this problem, which is common to nearly all countries. Consequently, it is not surprising that Recommendation No. 102 should contain a number of clauses on this subject (see Panel 38).

The urgency of appropriate measures is even greater in the developing countries than in industrialised countries, in view of the inadequacy of the public transport systems there and the inability of many workers to obtain even modest transport facilities of their own, such as a bicycle. Employers also have an interest from the efficiency point of view since workers who arrive tired at work owing to transport problems will certainly have poor productivity.

Certain undertakings in some developing countries have, in fact, without legal compulsion, introduced some of the measures contained in Recommendation No. 102, or other similar measures such as soft loans to help workers to purchase vehicles. Some undertakings provide transport either in the plant's own vehicles (often open lorries used also for goods transport), or through the hired services of a private transport firm. This may be the only solution where the worksite is, for example, in an isolated area with only poor public transport connections. In such cases, the employer's main aim is to ensure that workers arrive at the site on time, and his action forms part of a general policy of company efficiency.

The solutions proposed here cannot be implemented by the employer alone. The problems are now so serious and extensive, especially in large towns, that only intervention by ministries of labour, public transport, housing and industry, working jointly with employers' and workers' organisations, can introduce and co-ordinate the necessary action.

Welfare facilities to improve the living conditions of workers and their families

These may be divided into five groups:

- facilities to promote ownership or rental of accommodation matching the worker's space requirements and income;
- facilities to ensure regular supplies of adequate quantities and qualities of foodstuffs and other basic-needs articles at reasonable prices;

PANEL 38

Welfare Facilities Recommendation, 1956 (No. 102)

29. Where, in accordance with national or local custom, workers provide their own means of transport to and from work, suitable parking or storage facilities should be provided where necessary and practicable.

30. Where a substantial proportion of the workers experience special difficulties in travelling to and from work owing to the inadequacy of public transport services or unsuitability of transport timetables, the undertakings in which they are employed should endeavour to secure from the organisations providing public transport in the locality concerned the necessary adjustments or improvements in their services.

31. Where the workers' transport difficulties are primarily due to peak transport loads and traffic congestion at certain hours and where such difficulties cannot otherwise be overcome, the undertaking in which they are employed should, in consultation with the workers concerned and with the public transport and traffic authorities, and, where appropriate, with other undertakings in the same locality, endeavour to adjust or stagger times of starting and finishing work in the undertaking as a whole or in some of its departments.

32. Where adequate and practicable transport facilities for the workers are necessary and cannot be provided in any other way, the undertakings in which they are employed should themselves provide the transport.

33. In particular countries, areas or industries, where public transport facilities are inadequate or impracticable, and as an alternative to the provision of transport by the undertaking, transport allowances should, by agreement between the employer and the workers concerned, be paid to the workers by the undertaking.

34. Wherever necessary, undertakings should arrange for adequate transport facilities to be available, either through the services of public transport or otherwise, to meet the needs of shift workers at times of the day and night when ordinary public transport facilities are inadequate, impracticable or non-existent.

. . . .

- facilities to protect health, in particular by suitable access to medical or other forms of care needed in the event of sickness or accident;
- facilities to improve education and in particular to ensure literacy and compulsory education;
- facilities to promote access to useful leisure activities, in particular by the organisation of cultural, sporting and recreational activities and by measures permitting the suitable use of paid annual leave;
- miscellaneous facilities to deal with specific problems of the worker's family (e.g. if the head of the family dies or is disabled).

Welfare services in industrialised countries are nowadays of somewhat less importance, since it is increasingly considered that responsibility for improvements in this field falls on the public authorities or the workers themselves. However, in some countries, and particularly those with centrally planned economies, welfare services continue to be of especial importance since it is considered that improving the conditions of life of workers and their families requires action not only by national or local authorities but also by undertakings and workers' organisations; in fact, in several countries, undertakings continue to play a significant role in this field.

Welfare services are particularly valuable in the developing countries, where there may be serious deficiencies or inadequacies in social development policies. Undertakings with worksites in up-country areas (mines, oil wells, plantations, etc.) continue to provide workers with housing and other welfare services required for normal community life; since the national or local authorities seldom assume responsibility for these services, the undertakings are compelled to maintain them in order to keep their workers. Furthermore, in view of deficiencies in the general social infrastructure and low salary levels, employers have had to take significant action to improve workers' conditions of life even in urban areas.

Since action by employers to improve conditions of life is of particular significance in developing countries, some specific programmes of major importance in the developing countries, that is, in the field of housing, supplies and workers' health, will now be described.

Employers' action with respect to housing

The physical, moral and social significance of housing to the worker has already been considered in Chapter 1. Workers of modest means employed in hard physical work often have the poorest housing conditions, and shift or night workers need accommodation with particularly good insulation against noise and heat. The location of housing also plays an important role in conditions of life: the length and type of commuting may be a major cause of fatigue; the proximity of community services (schools, commercial centres, etc.) is a major factor;

foreign or migrant workers, agricultural workers, workers in building, civil engineering, plantations and forestry, for instance, are often housed in communal accommodation, and this should meet minimum standards to ensure that their health is not affected.

These problems have led to the adoption by the International Labour Conference of the Workers' Housing Recommendation, 1961 (No. 115), and various other instruments dealing with the housing of specific categories of workers, such as seamen and migrant workers.[5]

Problems over workers' housing are often related to urbanisation and aggravated by rapid industrialisation which has outpaced the provision of housing, roads and transport facilities. The situation and working conditions of workers should be taken into account at the planning stage so that conditions of life can be given adequate consideration.

Providing for the ownership or rental of adequate housing implies the existence of a national home financing scheme through which low-cost housing can be built for workers or loans provided at moderate rates of interest. In developing countries, the necessary large investments can be achieved only if national savings for housing have reached an adequate level: poor income distribution is often a barrier to substantial private capital formation. There is often reluctance to invest private capital in the construction of low-cost housing which is considered to offer high risk, low yield and a long pay-back time in comparison with other investments which give larger and more rapid profits. Public funds which might be available for low-cost housing are usually subject to serious restrictions, and governments have to reserve a large part of these funds for productive investment. Moreover, public expenditure on housing competes with investments that have just as high priority in other areas of social development policy, such as health and education.

Realising that public or private home financing schemes can scarcely meet their demand, certain workers in the formal sector of the economy in developing countries have turned more and more often to employers for assistance in housing. In some cases, workers' organisations have launched programmes in which they have attempted to channel their members' savings into house financing. Finally, in a number of developing countries, attempts have been made to expand the resources of institutions for low-cost home financing by compelling employers to make contributions to the institutions' budgets.

At the level of the undertaking

Employers' action on housing for their workers usually takes place within the framework of the undertaking; it is seen most clearly amongst firms which, for such reasons as the isolated location of their worksites, have constructed housing complexes for their employees.

Part IV of Recommendation No. 115 deals with housing provided by employers and lays down a number of principles in this context (Panel 39).

PANEL 39

Workers' Housing Recommendation, 1961 (No. 115)

12. (1) Employers should recognise the importance to them of the provision of housing for their workers on an equitable basis by public agencies or by autonomous private agencies, such as co-operative and other housing associations, separate from the employers' enterprises.

(2) It should be recognised that it is generally not desirable that employers should provide housing for their workers directly, with the exception of cases in which circumstances necessitate that employers provide housing for their workers, as, for instance, when an undertaking is located at a long distance from normal centres of population, or where the nature of the employment requires that the worker should be available at short notice.

. . . .

Housing complexes of this type may be built by employers as the result of an obligation under labour legislation or of a clause in a collective agreement. In India, for example, the obligations of plantation employers as regards the construction of housing are laid down by legislation and those for the steel industry form part of collective agreements. Although employers' efforts in many cases are laudable, it is often found that the number of dwellings that meet specified standards in these complexes is out of all proportion to the number of workers who are entitled to them by legislation or collective agreements. A striking feature of these complexes set up by employers is the extreme variety of housing conditions. With few exceptions, oil and mining firms have usually built better-quality housing than that provided in plantations and agricultural, forestry and animal husbandry undertakings. For example, the housing conditions in Sri Lankan plantations have been described in an ILO report published in the mid-1970s (Panel 40).

Undertakings in developing countries may sometimes contribute to the provision of workers' housing in ways other than direct construction. Housing loans may be made on better terms than those available from "low-cost" housing finance institutions (see Panel 41). This type of loan is relatively common in Latin America, in particular in Colombia and Mexico.

Individual employers may help to solve workers' housing problems by giving financial support to programmes launched by the trade union to which the workers belong. Such programmes are very rare in

PANEL 40

In Sri Lanka little progress seems to have been recorded in the field of housing for plantation workers, most of whom are Tamils who came originally from southern India and who have been in Sri Lanka for several generations. Although some of them were built practically a century ago, the back-to-back type line rooms with a common verandah running the length of the building and the single line rooms continue in fact to be the type of housing generally supplied for workers' families. Each family frequently occupies only one room which is used as kitchen, dining-room and bedroom combined. The slow rate of progress was illustrated by the socio-economic census carried out in Sri Lanka in 1969-70, which revealed that 225,720 out of 251,655 dwellings available (i.e. 89.7 per cent of them) were of the line room type and that these housed 1,170,700 out of the 1,316,050 persons (89 per cent) residing on plantations.

The case studies carried out in Sri Lanka in December 1975 in connection with this report show that the situation has improved little over the past six years. [. . .] line room housing is still predominant and in one case (Govinna rubber plantations) the only type of housing provided for the workers on the ten plantations visited. Several factors were put forward by the persons questioned to explain this situation: firstly, rumours of nationalisation which eventually proved correct but which in the intervening period put a brake on, or even halted completely, all action by the employers in regard to housing; secondly, poor prices for tea on the world market over a number of years, increased production costs and higher taxation, that were reported to have led to reduced profitability of the plantations; and finally, the difficulty in obtaining building materials whose distribution is strictly controlled by the Government. In this connection it should be noted that a scheme to reorganise the tea industry launched by the Government in 1973 with a view to checking the drop in production provided that subject to certain conditions planters might receive subsidies, part of which could be used to improve their workers' housing (up to 10 per cent of the sums spent for this purpose). Unfortunately, no advantage has been taken of this opportunity.

ILO: *Housing, medical and welfare facilities and occupational safety and health on plantations*, Report III, Committee on Work on Plantations, Seventh Session, Geneva, 1976, p. 14.

PANEL 41

In Colombia, the working capital fund of Acerías Paz del Río SA is intended to help workers to purchase housing of their own. Financed by periodic contributions by the employer (20 million pesos under the two-year collective agreement concluded in March 1977) and by repayments of loans, this fund is managed by a board composed of an equal number of company and trade union representatives. The loans are granted according to a points system based on length of service in the company, the number of dependants, attendance record during the last two years and location of the housing. To simplify the problem of workers' transport, which is especially complicated for the company because the staff's living quarters are spread out over such a large area, the terms are much stricter in respect of length of service required and the rate of interest when the distance between the workplace and the location proposed for the housing is over 30 kilometres. As it is unable to comply with all of the applications for housing loans it receives through its working capital fund alone, the company has also called upon the Territorial Credit Institute for its collaboration in this connection. Lastly, it has a special department of architects and legal advisers working full time on workers' housing problems. This department draws up standard housing plans in the light of the wide variety of the workers' financial resources, evaluating the amount of the loan required for each case, appraising the applications for loans made to the company's working capital fund, informing workers of the financing possibilities offered by public and private institutions, and helping them to obtain the legal documents required for the real estate transactions involved.

ILO: *The improvement of working conditions and working environment in the iron and steel industry*, Report III, Iron and Steel Committee, Tenth Session, Geneva, 1981, p. 43.

developing countries, but for the sake of interest two programmes currently under way in Latin American countries will be described.

In Honduras, the savings and home loans scheme set up in 1965 by the Workers' Union of the Tela Railroad Company, a large agro-industrial complex employing some 27,000 workers, shows clearly the potential for trade union action at the level of the undertaking. The scheme encourages company employees, unionised or not, to save regularly for housing purposes. The trade unions' savings and credit section offers workers loans which, together with personal savings, account for 50 per cent of the house value (the remaining 50 per cent is financed by the undertaking). After five years of operation, four workers' housing complexes totalling 858 single-family houses had already been built, and some 8,000 of the company's 27,000 employees had joined the scheme and were saving a minimum of two lempiras (US$1) each week.

In Venezuela, a similar scheme was set up by the National Telecommunications Workers' Confederation in collaboration with the national telephone company. Under the house-purchase programme run by the confederation for its members since 1963, several thousand homes have been constructed in various parts of the country, and trade unions affiliated to the confederation have set up savings and loans associations. Membership of the associations is voluntary, but workers who join undertake to make regular savings of not less than 3 per cent of their monthly wage. The company strongly encourages membership: it pays into individual savings accounts a sum equal to at least 50 per cent of savings, and takes over 20 per cent of housing costs. Workers selected to take part in the programme receive from their association a loan for the part of housing costs not covered by their savings or the employers' contribution; the loans are repayable over ten years at an interest rate of 8 per cent per annum, and there is life insurance protection for dependants in the event of the worker's death or total permanent disability. The operation of the programme has been entrusted to the Workers' Bank, a public body responsible for low-cost housing development in Venezuela.

Other methods used by undertakings in developing countries to assist workers with their housing problems include:

- regular help for entry into public or private low-cost house-building programmes by guaranteeing workers' debt repayments, assisting with initial down payments, or taking over part of the interest payable on the loan;
- the distribution in advance of part of the sums usually payable only on termination of employment (redundancy money, etc.) to help workers to finance housing;
- help for workers to rent accommodation by loans for guarantee down payments, by guaranteeing the regular payment of rent or, less commonly, by taking over part of rent payments;
- the provision of building materials (steel, cement, timber) at reduced prices and of transport, especially where the worker is himself building, equipping or repairing his home.

Beyond the level of the undertaking

In some Latin American and Asian countries, employers may help to improve workers' housing conditions, on the basis of legislation or collective agreements, at the level of an industrial sector. For example, the statutory labour welfare funds, set up in India for mica, iron, coal, limestone and other mining firms, are financed by an ore-production tax

PANEL 42

The Mexican Act of 21 April 1972, which set up the autonomous tripartite National Workers' Housing Fund Institute (INFONAVIT), also requires undertakings to open savings accounts with the Institute for each worker, and to credit these accounts each month with a sum equal to 5 per cent of the workers' wages. The Institute grants workers soft loans to buy, build, equip or repair housing, and also finances housing complexes throughout the country. It was estimated, at the time the Act was promulgated, that this employers' tax would permit the construction of around 1 million new housing units in ten years (there was an estimated housing deficit of 2 million units) and create 350,000 new jobs in the building industry and related branches.

paid by employers and have, as one of their functions, the improvement of housing conditions in mining communities.

The contribution made by employers is most significant when it is organised nationally, that is, where legislation requires all undertakings to help to fund institutes for financing the construction of low-cost housing; this is the case in several Latin American countries but occurs only rarely in Africa and Asia. In some countries (e.g. Argentina and Gabon), legislation requires employers to make monthly payments of a given percentage of their wage bill to the institute in question. In other cases (e.g. Chile), the contribution is a percentage of net annual profits. In still other countries, legislation requires undertakings to make annual payments to the competent institution of the sums that their workers have accumulated as seniority benefits (often one month's wage per year of service) and which would normally be paid out on termination of service. In Brazil, for example, the National Housing Bank receives each year 8 per cent of the country's total annual wage bill; this accounts for around 60 per cent of its total funds.

The system operating in Mexico since 1 May 1972 shows clearly the considerable impact that this type of employers' contribution can have (Panel 42).

Employers' action in the field of provisions

Employers' actions to help workers by providing regular supplies of food and other articles in adequate quantity, quality and variety and at a reasonable price are of particular value in developing countries, since the workers' purchasing power is low and (in many countries) declining, owing to rapid rises in the price of basic-needs goods and services. Inflation stems from such factors as the stagnation or slow rise in food production in the face of often rapid demographic growth, deficient distribution networks (resulting from the virtual monopoly position of a number of large wholesale firms, the proliferation of intermediaries and speculative practices), and inadequate governmental control of prices.

The irregular payment of wages may also aggravate the negative consequences of low wages on the nutrition of workers and their families.

Works stores

Works stores may be set up by employers under their own management or that of an agent to sell food and other essential commodities to meet the normal requirements of workers and their families. For undertakings in isolated regions with poor retail services, the works store was originally considered an essential service and aimed at preventing the monopoly of a few local traders who might exploit the workers. Gradually, workers in urban undertakings have come to recognise the value of the works store in protecting their purchasing power against rising prices, in particular.

The Protection of Wages Convention, 1949 (No. 95), specifies that, where works stores for the sale of commodities to workers are established or services are operated in connection with an undertaking, the workers concerned shall be free from any coercion to make use of such stores or services. It also stipulates that, where access to other stores or services is not possible, the competent authority shall take appropriate measures with the object of ensuring that goods are sold and services provided at fair and reasonable prices, or that stores established and services provided by the employer are not operated for the purpose of securing a profit but for the benefit of the workers concerned.

Works stores are relatively common in such Latin American countries as Argentina, Chile, Colombia, Mexico, Peru and Venezuela, where they have often been set up as the result of collective agreements, and are found in both rural and urban centres. In Asia, "fair price shops" may be found in large undertakings in India and Pakistan; however, in Africa, works stores are uncommon in urban undertakings, and in isolated regions only a few are to be found, in large plantations or mining operations. In principle, works stores should sell products at reasonable prices and not bring profit to the employer. To ensure this, labour legislation in some countries lays down price guide-lines limiting the profit margin that the undertaking can add to the supplier's price to cover transport and operating costs (Bolivia, Chile), or prohibiting works stores from selling at prices higher than official prices or local market prices (Mexico). In Latin America, collective agreements may also contain provisions about prices in works stores (prices to be clearly and permanently displayed, procedures for price increases, etc.).

PANEL 43

Many plantations allow the overseer or farm manager to operate a store catering mainly to the workers in the farm. The merchandise offered for sale is mainly basic commodities like rice, milk, salt, dried fish, canned goods and clothing. These are sold on credit but the prices are usually 25 to 60 per cent higher than what is available when purchased from stores in the towns or cities.

Most workers have no choice but to purchase from the stores because they do not have enough income to make cash purchases in town. Credit purchases on their part is difficult in town stores where the vast majority of them are unknown to the store owners.

In some plantations, even if workers could afford to purchase goods at lower prices elsewhere, they are compelled to buy from the company or farm manager operated stores because credit purchases from those stores are made a condition precedent for workers to obtain work. Some even require a quota for credit purchases before they give work. As a result, these workers are driven deep into debt, which suits the aims of such plantations. These workers feel a moral obligation to pay their debts and they remain in the plantation.

ILO: *Housing, medical and welfare facilities and occupational safety and health on plantations*, op. cit., p. 32.

The Welfare Facilities Recommendation, 1956 (No. 102), stipulates that, in cases where workers have to pay for welfare facilities, payment by instalment or delay in payment should not be permitted. Labour legislation in the French-speaking African countries, for example, confirms this principle of cash payment; however, other countries authorise credit purchases up to a specified percentage of the workers' wages (Bolivia, Pakistan). In many developing countries, it seems that works stores are attractive to workers only if they sell on credit. Credit sales have clear disadvantages since they complicate the payment of wages, and often lead to arguments and inevitably to the employer's managing his workers' family budgets. They may also result in clandestine trading by workers who profit by selling to persons outside the undertaking the products purchased on credit in the works store.

The management of the store by an agent may also be a source of difficulties. First, workers' representatives are not always involved in the store's supervision as they should be; second, unless the agent is adequately supervised by the employer, he may soon start to act like an ordinary trader looking for fast and high profits.

Panel 43, which reproduces comments supplied by the National Congress of Unions in the Sugar Industry of the Philippines, describes the way in which workers may be abused.

Consumer co-operative stores

There is abundant documentation on the virtues of the consumer co-operative movement in preserving the purchasing power of underprivileged sectors of the population. Unfortunately, there are not many co-operatives in developing countries, since they have been impeded by such obstacles as shortage of starting capital; difficult or irregular supplies; the lack of skilled managers and operating staff; and workers' lack of understanding of the meaning, advantages and demands of the co-operative movement.

More recently, co-operatives have been set up at the level of the undertaking, sometimes with government support and managed by workers or their trade unions; this has the advantage of establishing a homogeneous group of individuals with common interests and objectives. Developments of this type are to be found in Colombia, India, Mexico, the Philippines and the Sudan, to give but a few examples. In some cases, the consumer co-operative has replaced the undertaking's works store. The co-operative may receive help from the employer, usually in the form of premises, equipment and fittings that are provided free or at a nominal rent. Undertakings may also make donations, or interest-free or low-interest loans, to help the co-operative shop to build up its starting capital; this is particularly valuable, since the co-operators' initial personal contributions are rarely enough to launch trading on the scale envisaged. Once established, the co-operative store may also be able to count on the assistance of the employer, who may use his commercial contacts and transport facilities to ensure regular supplies. Loans may be granted on attractive conditions or financial liabilities to suppliers may be guaranteed. The employer may also take over the co-operative shop's administrative and overhead expenses (maintenance of premises, electricity bills, etc.) or contact the authorities to obtain technical assistance for the managerial staff.

Consumer co-operatives established within the framework of an undertaking have had varying levels of success. Some have worked well and have paid their members regular dividends; just as many have been poorly managed and have soon failed (as happened, for example, in a number of plantations in Costa Rica and Malaysia). It can be seen that the major impediment to the consumer co-operative movement in developing countries is that workers and their representatives cannot become experienced co-operators or managers overnight; training is needed, and this is mainly the responsibility of the public authorities.

Other forms of assistance

Employers may also assist in the area of smallholdings, the supply of daily food rations or the encouragement of private trading in the undertaking's facilities.

First, in plantations, large agricultural undertakings, and so on, workers may be given permission to undertake subsistence farming or animal husbandry on land belonging to the undertaking, in order to improve their nutritional standards. It has been traditional to permit workers to cultivate a vegetable garden and raise farm animals around service housing, but allowing workers to use larger tracts of land for cereal crops is a more recent practice. Rice growing by workers on low land unsuitable for tea and rubber crops is widespread on plantations in Bangladesh, India and Sri Lanka. In most cases, the land is provided free or a nominal rent may be requested. Planters may encourage workers to start subsistence farming by doing the heavy work for them (ploughing and harrowing), distributing free or low-cost seed and fertiliser, providing agricultural tools at cost price, giving prizes to workers with the best vegetable gardens or making advances or loans to help them to start farming. They may also help in husbandry, for example by providing selected animals at low cost, constructing cattle shelters, paying herdsmen or supplying veterinary assistance (dips, vaccinations, etc.). In several countries, workers have increased their earnings by farming and cattle raising whilst improving their families' nutrition. In other cases, use has not been made of the land available because of a tendency by some employers to take back arbitrarily the land the workers have been cultivating, once the price of the plantation products has improved on the world markets.

Second, the practice of the provision by employers of daily food rations is no longer widespread, even in those countries, mainly in Africa, where labour legislation still requires employers to continue this practice for those workers who were recruited outside the region of the plant and who cannot independently obtain regular food supplies for themselves and their families. Where this practice has been maintained, free or subsidised rations are usually granted for only a limited period immediately after recruitment and comprise only a small number of foods, such as rice, corn or wheat flour, and so on.

In some countries (Bangladesh, India and Sri Lanka, for example), the government sometimes uses undertakings to distribute rationed basic-needs articles to the population at official prices; the undertakings themselves are also required to supply certain basic foods to their workers. Finally, some plantations also carry out subsidiary subsistence farming and cattle raising, and sell the products (rice, meat, milk, etc.) to workers at prices lower than those on the local market.

Third, undertakings in regions with no adequate commercial infrastructure (plantations and large agricultural, mining and oil undertakings) have often encouraged private trading on the worksite by helping small traders to establish themselves or by periodically organising public markets. This gives workers a larger choice of articles or the opportunity to diversify their purchases. Unfortunately, the prices and hygienic practices of these traders are not always sufficiently supervised.

Finally, other measures taken by some undertakings in developing countries to facilitate the supply of foodstuffs and other basic-needs articles to workers and their dependants include:

- the free distribution of milk and other foods to children in crèches or schools run by the undertakings;
- the provision of transport facilities or the allocation of paid leave to allow workers and their dependants living on isolated worksites to go to nearby localities in order to obtain supplies;
- the distribution of coupons for the purchase of a given quantity of foodstuffs in a stipulated shop;
- the organisation of domestic economy, cooking, nutrition or food hygiene courses for the wives and daughters of workers; and
- collaboration with the authorities to improve dietary habits.

Employers' action in the field of health

In spite of the progress made during the past decade, there are still serious shortcomings in the health sector in the majority of developing countries. The public funds allocated to this sector are inadequate, the health infrastructure is particularly deficient in rural zones, and social security covers health hazards only partially. Under such circumstances, the services that employers can provide are clearly of great importance.

At the level of the undertaking

Activities by employers to protect the health of workers and their dependants are usually organised at the level of the undertaking (see Chapter 2), either to meet requirements under labour legislation on occupational health or to improve the conditions of work and life of persons on isolated worksites. In the first case, the employer may be required to do rather more than just purchasing a medicine chest or installing a first-aid post. For instance, the Mexican Federal Labour Act of 2 December 1969 requires undertakings employing more than 100 workers to set up an infirmary and those with more than 300 workers, a sick bay. If the staff agrees, these undertakings can discharge this obligation by concluding contracts with clinics or hospitals in the area, or at any rate near enough to allow a sick or injured person to be taken there easily and rapidly.[6] Collective bargaining has often led to agreements by which the undertaking's health services provide health care for workers and their dependants, whether the sickness or injury is occupational or not. This is particularly true of the sugar industry.

In the second case, isolated undertakings are usually required by law to set up health services for both workers and their dependants, no matter what the cause of the disease or accident. In Venezuela, for example, all oil and mining employers, and other employers with more than 300 workers and camps more than 10 kilometres from a locality where adequate medical assistance can be obtained, are required to employ a physician and a pharmacist if the number of workers does not exceed 400, and an additional physician for each additional 400 workers or fraction of this number over 200. They are also required to operate one or more hospitals with modern health-care equipment, including surgical services and an adequate stock of medicaments. Finally, they must have a laboratory with all the necessary diagnostic equipment.

Health services set up by undertakings on worksites in areas with no suitable health infrastructure often have severe deficiencies or inadequacies (dilapidated premises, insufficient medical and surgical equipment and medicaments, too few physicians, too few and inadequately trained paramedical staff, etc.).

In the urban zones of developing countries, fewer undertakings operate their own health services since access to public health services is much easier than in rural areas. However, in several countries urban undertakings have continued to be active in this domain, especially when the social security scheme does not cover health care for workers and their families suffering from non-occupational diseases.

Undertakings which operate their own health services for workers and their dependants usually also assume the total or partial cost of specialised health care when this cannot be provided by the undertaking's own medical staff. The employer then often meets the cost of transporting the patient and, where necessary, the person accompanying him, and may give a subsistence allowance for each day spent away from the worker's normal residence for this purpose.

Undertakings which do not run their own health services may nevertheless help workers and their dependants to obtain medical care where this is not adequately available through social security schemes or public health services. This may be done in the following ways:

- granting non-interest loans with long terms of repayment;
- granting advances on the sums which workers have accumulated during their years of service in the undertaking and which would normally be payable only at the end of their employment;
- the joint employer/worker financing of the services of a physician or dentist;
- the establishment of a special fund in the undertaking, financed by contributions from the employer and employees, to pay health-care costs for workers and their dependants;
- encouraging workers to join private health insurance schemes and covering a part of the premiums;
- encouraging workers to save regularly to meet unforeseen difficulties such as illness (in particular by matching or part-matching the sum saved by workers in individual savings accounts); and

PANEL 44

In India [. . .] it is interesting to note that the industry is in various ways financing health services in the mica and iron mines. Thus, a Mica Mines Labour Welfare Fund, financed by the mica-mine employers through a levy at present equivalent to 2.5 per cent of the value of their exports, in 1974 devoted nearly 55 per cent of its resources (some 2,769,000 rupees) to improving the health of the 48,000 beneficiaries (12,000 workers and their dependants) in the states of Andhra Pradesh, Bihar and Rajasthan. At that time the fund ran three hospitals (160 beds in all), two regional infirmaries with 40 beds each, 39 fixed dispensaries, three of them with a ward (15 beds in all), four mobile medical units and 12 mother and child welfare centres. As part of its campaign against tuberculosis it also ran a special 50-bed sanatorium, equipped a special ward in one of its central hospitals (20 beds) and reserved another 20 beds in a public hospital and two public sanatoria. Whenever a member of a worker's family was being treated in hospital for tuberculosis, the fund paid a subsistence allowance of 50 rupees a month, and whenever there was no room for a patient in a hospital or sanatorium, offered treatment by one of its mobile teams. It had likewise entered into an agreement with the Tetulmari leper hospital in Bihar State for the treatment of workers and their dependants suffering from leprosy. Lastly, it was running a school health programme for children attending the elementary schools run by it in Andhra Pradesh. At the same time, the Iron Ore Mines Labour Welfare Fund, financed by employers through a charge of 0.25 rupees per ton of ore extracted, was offering its 212,000 beneficiaries (53,000 workers and their dependants) the services of a central hospital, three branch centres and seven mobile dispensaries. Besides which, a number of extra beds were reserved in establishments specialising in the treatment of leprosy, tuberculosis and mental disorders. Its subsidies had helped 12 undertakings to create dispensaries meeting statutory requirements, while gifts had been made to five others, so that they might equip their hospitals with an X-ray unit and other special equipment.

ILO: *The welfare of workers in mines other than coal mines*, Report III, Third Tripartite Technical Meeting for Mines Other Than Coal Mines, Geneva, 1975, pp. 38-39.

- agreements between the employer and a public or private medical service to provide workers and their dependants with certain forms of medical care which would not otherwise be available, in return for a limited financial participation on the part of the worker.

Beyond the level of the undertaking

In some developing countries, the employer's work in the field of health care may be supplemented by programmes organised within a specific sector. The example in Panel 44 describes a programme of this type which has been in operation for several years in the mica and iron mines in India and which is financed jointly by all the employers in question.

Health services in a specific sector may also be financed and organised jointly.

Conclusion

It was decided that the establishment and development of welfare facilities for workers should be one of the main lines of action of the International Programme for the Improvement of Working Conditions and Environment (PIACT), because no artificial division can be made between working and living environments, which are closely inter-related, sometimes overlap and always have significant interactions. Improving workers' housing, feeding or commuting conditions will also improve their rest and recovery after work, increase their productivity and make it easier to improve their working and environmental conditions.

Measures to improve the workers' living environment are a valuable adjunct to action to promote better working conditions, and the latter are much more likely to succeed if they form part of a wider policy of improving the quality of life in general.

The need to act simultaneously to improve both working conditions and conditions of life is particularly apparent in the developing countries. However, "global" action is difficult in these countries, since the resources that are available to the public authorities to improve social infrastructure, in particular, are usually out of proportion to the enormous needs that exist. The special importance of action by employers to improve welfare facilities for workers in these countries will therefore be clearly understood.

Notes

[1] FAO: "Nutrition and food", in ILO: *Encyclopaedia of occupational health and safety* (Geneva, 3rd (revised) ed., 1983), p. 1485.

[2] FAO: *Report of the FAO/ILO/WHO Expert Consultation on Workers' Feeding, Rome, Italy, 10-15 May 1971*, Nutrition Miscellaneous Meetings Series, No. 2 (Rome, 1971), p. 3.

[3] W. Galenson and G. Pyatt: *The quality of labour and economic development in certain countries: A preliminary study*, Studies and Reports, New Series, No. 68 (Geneva, ILO, 1964).

[4] FAO: *Report of the FAO/ILO/WHO Expert Consultation on Workers' Feeding*, op. cit., p. 11.

[5] See also T. Umezawa and J.-M. Clerc: "Housing of workers", in *Encyclopaedia of occupational health and safety*, op. cit., pp. 1059-1061.

[6] ILO: *Legislative Series*, 1969 – Mex. 1, section 504, iii-iv.

WORKERS IN THE RURAL AND URBAN INFORMAL SECTORS IN DEVELOPING COUNTRIES

7

The improvement of working conditions and environment in most developing countries should be considered within the larger context of overall social and economic development rather than within the narrow confines of the workplace *per se*. This wider view is necessary owing to the combined effect on workers' welfare of many economic and social forces, such as deep and increasing poverty, low and uncertain incomes, unemployment and inadequate access to minimum standards of basic social services. This intertwined relationship between working conditions and the environment, on the one hand, and the socio-economic environment, on the other, and the need for a global approach to working conditions in developing countries are the main focus of this chapter.

Working conditions in rural areas of developing countries

A complex situation

Nowhere are improvements in working and living conditions more desperately needed and yet so complex and difficult to design and implement than in the rural areas of developing countries, where a large proportion, if not the overwhelming majority, of the population work and live under conditions even worse than those in the industrial or urban centres. In rural areas, poverty is more pervasive and acute than in urban areas, incomes are lower and more uncertain and social poverty (where income and wealth inequalities are reflected in dependence, exploitation and inferiority) is more pronounced. Rural populations, by producing food, cash crops and raw materials, generate the surplus, indeed the wherewithal for accelerated industrial development and capital accumulation; but they still have poorer access to adequate housing, health, educational and other social services, and this explains their higher illiteracy rates, shorter life expectancies and reduced productive capacities.[1]

Deep and growing impoverishment

Perhaps even more significant is the trend in the incidence of rural poverty. Although growth in total and per capita gross national product in the developing countries has been quite good during the past two or three decades, there is disturbing evidence of increasing impoverishment in many parts of Asia, Africa and Latin America. An ILO study of rural poverty in Asia found, in most of the countries studied, a clear increase in the proportion of the rural population living below the poverty line, a decline in the share of the lowest income groups in aggregate consumption and income and, even worse, an absolute decline in the real income of a significant proportion of low-income households.[2] Several studies also confirm the existence of substantial and increasing rural poverty in Africa and Latin America.[3] This demonstrates the enormity of the task to be accomplished and the need to ensure that the discussion of improvements in working conditions and the working environment is clearly related to the reality of deep and increasing mass poverty, underemployment and unemployment.

Need for a wider analytical perspective

Measures to improve working conditions and environment in the rural sector must often be wider in scope than those required in the industrial sector. In contrast with industry or the organised sector, where working conditions can be considered more or less in the context of the workplace *per se*, the rural sector in developing countries has problems which cannot be divorced from the special nature of agricultural work and the global forces that bear decisively on the rural population's living and working conditions.

There are at least four major determinants of the working environment in developing countries: the nature of agricultural work; the physical environment; the living environment; and the economic environment, which covers a wide range of factors such as agrarian structure, organisation of production, technology and prices.

Nature of agricultural work

In developing countries as a whole, agriculture remains the major occupation of the rural population, although non-agricultural activities may form a significant share of total rural output and employment. Agricultural work is arduous by its very nature, and the low level of technology and capital inputs mean that farmers are subjected to heavy manual labour and, in the peak seasons, to long hours of work as well.

These problems assume greater importance when looked at against the background of the poor nutritional conditions of agricultural labourers, the hostile physical environment and the extremely poor living and working conditions.

The close link between nutrition, health and productivity has been established empirically. Inadequate food consumption and nutritional deficiencies increase susceptibility to disease, diminish work capacity and reduce productivity. A total daily intake of 2,000 calories will maintain a calorific balance only if the effective working time does not exceed four or five hours per day (see also Chapter 1). Yet much of the rural population in Africa, Asia and Latin America has an even lower calorie consumption. Forestry work is heavy and hazardous and involves long hours of work; yet the daily calorie intake of an Indian forestry worker is 1,700-2,000 calories, compared with 3,000-4,000 for his European and North American counterpart.

The relationship between nutrition and work intensity is all the more significant in the case of working children and, more particularly, of housewives and mothers who, in addition to their domestic activities, often do heavy physical work for long hours in the fields.[4] In some parts of rural Asia and Africa, women have a consistently higher daily workload than men in every age group, but still may only eat what is left by their husbands and children. This may account in part for the observed decline in the average life expectancy of rural women in some countries. In one West African country, this decline is widely attributed to the extra strain imposed on an already undernourished female population by the cultivation of additional land.

Physical and living environment

In many developing countries, the effects of heavy agricultural work are aggravated by the natural rhythms of agricultural life and the rigours of an untamed physical environment. Climatic factors determine the production cycle, the pace of work and labour use; work is consequently not uniformly distributed over the year. Very long hours are worked during the peak season when food stocks are usually at their lowest – while male labour, in particular, is underemployed for the rest of the year. Rural people thus also suffer from irregular incomes and food supplies, and are subject to all the vagaries of the weather.

Disease is a further effect of the physical environment of rural populations in developing countries, and large numbers suffer from such debilitating and fatal diseases as malaria, trypanosomiasis (sleeping sickness), Chagas' disease, schistosomiasis (bilharzia) and onchocerciasis (river blindness).[5]

Poor living conditions make rural dwellers more vulnerable to disease, shorten their working life and diminish its quality. The common endemic diseases in the developing countries – intestinal parasitosis and

PANEL 45

In 1975, only about 28 per cent of the population in low-income countries had access to safe water, and the proportion is much less in rural areas. Further, rural populations in the poorer developing countries sometimes have almost no sewage disposal facilities. The link between sanitary conditions and health can be seen from the evidence that privy construction in Costa Rica helped to halve the death rate from diarrhoea and enteritis between 1942 and 1954, and that improved water supply and toilet facilities cut cholera incidence by about 70 per cent in the Philippines. Briefly, then, the poor physical and living conditions under which rural people work and live contribute to their ill-being and the poor quality of their working life.

infectious diarrhoea, typhoid, cholera, etc. — are generally faecal-related or faecally transmitted, and stem from the contamination of food, water or soil by human waste. This in turn can be attributed to the lack of community water supply and sewage disposal facilities and to poor housing (see Panel 45).[6]

Disease may also have high economic costs: it reduces the availability of labour and the productivity of workers and plant; an unhealthy natural environment also impedes the exploitation and development of natural resources and animal wealth. Sickness causes absenteeism which, in turn, disrupts the production process and causes loss of output.

Improvements in the physical and living environment can significantly reduce absenteeism and raise output. Where malaria is endemic, it may be necessary to recruit 30-40 per cent more workers than are normally required, in order to compensate for absence through illness; the eradication of malaria eliminates the need for this practice. For example, it is reported that, in Burma and Pakistan, rice production increased by 15 per cent in the first year after malaria eradication.[7] A study of tuberculosis control in the Republic of Korea showed that each US$1 invested in a disease-control programme resulting in an increase in work life and a decrease in sickness could yield a return of US$150.[8] Further, according to a study of construction and rubber plantation workers in Indonesia, 85 per cent of whom suffered from hookworm infestation and iron-deficiency anaemia, 60 days of anaemia therapy, costing a total of US$0.13 per worker, resulted in a productivity increase of some 19 per cent.[9]

Economic environment

The poor and unhealthy working conditions arising from the nature of agricultural work, and the poor physical and living environment in developing countries, spring essentially from the low levels of development. The problems and their solutions are intertwined with the tempo and pattern of economic development, and hence with the overall economic environment.

There are at least three economic issues central to the improvement of working conditions and environment, namely:

(a) the rate of growth of total output, that is, whether or not the economy is generating sufficient growth and employment opportunities to raise income per head;

(b) the pattern of growth, that is, first, whether this pattern results in the production and delivery of goods and services needed to meet the consumption requirements and basic needs of the poor and disadvantaged groups or the requirements of the rich and well-to-do; and second, the extent to which growth generates employment and income; and

(c) the extent to which employment and incomes are generated without creating adverse effects on working conditions and environment.

Growth, growth patterns and poverty are interlinked factors; rapid economic growth is essential for higher levels of welfare [10] and provides the wherewithal for sustained increases in consumption levels and for expanding the frontier for increased production.

However, although rapid economic growth is a necessary condition for improvements in the well-being of rural people, it is not a sufficient condition. Various studies in the ILO and elsewhere have shown that there has been an increase in the incidence of poverty in many developing countries in spite of respectable and sometimes rapid growth rates. For growth to benefit especially the poor, it must reach the poor in the form both of the goods and services they need to meet their basic needs, and of incomes required to purchase these goods and services. This therefore means that the growth pattern must be appropriate in terms of both its output composition and its input combination.

The pattern of growth (in the senses outlined above) in many developing countries – in both rural and urban areas – has been inappropriate, which explains the failure of rapid growth to reach or trickle down to poor people, especially the rural poor. To begin with, there is the well-known urban bias of development: health and educational services and most development activities are concentrated in urban areas; the consumption requirements of urban people and of those in the higher income brackets usually receive a disproportionate weight in planning and investment decisions. This has resulted, inter alia, in a substantial shift of resources from the rural to the urban sectors.

Within the rural sectors too, there have been serious imbalances in resource allocation. Cash crops, often for export, are favoured over food crops, a factor which partially accounts for the decline in food production per head in many developing countries and the deterioration of the nutritional status of a large proportion of their populations.[11] Further, notwithstanding the surplus labour available, the technologies used have been generally capital-intensive or labour-displacing. Where this has not been the case, structural bottle-necks such as inadequate access to land, credit and technical know-how often prevented poor or small farmers from taking advantage of available technologies or innovations, as has been seen from the experience with the Green Revolution.[12]

Agricultural modernisation and working conditions

The foregoing analysis underscores the need for rural development policies to be oriented towards the poor and small farmers who form the majority of the rural workforce, to tackle the problem of meeting these farmers' basic needs and to stimulate employment creation. This being said, however, it needs to be pointed out that agricultural modernisation in general and new technologies in particular can sometimes have adverse effects on working conditions,[13] as is illustrated by Panel 46. The case of the expansion of commercial agriculture sometimes leading to a decline in food production, and thus to a deterioration in the nutritional and health status of the rural population, has been already noted.

Another important point that needs to be made is the impact of new technologies and innovations on workers' health and workload. The expansion of land under cash-crop cultivation following mechanised clearing may result in an increased burden of work in weeding, harvesting and carrying operations. In Indonesia, the introduction of the dwarf HYV rice variety had detrimental effects on workers.[14] Farmers had to stoop excessively during harvesting and had to furnish extra effort because of the additional operation required for crushing the paddy stalks. Moreover, the new variety meant heavier loads. Whereas the head load for the traditional variety was 20 kg and 40 kg for female and male workers respectively, the head load for the new variety was nearly 100 kg, owing to the size of the standard jute sacks available at the market. Increases in the burden of work, especially of women, resulting from the introduction of new farming techniques or plant varieties have also been observed in other countries, such as Bangladesh and Egypt.

Technological choice and working conditions

Are there technological options which come close to having the ideal characteristics of providing employment, improving working conditions, enhancing safety and improving or protecting the environment? Limited

PANEL 46

Paradoxically, rural economic development may sometimes be a precursor to parasitic diseases: for example, an increase in population density may lead to a rising malaria transmission level and a consequent increase in the incidence of infection. Planned development projects such as the Tennessee Valley Programme, the Kariba Dam, the Aswan High Dam, the Mekong River Project and the Trans-Gabon Railway need thousands of labourers; if the project is carried out in a tropical malarious zone, the consequences for such a large concentration of labour are likely to assume disastrous proportions unless provision has been made in the scheme for an anti-malaria programme. The construction of the Panama Canal is an example that should not be forgotten. In eight years there were 50,000 deaths from malaria and yellow fever and the project had to be discontinued for a while. There are similar examples from India, such as the large railway construction project at Rangahapur (1942), which had to be closed down owing to malaria, and the Sarda canal construction project (1920-29), in which 96 men out of every 100 went down with fever and work had to be stopped until the conditions were ameliorated by the control of malaria.

E. P. Mach: "Selected issues on health and employment", in *International Labour Review*, Mar.-Apr. 1979, pp. 142-143, based on WHO: *Manual on personal and community protection against malaria in development areas and new settlements* (Geneva, 1974), pp. 9-10.

though our knowledge is, there are scattered examples of technologies which can sometimes meet these criteria. In Indonesia, for example, an experiment was undertaken to determine the most efficient among five different types of hoes available for cultivating rice terraces in Bali.[15] Using an efficiency index (defined as the amount of work done divided by the energy expended in performing the work), it was found that the efficiency of the hoes ranged from 8.42 for the worst to 12.42 for the best.

Perhaps more revealing are the results of a detailed study of technological choice in forestry in the Philippines. The forestry industry is one of the most hazardous and has always had a relatively high accident frequency rate. Most accidents occur during timber felling and conversion.

In the Philippine case study, three alternative methods of felling and cross-cutting trees of large diameter were examined: *(a)* with a large, heavyweight power chain saw of 9-13 horsepower; *(b)* with a smaller, lighter power chain saw of about 3-5 horsepower; and *(c)* with a traditional two-man cross-cut saw.[16] The small chain saw was more labour-intensive than the large chain saw, but less than the traditional two-man cross-cut saw. However, the productivity level under the small chain saw technique was three times that of the two-man cross-cut saw. Moreover, manual sawing placed great physical strain on the workers, especially those who suffered from dietary deficiencies. The use of the large power chain saw also had hidden ergonomic costs, since it was heavy and entailed a high risk of accidents and hearing loss. The study concluded that the small chain saw not only was the least costly method of production but also offered several ergonomic advantages: it was less likely to impair workers' hearing, was equipped with anti-noise and anti-vibration devices, and was light and well balanced. Similar choices and options were found for other operations also; but the important lesson from this particular example was that, at least in forestry operations, the scope for technological choice was relatively broad and that appropriate techniques existed. These techniques were economically justifiable, especially when the concept of "least-cost" covered ergonomic and environmental considerations as well as purely economic ones.

Some policy implications and issues

Need for rural development strategies directed against poverty [17]

The inter-relationship between the working environment and the overall living environment underlines the need for measures to improve working conditions within a larger framework of rural development strategies aimed at curbing poverty. For a large proportion of the rural population, the poor conditions under which they work derive largely from their poverty, that is, their inability to meet certain minimum requirements for private consumption as well as their limited access to essential services. This implies that rural development must embrace at least two central and specific objectives.

The first should be the provision of adequate employment opportunities at income levels that can enable rural people to meet their minimum consumption requirements. As land is the major source of livelihood in rural areas, a policy of land reform that can ensure access to land for those in need of it should be an integral part of any development strategy designed to curb rural poverty. This should be complemented with the provision of irrigation as well as new technologies and inputs, especially to poor and small farmers so as to enhance their productive capacity and to ensure full employment all year round. These may need complementing by labour-intensive public works schemes which – besides providing badly needed infrastructural facilities such as roads, dams, and so on – can also be a major source of employment, especially for low-income households and for those workers who may be seasonally unemployed.

The second objective of such a strategy should be the attainment of minimum standards in such areas as sanitation, the safety of drinking-water, public transport, and health and educational services. It was noted earlier that the poor quality of the labour force can often be attributed to poor sanitation and living conditions and to the inhospitable physical environment. Indeed, gains in welfare derived from the provision of sustained employment opportunities can be jeopardised if public health facilities are not available and if safe drinking-water and sanitation are not provided. Yet medical facilities and social services are often urban biased and hardly oriented towards providing, on a mass scale, minimal protection from commonly prevalent diseases or towards attacking the worst and commonest forms of poverty. In policy terms, therefore, a rural development strategy directed against poverty will require a reversal of the hitherto well-known urban and élitist bias in resource allocation and social services, and the design and delivery of the cost-effective services most needed by the poor.

Need to promote participation

The design of rural development strategies should not be conceived solely as a technocratic exercise. The participation of the rural poor and of rural workers in all decisions that affect their well-being is important, not only because it is an integral element in welfare, but also because it can contribute both to a more efficient utilisation and mobilisation of resources and to the initiation and consolidation of agrarian change.[18] Organisations of the rural poor, be they tenants or workers in private farms and plantations, can enable them to articulate and defend their interests in many domains, including working conditions and occupational safety and health. They can also serve as mechanisms for the delivery of inputs and services. That is why the ILO, through various international labour Conventions and Recommendations, has been urging countries to initiate and support such organisations (for instance, the Tenants and Share-croppers Recommendation, 1968 (No. 132), and the Rural Workers' Organisations Convention, 1975 (No. 141)). In practice, there are very many obstacles to the establishment and effective performance of such organisations: obstacles due to material factors (geographical scattering, size of work units) or cultural factors, due also to sometimes hard and, in some cases, violent opposition from landowners or political opponents. Both governments and established trade union organisations have therefore to assume a major responsibility – that of playing the role of catalyst and supporter in breaking down the obstacles and promoting independent grass-roots organisations in the rural sector.

Rural development, technological choice and working conditions

Although a concerted attack on the worst forms of poverty is a necessary condition for any improvement in the working environment, paradoxically, rural development can have adverse consequences on working conditions; technologies and innovations can also result in a greater work burden and in serious risks to health. Hence there is a need to apply ergonomic and environmental criteria in the design and implementation of rural development programmes and in the choice of technology.

It is conceivable that the inclusion of improvements in working conditions in the "appropriateness" or choice of technology may well pose certain conflicts. It may, for example, imply a trade-off between reducing the work burden and protecting employment opportunities. In Indonesia, for example, the introduction of rice mills is considered to have destroyed women's income-earning opportunities. In Java alone, 12 million women's work-days were lost, and this meant a loss in income of US$50 million for those women who earned their income through the hand-pounding of rice. On the other hand, grinding and hand-pounding are considered so laborious by East African village women that they are willing to go long distances carrying 24 kg of maize to have it ground. In such situations, it may be appropriate to ensure that those who could be potentially affected by the innovations are first consulted to determine their needs and priorities. However, such conflicts need not always arise. Limited though they are, some studies have shown that it is technically possible and economically feasible to arrive at technological options that come closest to the ideal characteristics of providing employment, improving working conditions, enhancing safety and protecting the environment. But knowledge of technological options in specific socio-economic conditions is still at an embryonic stage, and the need for research into this area can hardly be over-stressed.

The urban informal sector

Rapid urbanisation and the failure of modern industry and other organised sectors of the urban economy [19] to generate adequate employment opportunities, and the goods and services that most urban people need and can afford, have led to the emergence of a large unorganised sector, often referred to as the "informal sector", as a major source of income and output. Anywhere between 20 and 70 per cent of the urban labour force in developing countries are employed in this sector; the average level is around 40-50 per cent. It encompasses a wide range of activities such as those of petty traders, street hawkers, shoeshine boys, cooks, carpenters, masons, tailors, taxi-drivers and other tradesmen – all of which, besides providing employment, are also critical in the provision of goods and services needed by a large, though often poor, section of the urban population.[20] A vast subsector of the urban economy consists of marginally productive as well as economically efficient and adequately remunerative activities. Further, its importance derives not only from its current contribution but also from its future role in providing a livelihood for future additions to the already swollen urban labour pool.[21] Therefore, a strategy to improve work opportunities and work situations in urban areas will have to give a central place to the informal sector.

Characteristics and contribution

Informal sector activities are characterised by ease of entry, reliance on indigenous resources, family ownership, small scale of operation, labour-intensive and adapted technology, skills acquired outside the formal school system, and unregulated and competitive markets.[22]

There are at least two important factors that account for the special characteristics of informal sector enterprises.

The first is related to the origin and motivation of informal sector units or enterprises. These enterprises are established basically as a response to the lack of income-earning opportunities in the formal sector. They are created not by the classical entrepreneur eager to exploit investment opportunities and make profit, but by migrants and others who have neither capital nor skills but who none the less have to make a living and who therefore have had to generate their own employment opportunities.

A second critical factor is the policy environment within which informal sector units operate. Unlike formal sector enterprises, which are officially recognised and fostered by governments and therefore enjoy considerable advantages, informal sector activities are often ignored and in some respects even harassed by public authorities. They operate largely outside the system of government benefits and regulation, with no access to, for example, the formal credit institutions or the main sources of technology. Moreover,

> Many of the economic agents in this sector operate illegally, though often pursuing similar economic activities to those in the formal sector. [. . .] Illegality here is generally due not to the nature of the economic activity but to an official limitation of access to legitimate activity. Sometimes the limitations are flouted with virtual abandon [. . .]; sometimes the regulations are quite effective. The consequence is always twofold: the risk and uncertainty of earning a livelihood in this low-income sector are magnified, and the regulations ensure a high quality of services and commodities for the wealthy few at the expense of the impoverished many.[23]

These two factors help to explain why formal sector enterprises tend to be large and capital-intensive while informal sector enterprises adopt labour-intensive, often indigenous technology and tend to be small and economically vulnerable.

Earnings and conditions of work

By far the most dominant form of employment in the informal sector is self-employment, rather than wage employment. Such "peripheral" workers create their own jobs, taking advantage of demand opportunities. Several studies confirm a high concentration of younger and older workers and a higher proportion of female workers than in the formal sector. Short of capital and lacking in skills and know-how, informal sector participants, notably in tertiary activities, suffer from precariously low and fluctuating earnings. Average earnings may not reach even 40 per cent of the levels in the formal sector. And although not all the urban poor belong to the informal sector, most of them nevertheless depend on it for a livelihood. In numerous cities, as many as around 80 per cent of those earning below the legal minimum wage belong to this sector.

Although earnings are generally low and far below the level required to satisfy minimum basic needs, hours of work for most informal sector participants are generally long. In Freetown, Sierra Leone, for example, two-thirds of those surveyed worked between eight and 12 hours daily, and in Lagos, Nigeria, 92 per cent operated for more than nine hours per day.

Besides having low incomes and long hours of work, workers in this sector suffer from a lack of work premises and supporting infrastructural facilities, especially access to suitable locations within the city. As many as a third do not have a fixed location to carry on their business. Of those with fixed locations, a large proportion neither own nor rent their locations, which implies that they operate legally or illegally in public places, especially city centres. One consequence of this is that many operate in temporary sheds and structures which can be easily dismantled at the request of the public authorities. Many are therefore forced to operate in their residential premises. However, the premises are generally structurally weak and poorly served. It is reported that in Jakarta, Indonesia, for example, only 30 per cent of the households in the informal sector lived in "permanent" structures. Further, most were not provided with urban services; for example, only 25 per cent in Lagos, Nigeria, and virtually none in Colombo, Sri Lanka, were served with electricity and water. These scattered examples indicate the poor conditions under which informal sector particpants work and live.

Occupational and community health hazards

The inevitable recourse to primitive and improvised technology and work processes, as well as the inadequate premises and the use of unknown products, dangerous machinery and potentially risky tools – all compounded by overwork and a low level of preventive knowledge – and the failure to apply elementary preventive measures, such as good housekeeping, ventilation, and minimal hygiene and fire precautions, must take their high "unrecorded" toll in terms of occupationally related loss of these workers' lives and damage to their health. To complicate matters, as in the case of many rural workers in developing countries, these hardships at work are linked to and overshadowed by much community sickness due to malnutrition and to vector-borne, communicable gastro-enteric and respiratory diseases – themselves the result of poverty, squalor and ignorance.

Problems and policies

The development of the urban informal sector must not be viewed or considered in isolation. It should form an integral part of an overall national development strategy if regional or socio-economic disparities are not to worsen. This therefore means that the development of the informal sector must be linked to a national policy which stresses rural development and decentralised industrial and urban development, so as to ensure that the benefits of growth are diffused widely and equitably and the "push-and-pull" factors in the rural-urban migration process are weakened.

The formulation or design of policies and programmes for the urban informal sector is not an easy matter. Participants have varying characteristics, besides being organisationally diffuse and outside the orbit of formal policy and legislation. It is therefore difficult to attack the problems associated with poor working and living conditions through a single instrument, much less through legislation. One must consider at least two complementary sets of policies for this sector. The first set refers to employment policies, while the second is concerned with the provision of social services and physical infrastructure.

Increasing income-earning opportunities

The major employment problems in the informal sector are low incomes and productivity. These can be attributed to a variety of factors, of which two appear to be especially important. Generally, informal sector participants, because of their deprived background, have little or no capital and are mostly illiterate and unskilled. Because they have no property, they have little or no access to formal credit sources,[24] and therefore their ability to expand their activities is very much restricted to low levels of income or profitability. A major task of public policy, therefore, is the provision of greater access to credit, skills and other inputs for informal sector participants. This requires not only a greater infusion of resources into the sector but also the design of innovative organisational structures that can ensure the delivery of resources to workers at the lowest cost. Therefore, the possibilities for enhancing the economic and institutional capacity of informal sector participants through the establishment of co-operatives or other forms of organisation that can facilitate large-scale operations and deliver economies of scale need to be considered.

A second set of factors that have a bearing on incomes relate to activities and measures that fall, by and large, in the public domain. It was pointed out earlier that existing government regulations and modes of operation favour the large-scale undertakings, and directly or indirectly hinder the development of small-scale and informal sector enterprises. These regulations may take various forms, such as the rationing of production and trade licences and subsidised credit, special tax exemptions, and the provision of infrastructural facilities at low cost to large firms. Informal and, indeed, many small-scale enterprises operate largely outside the system of government benefits and regulations; and many operate even illegally, by being engaged in legitimate activities to which access is restricted by government regulations. It is therefore imperative that government fiscal, credit and licensing policies be reassessed in the light of their implications for overall efficiency in resource allocation and for expanding the market frontier and productive capacity of informal and small-scale enterprises.

Improved access to social services and physical infrastructure

As pointed out earlier, although not all the poor are in the informal sector, most of those who do live there belong to the poorest urban groups. Their poverty derives not only from inadequate income-earning opportunities but also from limited access to public services. This arises partly from their low incomes but also because of the heavy biases in the design, location, pricing and delivery of such services as education, health, transport, clean water and good sanitation. For example, in the field of health, one is often impressed by the expensive modern hospitals found in urban areas. Yet these generally benefit the rich minority more than the poor majority, notwithstanding the higher incidence of disease and mortality among the latter group. Public water-supply and sewerage facilities are usually also beyond their reach, and those to which they have access may cost relatively more. For example, low-income groups often have to pay 20 times more for water supplied by street vendors than middle and upper income groups pay for piped water supplied by public authorities.

Education presents a similar picture since, in poor households, not only are the parents often uneducated, but also their children suffer from unequal educational opportunities. Although the reasons for the limited access to education vary from country to country, the major explanation seems to lie in the fact that the poor often cannot afford the fees and generally live some distance from the nearest school. Although in a large number of countries most urban dwellers live in slum or near-slum areas, most of the schools are usually located in other areas of the city. This is an important issue, not only because education is important *per se* but also because it is a significant factor in raising incomes. Similar biases also prevent poor people from having easy access to public transport.

Their incomes are often too low to pay for transport; but, in any case, the services are often not available in the areas where they live, and they are forced to walk up to two hours each way to get to and from their work. The commuting distance, of course, increases with the size of the city: it has been estimated that the average journey to work of poor people is 3 miles in a city of 1 million, and 7 miles in a city of 5 million.

More dramatic, perhaps, is the extent of housing deprivation. In many cases, the overwhelming majority of urban dwellers live in slums and uncontrolled settlements, and many are totally homeless. Worse still, the dwellings of the poor may be razed to the ground by the public authorities, because they either endanger public hygiene, or do not meet unrealistic housing standards or simply mar the beauty of the city.

In almost all the above cases, the reason for this limited access of the poor to social services and housing is not only their low incomes but also the fact that most basic urban services are oriented to the middle and upper income groups. Housing and social service "standards" are set in terms of the middle and upper income groups' interests. Therefore, if the poor are to benefit, the approach to the design and delivery of such services must be radically altered and directed towards mass delivery at costs which the poor can afford. For example, the provision of stand-pipes (rather than piped water to individual houses) can have a significant impact on the health of poor people, since the provision of clean water and adequate sewerage is often the single most important factor in improving the health environment. The provision of small clinics in slum and squatter areas, where health problems are most severe, needs to be given priority in health policy, as should, in the field of education, small, inexpensive education units located in accessible areas. In housing, the provision of suitable land, equipped with basic infrastructural services such as access roads, drainage, water, sewerage and electricity, can bring significant cost reductions for the poor and provide them with increased possibilities for investment in housing construction.

Basic measures relating to safety and health

In such very depressed settings, where any vestiges of order or organisation are lacking, much has to be achieved on a wide, multi-sectoral front of socio-economic and health development, before any effective results can be expected from the isolated application of strictly sectoral occupational safety and health measures. Yet the challenge at least to contain the most obvious and prevalent man-made dangers cannot be entirely ignored. One possible ad hoc manner of bringing about some gradual change is through the establishment and systematic utilisation of a new category of specifically trained multi-purpose safety delegates. Their functions might include the identification and the recording of prevalent local hazards. In turn, they might advise on appropriate corrective action in respect of all potentially harmful work settings; these might include, inter alia, unsafe machinery and equipment, faulty work processes, harmful environmental pollution, potential dangers of fire, and shortcomings in protective measures for workers. These informal inspection and advisory functions could be enhanced if they were supported by training systems and specifically selected regulatory measures on the lines discussed above in connection with occupational safety and health in developing agricultural sectors. As regards harmful substances, the hazards in informal urban sectors derive less from pesticides than from a wide spectrum of chemicals, heavy metals and particulate matter.

Notes

[1] M. Lipton: *Why poor people stay poor: Urban bias in world development* (London, Temple Smith, 1977).

[2] ILO: *Poverty and landlessness in rural Asia* (Geneva, 1977).

[3] See, for example, idem: *Poverty and employment in rural areas of the developing countries*, Report II, Advisory Committee on Rural Development, Ninth Session, Geneva, 1979; O. Altimir: *The extent of poverty in Latin America* (Santiago, ECLA, 1978); A. Bequele and R. van der Hoeven: "Poverty and inequality in sub-Saharan Africa", in *International Labour Review*, May-June 1980, pp. 381-392; and World Bank: *Rural development*, Sector Policy Paper (Washington, DC, 1975).

[4] See, for example, E. Mendelievich (ed.): *Children at work* (Geneva, ILO, 1979); Z. Ahmad: "The plight of rural women: Alternatives for action", in *International Labour Review*, July-Aug. 1980, pp. 425-438; and I. Ahmed: *Technological change and the conditions of rural women: A preliminary assessment* (Geneva, ILO, 1978; mimeographed World Employment Programme research working paper; restricted).

[5] For an extended discussion of these health issues, see World Bank: *Health* (Washington, DC, 1975). Much of the discussion in this section is based on this study. See also E. P. Mach: "Selected issues on health and employment", in *International Labour Review*, Mar.-Apr. 1979, pp. 133-145.

[6] World Bank: *Health*, op. cit., p. 20.

[7] Mach, op. cit., pp. 142-143. For an analysis of the inter-relationship between health, nutrition and employment and the positive impact on output and employment of improvements in health and nutrition, see also A. Bequele and D. Freedman: "Employment and basic needs: An overview", in *International Labour Review*, May-June 1979, pp. 315-329.

[8] World Bank: *Health*, op. cit., p. 25, based on Felstein, Martin, Piot and Sunderesan: "Resource allocation model for public health planning: A case study of tuberculosis control", Supplement to *Bulletin of the World Health Organization* (Geneva, WHO), 1973, Vol. 48.

[9] S. S. Basta and A. Churchill: *Iron deficiency anaemia and the productivity of adult males in Indonesia*, World Bank Staff Working Paper No. 175 (Washington, DC, World Bank, Apr. 1974) cited in World Bank: *Health*, op. cit.

[10] For a succinct analysis of the links between growth, patterns of growth and poverty, see ILO: *Poverty and landlessness in rural Asia*, op. cit., and idem: *Poverty and employment in rural areas of the developing countries*, op. cit.

[11] This is especially true of sub-Saharan Africa. See FAO: *The Fourth World Food Survey*, FAO Statistics Series, No. 11; FAO Food and Nutrition Series, No. 10 (Rome, 1977).

[12] K. Griffin: *The political economy of agrarian change: An essay on the Green Revolution* (London, Macmillan, 1974), especially Chs. 7 and 8.

[13] For a discussion of the unexpected negative consequences of huge and capital-intensive development projects, see Mach, op. cit.

[14] A. Manuaba: "Choices of technology and working conditions in rural areas", in ILO: *Technology to improve working conditions in Asia* (Geneva, 1979).

[15] ibid.

[16] I. Ahmed: "Technologies for improved working conditions and environment in Philippine forestry", in ILO: *Technology to improve working conditions in Asia*, op. cit., pp. 98-128.

[17] For a discussion of the major features of anti-poverty or basic-needs-oriented development strategies, see ILO: *Employment, growth and basic needs* (Geneva, 1976).

[18] For an extended discussion of the issues and problems in promoting participation, see ILO: *Rural employers' and workers' organisations and participation* (Geneva, 1979); "Participation of the rural poor in development", in *Development* (Rome, Society for International Development), 1981, No. 1 special issue edited by Anisur Rahman; and P. Oakley and D. Marsden: *Approaches to participation in rural development* (Geneva, ILO, 1984).

[19] Between 1950 and 1975, the urban population in developing countries trebled and, as a proportion of the overall population, rose from 15.7 per cent in 1950 to 27.3 per cent in 1975. However, this rapid growth was not accompanied by corresponding employment growth in the formal sector. Thus, although industrial output increased by 5-10 per cent or more a year, employment rose by only 3-4 per cent, while the labour force was growing at a rate of 4-5 per cent.

[20] Products from the formal manufacturing sector are generally far too expensive for the poor. Informal sector activities, on the other hand, because they use little capital and skills, are able to reduce costs and sell goods more cheaply. In addition, by collecting and recycling waste material such as used bottles, waste paper, used clothing, plastic materials, tin cans, etc., they contribute both to economy in the use of resources and to the improvement of the environment. It is estimated that, in Nairobi, Kenya, for example, 77 per cent of the establishments use partly or exclusively recycled material. In Cali, Colombia, some 1,200 to 1,700 garbage pickers earn their livelihood through recycling waste materials collected from garbage dumps.

[21] A little simple arithmetic will illustrate the point. If, say, 20 per cent of the labour force is in the formal, organised industrial sector, the absorption of a 4-5 per cent increase in the labour force will require a 20-25 per cent increase in industrial employment. This, of course, is a difficult, if not impossible, task – especially since industrial output has to grow by far more than 20-25 per cent if, as experience has shown, labour productivity increases of 3-4 per cent per annum are taken into account. In the event, therefore, a large proportion of the urban labour force will continue to seek work opportunities in the urban informal sector.

[22] ILO: *Employment, incomes and equality: A strategy for increasing productive employment in Kenya* (Geneva, 1972), p. 6. See also ibid., pp. 503 ff.

[23] ibid., p. 504.

[24] See in this connection ILO: *Group-based savings and credit for the rural poor*, Papers and proceedings of a workshop, Bogra (Bangladesh), 6-10 November 1983 (Geneva, 1984).

HOW TO IMPROVE WORKING CONDITIONS AND ENVIRONMENT 8

In preceding chapters we have surveyed the factors that influence working conditions and environment, examined the various problems that arise, and described both the general improvements that are needed and, as far as possible, the specific measures that should be taken.

In this chapter we shall not go over again what has already been said. However, in order that the reader may draw all the many threads together, devise a policy for working conditions and environment, or set the action required in the appropriate context, we need to:

- consider briefly the men (and, of course, the women), institutions and means that contribute towards the elaboration of a policy of improving working conditions and environment (see "The men and the means", below);
- review the design and implementation of measures at the level of the undertaking (see "Action at the level of the work unit", p. 264) and at the national level (see "Action at the national level", p. 272), bearing in mind certain guide-lines; and
- study two particularly important resources: technology, its effects and use (see "Technological choice", p. 281); and education (see "Education and training", p. 287).

The men and the means

On whom and on what does the improvement of working conditions and environment depend? What can each one of us do to contribute towards that improvement? The first question is a general one that cannot be answered merely with a list of measures to be taken. It leads directly to the second, more specific question, since if concerted action is to be taken, trained and willing people are needed to undertake it: action must be matched to the circumstances and be logically planned. The labour inspectorate has the responsibility of applying a policy of which it is the tool; by its reports to the central authority and its work at the level of the undertaking, it plays an active role in the application of policy on working conditions and environment. Employers' and workers' organisations play a creative role (i.e. in their local or national bargaining); but they are also tools for transmitting legislation and theoretical and practical knowledge.

What does the improvement of working conditions and environment depend on?

Improvements in working conditions and environment are the outcome of overall development measures, specific measures and a particular approach to a country's economic and social policy; moreover, improvements require the right conditions for their implementation. Often, working conditions and environment result, in part, from action taken to deal with such phenomena as nutritional deficiencies, economic vulnerability, the technological dependence of developing countries, a deterioration in the terms of trade, urban growth and rural exodus, unemployment, getting and keeping a job, investment priorities, and so on. This fact cannot be overlooked even though the relation between general problems and working conditions varies from case to case: the employment of children is an example of a very close link, because it is mainly the result of family poverty which cannot be eliminated without a massive anti-poverty programme (see "Specific categories of workers and their problems", pp. 282-287). Extreme poverty will certainly be the major social problem in the years to come.

Improvements in working conditions and environment are the outcome of specific measures, in such fields as machinery guarding, guaranteed minimum wage, reduction of working time, and so forth. In each of the preceding chapters the reasons for these measures and the principles to be followed in their implementation were explained and, to this extent, the whole book has attempted to determine the ways in which working conditions and environment may be improved.

Improvements in working conditions and environment are, finally, the outcome of a particular approach to national economic and social policy. On the one hand, some improvements are related to human rights: social justice, equality, suppression of discrimination (race, sex or religion), freedom of association, collective bargaining, and so on (see Chapter 1 and "Specialists and educational, research and advisory bodies", p. 264). A prosperous economy is not a prerequisite for improving the workers' condition. Governments, employers and workers, working together within the ILO, are united by the philosophy that measures on basic rights should not be subservient to economic policy, since these very measures are themselves the proof of real progress. On the other hand, to be effective, action and policies to improve working conditions and environment must be an integral part of a country's economic and social policy. This basic principle will be dealt with in the section on "Action at the national level", p. 272.

Before measures to achieve improvements can be decided upon, implemented and have their effect, certain conditions must be met. Clearly, the means, the people and the institutions are needed, but so are also objectives and strategies, and this raises a number of questions. Who should carry out the action? On the basis of which priorities? Should action be isolated or co-ordinated; spontaneous or controlled; local and specific or general and common to the whole country?

The factors mentioned in the preceding two paragraphs are those which are of interest in this chapter, although space is lacking to consider them all in depth.

What can each of us do?

Programmes for improving working conditions and environment and the way in which such programmes are carried out will vary depending on the circumstances (region, country, undertaking, type of work, etc.) and also on what can be expected of those involved in the programmes.

The improvement of working conditions and environment entails changing an existing state of affairs in line with a set objective; however, both the existing state of affairs and the objective to be achieved will vary depending on the position, role and responsibilities of those instigating the change and the resources they can call upon.

To be effective, the action must be:

- suited to the objective, i.e. to ensure that progress is made in the right direction;
- suited to the role of the person involved, i.e. falling within his sphere of competence;
- suited to the target situation, i.e. capable of effectively modifying a situation that is well understood and for which the correct means are being used.

Since this book is aimed at a wide range of people with different functions in equally different contexts, readers will have to adapt the information it gives to their own functions and, in the case of those involved in training, to the functions of their trainees.

If a programme is to be effective, those responsible for it must have a positive attitude to working conditions and environment – and not just a morally positive attitude, since good intentions need to be accompanied by an insight into the problems and by the ability to maintain organised action over the long term. (Generous sentiments should not, however, be underestimated: they may have a "trigger" effect. Labour history is strewn with initiatives and discoveries stemming from people's sensitivity to the suffering of their fellows. But sentiment alone is not sufficient.) Selecting the action that suits the circumstances requires conviction and competence – both of which will be developed through education and training.

How can we improve working conditions and environment? What emerges above all is the size of the task, the great effectiveness of simple

PANEL 47

Labour Administration Convention, 1978 (No. 150)

Article 4

Each Member which ratifies this Convention shall, in a manner appropriate to national conditions, ensure the organisation and effective co-operation in its territory of a system of labour administration, the functions and responsibilities of which are properly co-ordinated.

. . . .

Article 6

1. The competent bodies within the system of labour administration shall, as appropriate, be responsible for or contribute to the preparation, administration, co-ordination, checking and review of national labour policy, and be the instrument within the ambit of public administration for the preparation and implementation of laws and regulations giving effect thereto.

measures skilfully applied, and the need for individual action by a wide range of people – not just the experts, but also people whose daily actions have a direct or indirect effect on other workers.

Public institutions

Legislation and the public authorities have a clear role to play, and a brief review will be given of the principles of labour administration, labour legislation and labour inspection.

Labour administration

The ministry responsible for labour affairs, usually called the Ministry of Labour, has a central role to play in matters connected with working conditions and environment. The Labour Administration Convention, 1978 (No. 150), specifies its main responsibilities (see Panel 47).

The Ministry of Labour will be responsible for designing a policy on working conditions and environment, drafting the necessary legislation and regulations, and expressing an opinion on labour matters within the government (see "Selecting priorities" (p. 274); "Guiding principles" (p. 276) and "Diversified action" (p. 278)). Dialogue with employers' and workers' organisations is both the source and the expression of this social policy since, in gathering information from them, the Ministry will develop a style of social policy that gives due weight – which will vary depending on the case – to concerted action. Finally, the Ministry "has to be alert to labour problems and to changes which might bring on problems; it has to investigate and keep up to date on facts and trends and to devise solutions and policy proposals, and it has to keep in touch with many other bodies." [1]

In implementing this policy (the section on "Action at the national level" (p. 272) deals solely with this aspect), the Ministry of Labour gives a lead to the labour inspectors, giving instructions, specifying their role, and supervising, advising and co-ordinating their activities to ensure a uniform policy; it also has the responsibility of ensuring that these services have the necessary resources to carry out their tasks.

Social policy is closely intertwined with other aspects of a country's policies, and consequently, by proposing lines of action and activities in fields of which it does not have direct control, the Ministry of Labour can exert considerable influence: for example, it has the responsibility to promote the concept of teaching about working conditions and environment in educational establishments. More generally, it should maintain close contacts with other ministries and ensure, by adopting a positive and resolute position, that the whole government has a forward-looking policy in social affairs.

Sometimes, responsibilities which would normally fall to the Ministry of Labour are handled by other ministries involved in economic or technical matters. For example, labour administration and the inspection of working conditions in mines, transport, docks or civil engineering may be within the purview of the relevant ministries. This dispersion of responsibilities is clearly not conducive to a uniform social policy.

Although within the government the Ministry of Labour has the major role in respect of working conditions and environment, many, if not most, other ministries also have a part to play:

- some are involved because the resources of other ministries depend on them (e.g. the Ministry of Labour's manpower policy cannot be implemented without collaboration with the Ministry of Finance and the Ministry of the Civil Service);
- others are concerned because they draw up the country's major lines of policy and priorities (Ministry of Planning, Ministry of National Economy or Finance);
- others are involved because they are responsible for one aspect of working conditions and environment (Ministry of Health, Ministry of Housing) or because their operations constitute an essential follow-through to the activities of the Ministry of Labour (Ministry of Justice, which arranges the prosecution and judgement of labour legislation infractions);

– still others play a part because they have the task of overseeing or stimulating certain sectors (Ministries of Industry, Agriculture, Transport and Communications, Handicrafts, Energy, Mines, and so on), and no policy to improve working conditions and environment could be effective without their involvement. Public undertakings have a special advantage in implementing government guide-lines and can play a pilot role. Ministries of Industry and Finance, in particular, and bodies under their control that promote small enterprises can, through incentive measures and advice to employers, take effective action in improving working conditions and environment (see "Coercive and incentive measures", p. 279);
– the Ministry of Education and Training can have a decisive long-term influence by integrating concepts of working conditions and environment into its teaching programmes;
– last but not least, the Prime Minister and the Head of State themselves have a capital role to play through the general orientation they prescribe, through the choices of policies they make, through the resources they allocate to implement these policies and through their overall stimulus and co-ordination.

Labour legislation

Although it is too often seen as coercive, legislation is the very foundation of social order and justice and, when it is lacking or not enforced, the way is open to all forms of abuse. Legislation on working conditions and environment specifies minimum but essential standards; these may be raised by collective agreements and may be adapted and made more specific at the shop-floor level. The law directly regulates certain components of working conditions and environment (occupational safety, hours of work, etc.), and, in its provisions relating to trade unions and collective bargaining machinery, establishes conditions for negotiations between employers and workers.

To ensure that this basic and irreplaceable legislative function is truly effective, legislation must be clear and precise, it must be neither fragmentary nor unnecessarily complicated, it must cover all workers without exception, it must be widely disseminated and understood, and it must have access to the resources necessary for its enforcement.

How can the gap between legislation and practice be spanned? As legislation becomes better and more comprehensive, one of the greatest problems in many countries is that of its effective implementation. Why is there this gap between legislation and practice? The enforcement of labour legislation is a subject that goes beyond the scope of this book. However, we may state that there are usually several factors involved: labour legislation may not always be sufficiently realistic and labour inspectorates may be too weak; the courts may be overloaded and

PANEL 48

Labour Inspection Convention, 1947 (No. 81)

Article 3

1. The functions of the system of labour inspection shall be:

(a) to secure the enforcement of the legal provisions relating to conditions of work and the protection of workers while engaged in their work, such as provisions relating to hours, wages, safety, health and welfare, the employment of children and young persons, and other connected matters, in so far as such provisions are enforceable by labour inspectors;

(b) to supply technical information and advice to employers and workers concerning the most effective means of complying with the legal provisions;

(c) to bring to the notice of the competent authority defects or abuses not specifically covered by existing legal provisions.

2. Any further duties which may be entrusted to labour inspectors shall not be such as to interfere with the effective discharge of their primary duties or to prejudice in any way the authority and impartiality which are necessary to inspectors in their relations with employers and workers.

sometimes too legalistic, and procedures may be burdensome. There are three basic reasons for this state of affairs: the population's poor living conditions; the priority given to economic objectives; and the lack of popular support for legislation on working conditions and environment.[2]

It is therefore necessary to upgrade the labour activities of the State, involve employers and workers more actively, and make greater efforts in the field of training.[3]

Labour inspection

Article 3 of the Labour Inspection Convention, 1947 (No. 81), states that the main function of the labour inspection system is to "secure the enforcement of the legal provisions relating to conditions of work and the protection of workers while engaged in their work". Consequently, labour inspectorates have a prime role in implementing a policy of improving working conditions and environment (see Panel 48, and also the Labour Inspection (Agriculture) Convention, 1969 (No. 129), Article 6). Provided that the inspectorate has adequate and well-trained staff, is not side-tracked from its main function by subsidiary tasks and is not paralysed by lack of resources, it can have an effective presence at the workplace and take decisive action by being severe, persuasive or explanatory, depending on the case.

In both theory and practice, the labour inspector must be independent of all pressures, and have the power required to sustain his authority. Penalties inflicted by the courts should be sufficiently severe to be dissuasive, and judgements should be made within a reasonable period if the inspector is not to lose his credibility and the law its authority. Labour inspectors should not be reluctant to deal with working conditions and environment in general, and occupational safety and health in particular. Such reluctance may often be ascribed to secondary and university education, which may provoke the feeling that these are technical and complicated matters. Negative attitudes of this kind should be avoided or corrected by suitable training. These considerations show the extent to which the effectiveness of labour inspection depends directly on decisions taken within the Ministry of Labour or the government. It is here too that choices are made on labour inspection procedures (specialist inspections by field of activity, or general inspectors only) and on the inspectorate's structure and composition (use of outside organisations for specific technical questions or integration of specialists into the inspectorate itself), and decisions taken on programmes, priorities and balance (programme areas, economic sectors and size of undertakings).[4]

Role of employers and workers

This subject will be dealt with only briefly since it has been mentioned repeatedly throughout the book. Employers must take a leading role in preventing occupational accidents and diseases and in improving conditions of work in undertakings of all sizes. Workers – acting either by themselves (where general working conditions and, in particular, job security permit) or through their organisations – should take an active part in measures aiming at improvement (see also "Key figures in the undertaking", pp. 270-272).

Employers' and workers' organisations

Active and independent employers' and workers' organisations have a decisive role to play in working conditions and environment, a role which may cover the information, education and stimulation of their members, the inclusion of improvement measures in collective bargaining, advising and lobbying at the government level and in committees or in research and training institutes, and so on. The value of discussion at the plant and national level is unanimously recognised, even though the dialogue is sometimes difficult; it is invaluable in understanding problems and in bringing about mutual comprehension and a positive approach to finding concrete solutions.

The improvement of working conditions and environment requires sound machinery that guarantees freedom of association and participation by both employers and workers.

The principles of freedom of association and the protection of the right to organise are fundamental here. Both are clearly laid down in the Freedom of Association and Protection of the Right to Organise Convention, 1948 (No. 87), which at the time of writing had been ratified by 97 countries. The Right to Organise and Collective Bargaining Convention, 1949 (No. 98), which has so far been ratified by 113 member States, requires that workers shall enjoy adequate protection against acts of anti-union discrimination in respect of their employment and that "measures appropriate to national conditions shall be taken, where necessary, to encourage and promote the full development and utilisation of machinery for voluntary negotiation between employers or employers' organisations and workers' organisations, with a view to the regulation of terms and conditions of employment by means of collective agreement" (Article 4).

The Workers' Representatives Convention, 1971 (No. 135), confirms the need for workers' representatives in the undertaking to "enjoy effective protection against any act prejudicial to them, including dismissal, based on their status or activities as a workers' representative or on union membership or participation in union activities, in so far as they act in conformity with existing laws or collective agreements or other jointly agreed agreements". It also specifies that "such facilities in the undertaking shall be afforded to the workers' representatives as may be appropriate in order to enable them to carry out their functions promptly and efficiently", taking into account the needs, size and capabilities of the undertaking concerned and provided that the efficient operation of the undertaking is not impaired (Article 2).

The Rural Workers' Organisations Convention, 1975 (No. 141), affirms the principle of freedom of association and establishes the right for all rural workers to join organisations of their own choosing without previous authorisation.

Collective bargaining

Collective agreements [5] are particularly suitable for laying down requirements with respect to working conditions and environment in a particular trade, geographical area or undertaking, and agreements usually deal essentially with these subjects. Negotiations between employers' and workers' representatives can review and resolve problems which, if neglected or overlooked, may be aggravated by time and lead to conflicts or management and production problems that are even more difficult to settle (absenteeism, low product or service quality, falls in output, etc.). Collective agreements are therefore more flexible than legislation and are better adapted to local problems of working conditions and environment or the technical and economic problems of a given sector. Collective agreements also stipulate flexible procedures to resolve conflicts arising out of their application, as well as the time-limits agreed upon for their revision. In matters of occupational safety and health, hours of work, wages and welfare facilities, collective agreements may bring about genuine progress and a tangible improvement in the workers' conditions.

Works committees (or other similar bodies) may also provide a forum for the discussion of working conditions and environment. Depending on the country and their terms of reference, they may have various names, and the problems they deal with are either general (works committees, works councils) or specific (occupational safety and health committees). They may be either bilateral (composed of a variable number of workers' and management representatives) or consist of delegates elected by the workers or nominated by trade unions.

Specialists and educational, research and advisory bodies

Many individuals or institutions can contribute to improving working conditions and environment through:

- the collection or sifting of knowledge;
- training and information;
- the provision of advice, encouragement and, where necessary, supervision.

These people or institutions may be active in various disciplines related to man and his work (occupational safety and health, ergonomics, occupational physiology and sociology, work organisation, etc.); they may come from universities and colleges (schools of engineering, national administration colleges, teacher training institutes), university research centres, bipartite and tripartite organisations (in many countries social security funds play an important role in training, in consultancy and even occupational safety and health inspection), private associations or semi-public organisations. If they are consulted and involved at an early stage and adequate information on their activities is ensured, they may make a valuable contribution to improving working conditions and environment.[6]

Action at the level of the work unit

The work unit [7] is the place where good or bad working conditions are created and it should therefore be at the centre of any examination of the causes of unsatisfactory situations and of the measures needed to rectify them.

No undertaking of any type can concern itself only with the production of goods or the supply of services, without regard for working conditions and the safety, health and welfare of its workers – and this not only for social reasons but also in view of the direct or indirect effects of working conditions and environment on productivity and on the satisfactory operation of the undertaking.

The undertaking does not operate in isolation since it interacts in a multitude of ways with the environment. In both the industrialised and the developing countries, growing emphasis is being placed on the undertaking's economic and social role in the national community (creation of wealth, provision of jobs, but also the cost of occupational accidents and bad working conditions, etc., to the community as a whole).

Need for a global approach to the problem

A multidisciplinary method required

As was pointed out in Chapter 1, the improvement of the working environment "should be considered as a global problem" (resolution of the International Labour Conference, adopted on 25 June 1974). In preceding chapters we have seen that even the most clearly identified and delineated problems can be fully explained only when they are dealt with in the context of people's working and living environments; consequently, any action that is envisaged should take this situation into account. Certain general rules applicable to all fields and to all workplaces do exist, and consequently the observance of legislation which lays down the minimum requirements is a basic principle in improving working conditions and environment. However, each workplace has its own peculiarities and no two are identical. Each one combines a number of interacting factors that constitute a unique system of inter-related working conditions and environment. These various considerations are dealt with in greater detail in "Two basic principles" (p. 268).

Working conditions and productivity

No work unit can have an exclusively social objective. It would be unrealistic to suggest that goods must be manufactured and services provided in other than a profitable way. The work unit must thrive and prosper. The search for acceptable working conditions and environment must be reconciled and integrated with these two objectives. Otherwise the improvement of working conditions would certainly be overlooked or set aside (for example, when safety instructions are not enforced because management has not integrated them into the basic operating instructions). The surest way of having working conditions taken

PANEL 49

It has taken a long time for the full extent of the interdependence between working conditions and productivity to be properly recognised. The first move in this direction came when people began to realise that occupational accidents had economic as well as physical consequences, although at first only their direct costs (medical care, compensation) were perceived. Subsequently, attention was paid to occupational diseases as well; and as a final step it was realised that the indirect costs of occupational accidents (working time lost by the injured person, the witnesses and the accident investigators, production stoppages, material damage, work delays, possible legal and other costs, reduced output when the injured person is replaced and subsequently when he returns to work, and so on) are usually far higher — as much as four times higher in some cases — than the direct costs.

The reduction in productivity and the increase in production rejects and manufacturing waste that result from fatigue due to excessively long working hours and bad working conditions — in particular, lighting and ventilation — have shown that the human body, in spite of its immense capacity for adaptation, is far more productive when working under optimal conditions. Indeed, in certain developing countries it has been found that productivity can be improved merely by improving the conditions under which people work.

Generally speaking, occupational safety and health and ergonomics have not been given sufficient consideration in modern management techniques, in spite of the modern tendency to consider an industrial undertaking as a total system or a combination of subsystems.

These problems have been seen in a different light since public opinion and, in particular, the trade unions became aware of them. It has been possible to detect in the stresses imposed by modern industrial technology the source of those forms of dissatisfaction which occur, in particular, amongst workers employed on the most elementary type of repetitive and monotonous jobs which are lacking in any interest whatsoever.

Thus, not only may a hazardous working environment be a direct cause of occupational accidents and diseases, but the worker's dissatisfaction with working conditions which are not in line with his current cultural and social level may also be at the root of a decline in production quality and quantity, excessive labour turnover and increased absenteeism.

In the developing countries the widespread lack of statistical data on industrial injuries and on absenteeism makes a detailed study of working conditions impossible; moreover, for workers in these

▷

countries working conditions may be only a secondary consideration, to be placed after the employment itself and the wages that accompany it. However, if one wishes to avoid, in the short term, the wastage of human and material resources — which is all the more serious in a developing country — and, in the long term, socio-political tension, great attention must be devoted to working conditions, and it must be recognised that nowadays the undertaking has an important social role to play in addition to its technical and economic function.

ILO: *Introduction to work study*, op. cit., pp. 47-48.

seriously into account is to relate them to production requirements. The opening paragraphs of the chapter on "Working conditions and environment" in the ILO's *Introduction to work study*,[8] reproduced in Panel 49, are particularly appropriate in this context.

The global approach and specific programmes

Since the individual measures to be taken in the work unit have already been dealt with in preceding chapters, this section will provide some general guide-lines for action programmes within the undertaking. Thus:

- each work unit should have a policy for improving working conditions and the environment in the same way as it has a policy for product or service quality; and this policy should be preventive;
- the improvement of working conditions and environment should form an integral part of the undertaking's general policy and of the daily implementation of that policy;
- the policy on working conditions and environment should be based on the rigorous enforcement of legislation, regulations and collective agreements;
- the policy should be backed up by consultation between employers' and workers' representatives; and
- the improvements that can be achieved by simple common-sense measures should be constantly re-emphasised.

The "global approach" may give rise to a number of objections. Is not a concrete programme within the undertaking no more than a series of specific and limited actions? Are these actions not usually related to safety and health matters and, consequently, should it be concluded that improvements in working conditions and environment can be given a rating, with occupational safety and health measures coming at the top of the list? These objections call for a number of comments.

- The global approach is essential to an understanding of the situation and may prove the only way of avoiding a wrong diagnosis and the ineffective or even harmful measures that may stem from such a diagnosis (see Chapter 1: "Working conditions and environment form a whole", p. 8, especially the section "The comprehension of work situations", p. 8; and "Aims and scope of the global approach", p. 20). Although it may seem that priority should be given to occupational safety and health measures that affect the physical environment, other measures are certainly important too, even though they may be less apparent or applied less frequently. For example, a construction worker working a 12-hour day for several weeks without a weekly rest would become accident-prone. In the same way, it would be quite legitimate for a recently established safety and health committee in an undertaking in a developing country to look at the composition and prices of meals served in the works canteen. Safety, health and welfare are affected by various factors which act directly or indirectly; all of them must be taken into account, and they demand a global approach since they are all inter-related.
- The improvement of working conditions and environment calls for separate, specific actions which may vary considerably in form and periodicity. Those relating to hours of work or welfare facilities, for example, may be taken by top management or by the personnel manager and may result from decisions taken at different intervals: those affecting safety and health, for example, are the result of day-to-day behaviour rather than of decisions taken at a given point in time. It is impossible to do everything at once; but all measures should form part of a coherent, integrated approach. With the global approach, the most important or urgent specific measures can be selected from amongst the various desirable courses of action. If they are to be effective, specific measures should be co-ordinated and form a coherent sequence (see also Chapter 2, "Overall occupational safety and health organisation", pp. 77-84).

Two basic principles

Enforcement of labour legislation

The scrupulous enforcement of labour legislation is a basic minimum in the improvement of working conditions and environment, since:

- labour legislation lays down compulsory requirements, applies to everybody and is backed up by the authority of the State;
- it lays down minimum standards;
- it is necessary to have compulsory requirements to counteract the effect of opposing demands (to make short-term economies,

complete urgent work, etc.) or of mere negligence or ignorance of workers' protection;

- it specifies the main problems and points out the need to solve them (the hazards of a particular product, the dangers of unduly long working hours etc.);
- it gives guidance in the solution of problems (calculation of overtime, information to be contained on wage slips, avoidance of contact with moving machine parts, etc.);
- where legal texts are distributed together with explanatory information, legislation may form the starting-point for a programme of improved training on social matters and consequently have a locomotive effect in improving working conditions and environment (this may be undertaken by the Ministry of Labour, employers' or workers' organisations or organisations responsible for basic and advanced vocational training).

A vigilant approach to the problems of man at work

The head of the work unit and those that assist him *must* adopt an alert and step-by-step approach to problem solving: otherwise, there is little hope that a policy of improving working conditions and environment will be implemented, for no matter how effective the labour inspectors and how powerful the deterrent effect of sanctions, it is impossible to have a labour inspector permanently at each workplace.

Too often working conditions and environment – especially those of a general nature, such as hours of work and welfare facilities – are approached from a purely legalistic point of view, as a set of rules imposed without justification by an external authority. Consequently, they are often dealt with solely from the operational point of view that is considered to be optimal (in the short term) for production requirements, since legal provisions and collective agreements are looked upon as constraints.

The correct approach should result in:

- questions being asked about the hazards that may exist in the use of a product or process, hours of work, etc.;
- questions being asked about the workers' personal problems; and
- lessons being drawn from any accidents, incidents or management problems that may occur.

Such an attitude does not derive merely from good intentions nor even merely from management's moral concern over the safety and health of the workers it employs. This is not to say that such motivation is of no importance; it should, however, be backed up by a minimum of information (i.e. a knowledge of the hazards and their consequences), and the opportunity should exist for each person to act effectively. Here can be seen the importance of all types of training.

Among the methods used to help to diagnose problems and to supervise the effective implementation of decisions are the "check-lists" used mainly in safety and health but which may also be applied to hours of work, work organisation and job content, for example. Some large undertakings also publish "social balance sheets" or "social reports"; these not only may be used for diagnosing working conditions and environment and for planning their improvement but are often reused for purposes of public relations and designed with this aim in mind.

This mental attitude must be permanent and must be shared – in an undertaking of any size – by all who have authority; it should be behind all decisions and their correct application and monitoring. It does in fact constitute a "policy", even though this term may seem pretentious for a small work unit. From time to time it may give rise to decisions of a general nature (on hours of work, pay, work organisation) or to specific measures (installation of a machine guard); however, it implies a sustained effort to create the correct approach to day-to-day work and in particular to the formation of safety consciousness.

Vigilance and sustained action are essential for maintaining (and even more so for improving) working conditions and environment, for a number of reasons:

- situations – of individuals or work or of the cost of living, housing, for example – change;
- the best devices and solutions tend to become less effective with time: tools wear out, plant requires maintenance, guide-lines are forgotten, and so on;
- the constant pressure of urgent production problems tends to diminish the priority given to safety and health if this priority is not constantly restated. The undertaking will need to mobilise all its efforts to implement such a policy (see, in particular, Chapter 2).

Key figures in the undertaking

1. *The chief executive officer.* Whether at the head of a small or a large agricultural, industrial or other undertaking, it is the employer who "gives the orders", and organises, manages, directs and supervises, with the aid of managerial staff when necessary. He is responsible for ensuring that the plant and its organisation do not imperil the workers' safety and health. He should therefore keep abreast of developments, issue guide-lines and receive reports on plant and premises, working conditions and the implementation of instructions; he should check that inspections are carried out, and, when assessing the performance of his managers, take into consideration the efforts they have made in the field of safety, health and working conditions.

The various means available to him in carrying out this task include the undertaking's work rules and operating instructions, which should contain provisions on working conditions and environment.

2. *The personnel director.* In all but the smallest undertakings, the chief executive officer will be assisted by a personnel director whose role, in view of his experience of the relevant legislation and his dealings with the authorities, is too often confined to legal matters or to the maintenance of order and discipline. His real task should be the dynamic and positive one of acquiring knowledge about the workers and about relevant factors in their lives both on and off the job, assisting in job assignment, making surveys and advising the chief executive officer about human factors, interpreting and enforcing labour legislation with the workers' welfare in mind and helping the chief executive officer in his dealings with workers' delegates. He should view welfare facilities and any responsibility he has for them as one of the undertaking's essential functions and not as a subsidiary or extraneous activity.

3. *Managers (or other terms to describe this leadership function).* The chief executive officer delegates authority to his managers, who should enforce not only production guide-lines but also rules on working conditions and environment in general and on safety and health in particular (see Chapter 2). Managers have a key role to play in the dissemination of directives and in ensuring their integration into day-by-day operations. Their task too is more than just giving orders, and should extend to exercising responsibility and leadership and encouraging involvement.

4. *Design and methods office workers* responsible for work organisation and planning deserve especial mention, since it is essential that they take safety, health and conditions of work into account at the design and planning stage and thus avoid deficiencies which might otherwise be very difficult to correct at a later date (see Chapter 2, "Hazards related to working procedures and work organisation", p. 50).

5. *The occupational physician, occupational medical service* (see Chapter 2, "Measures involving the worker", p. 72) *and the welfare service* (see Chapter 6) also come into this list but are dealt with elsewhere.

6. *Workers and their trade union representatives* should be consulted on working conditions and environment (a key component of the undertaking's policy) for a number of reasons:

- first, and justification in itself, the workers are the most directly affected by working conditions and environment;
- second, over the short term, concern for safety and working conditions often tends to take second place to economic requirements and urgent operational demands and, consequently, should be frequently re-emphasised;
- third, because only concerted action guarantees the correct implementation of decisions; and

– finally, because consultation improves the quality of the decisions taken; discussion with the workers involved will eliminate the type of error that is due to the omission of facts that would never be overlooked by the worker directly concerned; furthermore, ergonomists emphasise that "hands-on" experience adds an indispensable further dimension to objective measurements of noise, workload, and so on.

Dialogue, whether institutionalised or not, may be difficult or even produce conflict, since attitudes derived from different functions and responsibilities may be discordant; however, often only an objective and dispassionate global analysis will reveal the convergent interests of working conditions and good management.

Action at the national level

Policies are established to achieve goals within the constraints of perceived reality. Whether goal-setting should precede an assessment of the situation is a theoretical consideration, but the decision as to priorities among national objectives must stem from a knowledge – however elementary or inadequate – of the situation. If the themes that concern us in this volume are not given priority, the existing state of affairs may long be overlooked or misunderstood. Be this as it may, an essential prerequisite in all cases is that decision-makers, through their training, are aware of the importance of working conditions and environment.

Assessing the situation

Knowledge of the problems

A policy for improving working conditions and environment usually stems from an awareness of a situation and from the information that instigated this awareness. It must be preceded by a problem-identification phase of sufficient detail to allow definition of the final and intermediary goals and of the means to achieve them.

As a first step, data should be collected, and also compared and contrasted wherever possible, since statistics on occupational accidents and hours of work in particular may be misleading. It may be more reliable and important to realise, if only vaguely, that a problem exists than to have apparently accurate information which, because it is badly compiled, gives incorrect assurances and may consequently misguide social policy or inspire unsuitable measures.

It is not always easy, or even possible, to check the data systematically. Meetings with representatives of the authorities, and of employers' and workers' organisations, are an excellent means of bringing together and comparing existing data; they may also be a means of generating interest among participants and of mobilising energy for programmes that an analysis has shown to be necessary.

It may be asked whether it is essential at this stage to set up data collection facilities, or offices to design and produce data, and whether this is a first indispensable step in launching a policy for improving working conditions and environment. For example, it may seem necessary to have data from physiological or ergonomic studies, measurements of the working environment, reliable statistics on occupational accidents and diseases, surveys of public or plant welfare facilities, and so forth. No single answer can be given to this question, since so much depends on the urgency of the situation, the type and quality of available data and the goals that have been set. However, the following comments may be made.

First, no matter how valuable data may be for directing future action, it may be dangerous to specify that a programme cannot be started without adequate data being available, since if the data that are available are considered to be insufficient, action may be delayed or even paralysed. This does not mean that countries which have the resources and which are already operating programmes for improving working conditions and environment should not set up statistics research institutes and information dissemination centres (of which there are all too few). The intention is simply to draw attention to the danger of delaying measures for improving working conditions and environment under the pretext of inadequate information.

Second, high-quality data generally result from the implementation of programmes for improvement. Effective labour inspection will usually generate useful data on working conditions, and employers who are well informed and aware of the importance of safety measures, and trade union officials who are trained in this field, will produce accident statistics that are more reliable and generate greater safety consciousness.

Obstacles and means of overcoming them

An objective review of the obstacles and of the resources for overcoming them will guarantee the success of any working conditions and environment policy.

In many countries, the main obstacle to policy implementation is the fact that the responsible institutions are short of resources and are few and far between.

In developing countries, ministries of social affairs and their services (labour or medical inspection) and the employers' and workers' organisations are often weak, impoverished and, sometimes, powerless. Trade unions and employers' organisations may not be sufficiently interested in working conditions and environment, for reasons that have already been noted, and they lack a receptive audience. Labour administrations are short of staff and overburdened, and labour inspectors, whose work may be blocked for various reasons (see "Labour inspection", p. 261), do not carry out sufficiently frequent inspections. Weak support and limited resources are scarcely the right background for prolonged, energetic action to improve working conditions and environment. The situation in industrialised countries is usually very different, but they may still be relatively short of state services and lack adequate co-ordination between the various competent authorities.

Various other bodies and instruments (see "Public institutions", p. 258, and "Role of employers and workers", p. 262) are frequently overlooked, as though working conditions and environment were a matter of concern to only a small group of initiates; it might in fact be said, without denying the shortage of social structures in many countries, that another obstacle is the absence of wide community involvement in efforts to improve working conditions.

Other obstacles, more difficult to define but just as difficult to overcome, are the country's administrative organisation and the socio-economic and socio-political context. Pressure groups or factional or party rivalries may partly paralyse the state administration by wasting energy or by imposing a "wait-and-see" attitude and over-prudence in the implementation of policies, in order to maintain the status quo or for fear of encountering unforeseen resistance. A clear definition of government policy (see also "A coherent policy", p. 276) and of the functions and powers of those that implement it, and confidence in the will to pursue objectives are essential in mobilising those with the power to act in administrations, undertakings and economic and social spheres.

Existing resources, such as traditional culture and the population's physical, social or human qualities (social solidarity, team spirit, traditional skills, etc.), should be put to the best possible use in deciding upon hours of work, work organisation, choice of technologies, and so on, in developing strategies to implement decisions taken at the national level, and in issuing instructions to those in the field who are responsible for implementation (e.g. in the balance between constraint and sanctions, and dialogue, persuasion and appeals to solidarity). Neglect of these resources may generate resistance and retard action.

Selecting priorities

Is the improvement of working conditions and environment a national priority; if not, can it become one? What priorities should be set in this field?

Establishing an order of priorities for the improvement of working conditions and environment depends on the situation within the country and is the sole responsibility of the sovereign State. Previous chapters have indicated factors that may help in assessing the significance of individual problems, and an attempt will be made below (see "Guiding principles", p. 276) to list a number of characteristics that should be common to any policy for improving working conditions and environment.

The determination of priorities is a considerable problem for any government, but it is all the more acute in developing countries where everything is urgent and seems to demand top priority. In this regard, it may be considered that working conditions and environment have lower priority than malnutrition and hunger, the major endemic diseases, unemployment, low standards of living, inflation, the cost of basic commodities, and so on; also, it may be felt that working conditions are of concern only to wage earners, who are looked upon as "privileged" workers.

Here too, there is no single answer but a number of points should be made.

- The privileged position of one group over another is often relative and may not take all factors into account. For example, industry has far too many casual workers who do not enjoy the same protection as "normal" wage earners. Any relative advantages that these workers may have in no way compensates for the constant threat of fatal accidents and numerous other hazards.
- Working conditions and environment apply to all groups of workers – wage earners or not; urban or rural; industrial, agricultural or tertiary; or in the formal or informal sectors. Improvement programmes are needed everywhere.
- Measures to protect one group of workers need not work to the disadvantage of other groups of workers and may be of no cost to the national budget (e.g. regulations guaranteeing the regular payment of wages on a fixed date).
- Some measures have several aims or advantages, and reconcile priorities which may, at first sight, seem conflicting or concurrent. For example, a house-building policy may both improve workers' welfare and have positive effects on employment and productivity; vocational training may improve both output and occupational safety; the same applies to welfare and public transport facilities, to public and private health services and to measures to improve the conditions of work and life of rural workers, and so on.
- The improvement of working conditions and environment is part of a wider policy. However, working conditions are often forgotten, neglected or subordinated to production priorities and financial gain, which leads to dissatisfaction and even to serious and irreversible social or economic phenomena: degradation of the environment and impoverishment of rural populations as a result of policies that give sole priority to increased production at any price; rising health expenditure owing to poor working conditions; absenteeism and falling productivity; and negative attitudes to

factory work where workers' participation and conditions of life have been neglected; and so on. These and other examples given in earlier chapters are just a few of the long-term disadvantages of giving inadequate priorities to workers' needs.

The economic aspect of working conditions and environment, which has already been mentioned on several occasions elsewhere in this book (see, in particular, Chapter 1, "Economic constraints", p. 15), would require a much more detailed examination to correct the superficial treatment it has often received, which either contrasts matters of economic and social concern or subordinates the latter to the former in the name of realism. There is now a greater awareness of the inter-relationship between economic and social factors. Too often, analytical tools are still lacking, but a number are being developed to quantify and assess the cost of bad working conditions and the advantages of improving them. This endeavour is being pursued not only at the individual and plant level but also at the community level, since bad working conditions affect conditions of life, place a heavy burden on the community's resources and, finally, upset society's economy and equilibrium. Continuing research will help to strengthen the means of improving working conditions and environment. Over and above the economic gain to be achieved through such improvements, there is a social objective which is valid in itself and which should not be subordinated to profitability.

Guiding principles

The history of social progress in general and the improvement of working conditions and environment in particular has been marked with solemn declarations that have produced no effects, intentions that have achieved nothing and clearly deficient policies that have subsequently taken a U-turn. A coherent, co-ordinated and concerted policy is therefore important.

A coherent policy

First, there must be coherence between declared intentions or decisions and the measures that follow them. If decisions are to be effective, they should be translated into action in a minimum of time, and this implies that those responsible for implementing them – labour inspectors in particular – should be given clear instructions and the necessary resources.

Coherence should also be displayed through continuity of action, in order to demonstrate the authority's resolve. For example, any hesitation or policy changes affecting the type or role of workers' representation in the undertaking or the competence of labour inspectors in a given area would be counterproductive.

Finally, there must be coherence between the various decisions taken by an authority, and between those taken by different authorities. Not only formal but also *de facto* contradictions should be avoided. For example, measures encouraging undertakings to improve working conditions and environment will be fully effective only if the regulations are enforced; laws or regulations which are inadequately disseminated will not be enforced, and this will demonstrate that there is no real determination to put policies into effect. Ministerial authority and the workers themselves would be better served by legislation which was less advanced but which could really be enforced.

A co-ordinated policy

A number of ministries are involved in policies for improving working conditions and environment, and co-ordination at ministerial level is therefore essential.

Several ministries will nearly always be concerned when decisions of principle about structure have to be taken (strengthening of the labour inspectorate, creation of a new service, etc.).

Ministerial co-ordination is also necessary if the structures are to function properly, since certain measures, which are in themselves well conceived, may prove ineffective if they are carried out in isolation. For example, a Minister of Labour would be ill-advised to increase the number of labour inspectors unless he had the facilities for training them or if these inspectors, once trained, did not have the power or the resources to carry out their inspections, were burdened with other tasks or did not receive the necessary instructions, support or supervision. Consequently, the Ministries of Justice, Finance and, perhaps, the Civil Service need to be involved and a common policy established.

Co-ordination is also essential in establishing general lines of policy. For example, any decision to involve workers in occupational safety and health committees, works committees, and so on, would be doomed to failure if social dialogue were not encouraged and freedom of association not guaranteed. More generally, an effective and lasting social policy would be even less feasible if it were designed and launched by the Minister of Labour alone and aroused the suspicion or hostility of some or all of the other ministers.

Many complex problems need to be tackled simultaneously from a number of standpoints. For example, improving the conditions of life of rural workers (see Chapter 7) is an essential component of a rural development policy and not a separate, distinct policy of its own. If measures to improve conditions of life, decisions to involve rural populations and measures on agricultural technology are taken in isolation and without co-ordination, they are unlikely to succeed; and the same applies to action in the industrial sector.

A co-ordinated policy can, and must in certain cases, lead to joint action. For example, training in working conditions and environment presupposes close agreement between the Ministry of Labour and the Ministry of Education. An occupational medical service also implies a common purpose for the Ministry of Health and the Ministry of Labour; and the import of machinery that meets safety regulations entails liaison between the Ministry of Labour and the Ministry of Industry.

A concerted policy

Concerted action will guarantee that reliable information about the situation will be obtained, for no one is better aware of labour problems, hazards, workloads, fatigue and production constraints than the workers – be they blue collar, white collar or managerial – and the employer. Dialogue is clearly the best guide.

Concerted action will also help in the practical and long-term implementation of measures for improving working conditions and environment. If a central authority takes measures unilaterally and without consultation, these are likely to be incomplete and defective; even if they are sound and sensible, they may prove less credible and less readily accepted and may even be openly or secretly criticised, and there is less likelihood that they will be intelligently and effectively enforced.

Diversified action

Depending on priorities, general lines of policy, available resources, and so forth, a wide range of actions can be taken at the national level. Some of these will be mentioned here, but no classification on the basis of their nature or value will be attempted, in view of the equally wide range of situations which may exist. What is important is that they be drawn together by a central logical thread so that they complement each other and help to modify existing situations in the desired way.

Strengthening the institutions

The strengthening of existing institutions, or perhaps the creation of new ones, is usually a priority matter if the objectives are to be attained; this applies to labour inspection, in particular.

- The cost of recruiting and training new employees for administrative work may seem high for a poor country; however, this expenditure should be considered as an investment, and its medium- and long-term effects will be such as to make the financial sacrifice seem relatively small.
- However, the strengthening of existing structures often entails non-financial commitments which are just as essential as financial

expenditure. For example, the authority and efficiency of a labour inspectorate depend not only on the number and the quality of training of the employees but also on the penalties meted out for infractions and on the support and encouragement given by the Ministry. Similarly, the role that employers' and workers' organisations can play depends not only on their human and material resources but also on their interest in working conditions and environment and their ability to convince their members in this respect.

General and specific measures

General measures to improve conditions of life (construction of low-rent housing, improvement of public transport, provision of more vocational training, etc.) will have positive effects on the workers' situation. However, the State may also decide to carry out specific measures – for example, to improve the conditions of the poorest workers or to strengthen occupational safety measures in a nationalised industry, such as the mines.

Direct and indirect measures

Governments may introduce indirect measures to strengthen their means of action, for example by increasing the labour inspectorate staff, creating a safety and health service in a social security institute or providing training on working conditions and environment. More direct measures may relate to increasing the guaranteed minimum wage, setting up a factory or appointing persons with a known concern for workers' welfare to the top positions in national undertakings.

Coercive and incentive measures

The State has at its disposal a range of measures that can be implemented individually or in combination. It may instruct civil servants to enforce the law strictly and rigorously and may increase penalties for infractions of social legislation. It may offer financial incentives (such as tax reductions or exemptions, soft loans, preference in the distribution of materials or equipment), or encourage or officially and publicly support action in a given undertaking or sector.[9] These may usefully be combined with measures aimed at promoting the development and efficiency of small undertakings: financial incentives, and advice from consultants on the choice of technology and the design of equipment and organisational structures in small undertakings, are effective ways of simultaneously promoting improved working conditions and environment. This is a valuable policy measure from all points of view, but it does require clear delineation and adequate personnel training.[10]

The State may put pressure on undertakings or sectors in which it is a major client or has a majority shareholding, and insert clauses on working conditions in contracts that it signs.[11] It may also encourage, harmonise and co-ordinate action by virtue of its central position and authority. The active involvement of universities, employers' and workers' organisations, local communities and societies in research or the collection of information about working conditions and environment is essential to ensure that the work done is effective and is directed towards specific situations. These bodies may play a role at various levels of training, by providing information and creating public awareness (preparation of documents for the mass media), by designing industrial equipment, documenting life-styles, cultural technology and methods, and by studying the advantages and potential of modern technology, and so on.

Measures at the design and planning stage and corrective action

Although much can be done to improve existing workplaces, prevention is better than cure, and the best time to influence working conditions and environment is when plant and equipment are being designed or installed.

This is the least expensive way to adapt work to the worker and to integrate the worker into the work process, and is the best time to ensure, for instance, that workers are protected against hazards, that equipment and controls are anthropometrically adapted to the worker, and that machine parts are accessible for maintenance.

The provision of optimal working conditions and environment should be dealt with at the earliest planning stage of production plant or offices, as regards:

- the definition of requirements (programming);
- the siting or installation: available labour, the socio-cultural characteristics of the population, transport and housing facilities, integration of the work into the town or region.

Design specifications should include precise requirements as to safety and working conditions and incorporate a system of work organisation which matches the characteristics of the available manpower. When plans are being drawn up, the choice of technical solutions should be guided not only by reference to studies of similar work using the same technology (workers will be able to indicate what is arduous or dangerous and what must be corrected or redesigned) but also, to the extent possible, by general ergonomic or socio-technical data and information about the physical, climatic and socio-cultural environment.

Technological choice

The choice of the technology to be installed in a factory or office may have direct effects on working conditions and environment: hazardous machinery and plant; substances and microclimates; working procedures, work organisation, job content and work skills; hours of work; and wages.[12] The adoption of a technology involves more than merely purchasing plant, equipment and machinery and the transfer of technical data; it also entails transplanting a whole network of institutions, values, working methods and infrastructure.

The technologies chosen may not always be compatible with local conditions (climate, physical, social and cultural characteristics of the population, etc.), and the related work methods and organisation may reflect the concepts and ideas of the society in which the technologies were developed. If technology is transferred without precaution, it may have the disadvantage of totally separating the design process from the workers' socio-cultural environment. In rural areas and in small and medium-sized undertakings especially, the socio-cultural context plays a decisive role in all aspects of life and work, and there is a more intense social and communal life centred on the family, the extended family and neighbourhood relations (see Chapter 1, "Aims and scope of the global approach", pp. 20-21). Contrary to widely held opinions, however, technology is not predetermined, and case studies have shown a surprisingly large choice of economically viable production methods for a given situation. This choice makes it easier to plan acceptable working conditions and environment.

Whereas previously economic considerations were the sole criteria, the choice and import of technologies by developing countries are now more commonly examined from the viewpoint of their effects on employment. In fact, the whole range of social factors should be considered when choosing either local or imported technology. This was recognised by the 1976 World Employment Conference in the statement that "there is also a need to pay due attention to social aspects, working conditions and the safety of workers when introducing new technologies".[13] This new trend aims at achieving optimal results by giving adequate consideration to economic objectives, employment requirements and other social aims. It is being more and more realised that respect for the worker tends to facilitate rather than to impede the achievement of other objectives.

Only too frequently, a decision is taken to import a particular form of technology without due study, with short-term results alone being taken into consideration and with no thought being given to the need to avoid errors that have been made in other countries. Yet the rapid pace of technological change throughout the world offers numerous opportunities for improving working conditions and environment; if the maximum benefit is to be obtained, action must be taken at all decision-making levels. How can these concepts be put into practice? This was the question dealt with by the Inter-regional Tripartite Symposium on Occupational Safety, Health and Working Conditions Specifications in Relation to Transfer of Technology to the Developing Countries, held in Geneva in November 1981 (see Panel 50).

Although the need for measures in the field of occupational safety and health may seem more urgent and specific, and therefore attract more attention, the choice of technology affects all aspects of working conditions and environment. Existing technologies should be evaluated, selected and, where necessary, adapted, or new technologies should be created to resolve priority social problems. Many improvements that are feasible at the design or import stage are more difficult to correct at a subsequent date (see "Measures at the design and planning stage and corrective action", p. 280). It is, in short, necessary to master the technology and extract the maximum of advantages whilst creating the minimum of harmful effects for the workers and the population as a whole.

Specific categories of workers and their problems

Certain categories of workers are more vulnerable to bad working conditions and environment than others. General measures to improve working conditions and environment will therefore be of only limited effectiveness for these workers, as their specific problems call for special measures.

Limitations of space preclude the detailed examination of each of these specific categories, and only a few general remarks can be made. The reader is referred to Appendix D, "Guide to further reading", if he wishes to pursue the subject in greater depth.

Workers underprivileged or vulnerable owing to personal characteristics

Child workers

It has been estimated, and probably underestimated, that there were some 56 million child workers throughout the world in 1976. Children are not "small adults" but fragile human beings who must be trained to assume their future roles as citizens and workers. Most countries have followed the ILO's Minimum Age Conventions and Recommendations and adopted legislation fixing the minimum age of employment. However, the causes of child labour are largely unaffected by measures of this type. What is needed is enforcement of the law by a strengthened labour inspectorate, together with intensified efforts to control poverty, assist families and inform people of the dangers that work has for the physical development of children, and the adaptation of compulsory general education and vocational training suited to local conditions.

PANEL 50

To avoid or minimise the negative consequences of technology transfer, the following measures were suggested:

— ensure consideration of occupational safety, health and working conditions at the planning and design stage of projects;
— provision for approval by national authorities of occupational safety and related issues at the planning stage of technology transfer;
— the design team should include a representative from the technology-receiving country who will be involved in the operation of the plant and/or equipment;
— compulsory consultation with technology-receiving countries when designing projects;
— as far as practicable involvement of employers' and workers' organisations of technology-receiving countries at all stages from the planning phase for the introduction of the technology;
— prohibition of the importation of machinery, processes and equipment which do not meet the safety and health standards of the country of origin, with due consideration for local conditions;
— technology-exporting countries should provide all information they have available concerning occupational safety and health to the technology recipient. The technology-receiving countries should provide to the exporter all available information on technical conditions and on economic and social objectives as well as legislation relevant to the particular transfer;
— consideration of occupational safety and health in licensing arrangements;
— provision of welfare facilities such as adequate housing and transport;
— extension of medical facilities to cover the families of workers;
— once a new project has been decided upon, the licensor, contractor and consultants responsible for the design and implementation should be requested to state to governments the potential hazards of the project when in operation; the protective and remedial measures proposed including their cost; and the codes and standards used in the design. Governments of the importing country should employ suitably qualified consultants to check and verify these statements and to make an independent analysis of the plant or project throughout design and construction;
— the contractor or consultants should also be requested to design their plants on ergonomic principles appropriate to the climatic

▷

conditions, culture and social patterns and the physical capacities of the workers in the receiving country;

- tender documents for new projects should adequately specify safety features and standards, anthropometric and climatic conditions, and the required measures for monitoring and controlling the working environment. Contracts should not be awarded to those who sacrifice safety and health of workers to minimise cost;
- train negotiators for the transfer of technology in occupational safety and health and working conditions in order to ensure its inclusion in the technology transfer process.

ILO: *Inter-regional Tripartite Symposium on Occupational Safety, Health and Working Conditions Specifications in Relation to Transfer of Technology to the Developing Countries, Geneva, 23-27 November 1981* (Geneva, 1982), pp. 16-18 (extracts).

Young persons

Although not necessarily underprivileged workers, young persons form a heterogeneous high-risk group. Their needs are greater than and different from those of adults (nutrition, sleep, physical exercise); [14] they are more exposed than adults to occupational hazards because of their inexperience and susceptibility to specific risks, and their working conditions are often poor (arduous jobs, long working days, night work, low wages). They may also be subject to discrimination (exploitation under the guise of apprenticeship), and their working conditions may be unstable (temporary work, no integration into trade or trade union groups). They must be protected from hazards, their growth and the development of their personality should not be impeded, and they should be allowed to learn a trade, safe working habits and the means of protecting themselves against occupational hazards. Special medical supervision and the strict control of job placement procedures and conditions of work are all essential.

Migrant workers

Whether of foreign or national origin (often they come from rural communities), migrant workers are at particular risk in view of their uncertain legal status, their unstable employment, their frequently arduous or dangerous working conditions, their poor living conditions, the weakness of their trade union defence, their linguistic or cultural shortcomings, their low resistance to disease, their desire to make large savings on meagre wages, the relatively common discrimination they encounter and the difficulties of information and adaptation that they experience when transplanted into a foreign environment. The principle of equality of treatment expressed in international labour Conventions should be observed in all aspects of migrant workers' working and living conditions, and this implies the provision of well-organised official recruitment services, the existence and enforcement of bilateral agreements between States, and a major role for trade unions and labour inspectorates.

Disabled workers

These workers have specific employment problems and, throughout their working lives, will have difficulty in obtaining working conditions that are suited to their situation. Their handicaps may vary not only in origin but also in relation to the professions and jobs they are doing, and the concept of occupational handicap is relative to their specific occupation. A charitable attitude to the employment of disabled persons has been replaced by a more positive attitude which adapts the disabled person to the working environment, adapts the work to the person's handicap and emphasises the abilities and occupational capacities of disabled persons rather than their disabilities, as is indicated in the Vocational Rehabilitation (Disabled) Recommendation, 1955 (No. 99). In the event of disability resulting from an accident, preference should be given to adjusting and modifying the job to allow reintegration, rather than to retraining for a new job. Measures to reduce the physical effort demanded by work are in general helpful in the employment of disabled persons. On the other hand, the severe demands of work usually designed for healthy and well-trained persons increases the number of people unfit for work. (This raises the whole problem of jobs designed for "average, well-trained" workers since, by definition, the "average" worker has greater capacities and is more skilled than a large number of workers in general. This problem is considered in Chapter 5, "Work organisation and job content".)

Working women

Many women have unstable employment and arduous and low-skilled jobs and receive scarcely equitable wages, as a result of either their status as mothers, their dual family and professional function or cultural traditions. Are protection measures desirable? In fields such as night work (which is dealt with in greater detail in Chapter 3), the matter is controversial. The man's role in the family is changing more and more, and the Workers with Family Responsibilities Convention (No. 156) and Recommendation (No. 165) were adopted in 1981 with this in mind.

Workers underprivileged by the *de jure* or *de facto* absence of social protection

Homeworkers

The practice of home work is spreading and involving new areas of activity. Working conditions are often bad, with low wages, unduly long hours of work, job insecurity and ill-defined occupational status. The absence of social protection and the poor enforcement of regulations means that these isolated and defenceless workers are at high risk, underprivileged and even exploited.

Workers in temporary employment agencies

The practice of providing temporary workers, which was originally intended to help undertakings to find temporary replacement workers in the event of the unexpected absence of a permanent worker, has spread rapidly in recent years and is the subject of heated controversy. These workers are employed by agencies who place them with "user" undertakings to carry out work which is, in principle, temporary. The workers have low security, may be exposed to dishonest practices and require special protection (which should be provided by clear legislation) and close supervision, in view of the insecurity inherent in such unstable work and the poorly defined rights and obligations of the three parties to this triangular work relationship. Undertakings are attracted by the flexibility offered by the employment of workers with whom they have no bond, and workers are attracted by temporary work in periods of underemployment.

Moonlighters

The extent and, even more, the growth of moonlighting, a common form of clandestine employment, is indeed a worrying phenomenon. Moonlighting is prejudicial to the worker both when it is his sole job and when it involves double-jobbing. Various categories of workers are involved, but migrant workers form the most important group. Although some workers may turn the situation to good account, the majority are isolated and defenceless, lack social protection and may be exploited and exposed to very poor working conditions.[15]

Other forms of unstable work

Various types of work may be grouped together with moonlighting, since the workers involved tend to be situated outside the normal wage-earning structures and the protection they provide: young workers falsely classed as apprentices or trainees, wage earners wrongly called casual or temporary workers, freelance workers, workers artificially placed on fixed-term contracts. Detachments, employees on loan and labour subcontracting are all related schemes that are dangerous not only for the workers in question (often exposed to arduous, high-risk work with no supervision or protection) but also for other workers whose job stability is ever more threatened and for the very principle of social protection.

Education and training

It is no exaggeration to say that there is a widespread shortage of training in the field of working conditions and environment. The more flagrant aspects of this shortage are well known: ignorance of even the most common occupational hazards; unawareness of new hazards stemming from the use of new products or processes; poor appreciation of the consequences of a given form of work organisation, machine design or job instructions. The millions of young, migrant or "green" workers who receive no instruction about the work they are to do are living examples of the risks incurred, and passivity in the face of unacceptable working conditions and environment is the direct consequence of this widespread ignorance. The phrase "information starvation" is well coined, and the situation is all the more pernicious in that the training would, by its very nature, have advantages other than those of purely raising the workers' level of knowledge; well-designed training is the first step in a process because it gives workers the ability to acquire knowledge and raise their standard of skills.

Everybody has the right to be trained, in particular the right to be informed about and trained in the hazards of their job to themselves and their colleagues. Those with supervisory responsibilities have the right to understand the effects of their decisions on the workers who report to them or on those whose conditions of work depend directly or indirectly on them (engineers, planners, etc.). Inadequate or non-existent training in working conditions and environment effectively implies that the worker's training in his job is incomplete, since manual workers, engineers, planners, chief executive officers, and so on, can be fully competent only if they have received adequate training in all aspects of their work. The accumulation of these deficiencies is what constitutes bad working conditions. This demonstrates how great is the need for training: for training has an essential place in improving working conditions and environment.[16]

Role and objectives

Before determining the action to be taken, we must clearly define the role and objectives of training in a strategy of improving working conditions and environment.

Importance

In such a strategy, training is of importance for two reasons.

The continuous integration of improvements into the work process is vital, but it is possible only if everyone involved is properly trained. The work of labour inspectors in all parts of the world is easier and more effective in those undertakings where management and workers' representatives are aware and informed of the problems that exist. The role of training is therefore essential, and complements the inspection and constraints exercised by public authorities and the dialogue between employers and workers. Training, participation and inspection are three basic forms of mutually supportive action. And education is a source of inspiration for all other programmes, whether they take place within the workplace or have an indirect effect from the outside.

Training is the key to lasting change in those who receive it, and therefore, when it is applied to working conditions and environment and widely disseminated, it is a basic factor in the concern that our civilisations show for social matters and the worker. Without training, society would probably tend to pay far too much attention to the production of goods, which, in spite of its undeniable benefits, may still lead to profound suffering for the worker and, over the long term, result in imbalances in society itself.

A number of specific factors should be added to these general principles. In many developing countries and in various sectors in industrialised countries (particularly in small and medium-sized undertakings and in activities such as moonlighting, temporary work, home work and other precarious forms of employment – see "Specific categories of workers and their problems", p. 282), labour inspectorates, employers' and workers' organisations and other institutions contributing to the improvement of working conditions and environment are active in this field only to a limited or insignificant degree. Under these circumstances, of the three major prongs of action indicated above – training, participation and inspection – education is virtually the only one that survives. Education is the main approach for resisting a deterioration in working conditions and environment and for achieving improvements in the short, medium and long term.[17]

Function

The primordial role of training in working conditions and environment is to promote action. It must therefore stimulate awareness, impart knowledge and help recipients to adapt to their own roles.

Stimulating awareness

To stimulate awareness, it is first necessary to match the content and level of the curriculum to the trainees' level of comprehension and knowledge. If the programme is overburdened or beyond the trainees' comprehension (as frequently happens with training in the legal and technical aspects of occupational safety and health), it may be not only ineffective but actually harmful, since it may provoke a negative reaction to the subject both during the course and for long afterwards. The level of achievement at the end of the course should not be the only factor governing course design; more important is the amount of information the student can understand and assimilate. The subject is too serious for it to be consciously or unconsciously neglected or avoided by labour inspectors, personnel directors, trainers and instructors, employers and trade union officials, as a result of their having attended courses that were badly planned or too technical.

Awareness presupposes a consciousness of what is at stake: that is, of the risk (e.g. of handling a given product) or the problem (e.g. a schedule of hours of work) and of their repercussions on the workers themselves, their families, the undertaking and the community. The way in which the facts are presented and explained will influence the trainees' feelings and concern, and will also give a first indication of the measures to be taken; this will then lead to reflection on what can be done and provide an incentive for obtaining further information, for stimulating an understanding of the situation and for taking the necessary action. The correct appreciation of what is at stake is essential if trainees are to be convinced of the importance of what they are being taught, so that they then apply the regulations intelligently – and not just blindly and mechanically – and use their own initiative whenever circumstances require. This reaction will be deeply ingrained and will produce positive results long afterwards.

Finally, stimulating awareness also means demonstrating opportunities for change. The achievements of others can be used here to show that the goals are not utopian; if well selected, they will stimulate the trainee's imagination in seeking a suitable solution to his own future problems. A major requirement is to persuade all trainees that they are themselves capable of action and that they can act on the reality around them. Consequently, it is important to emphasise simple measures that can be carried out at low cost and without advanced technical know-how. Similarly, it is important to show the contribution that improved working conditions and environment can make to the operation of the undertaking and to the reduction of community expenditure.

Increased awareness is not only the starting-point for training, by conditioning the trainee to make the effort to pay attention and to understand what he is being told; it is also a major part of training itself, since it is the key to the trainee's personal adaptation to new situations and to personal improvement.

Imparting knowledge

Imparting knowledge is a second function of training; thus, when deciding upon the content of the curriculum, one should not lose sight of the goal of providing the trainee with the wherewithal for action. The knowledge imparted will depend on the trainee's needs and functions, but in all cases (since little training time will be available) emphasis should be placed on key concepts.

It is preferable to teach a limited number of basic methods and principles which will guide trainees in their action – for instance, the principles that group protection should be given preference over personal protection; that hazards should be eliminated at source rather than by multiplying the number of safety instructions; that risk analysis should be carried out before any safety measures are taken. A sound knowledge of these basic principles will be invaluable in helping trainees to deal with situations that are new to them. Training is often a matter of correcting errors and of combating misconceptions. For example, a typical misconception is that safety and working conditions are fields reserved for specialists; on the contrary, the teachings of the pioneers in accident prevention – that working conditions and environment are everybody's concern – should be reiterated and emphasised. Whenever possible, reference should be made to the trainees' own areas of knowledge and specific work situations.

The measures that can be most widely applied in a broad range of situations and activities, by people with no special technical competence, are simple measures. This should be emphasised not just for the trainees' own benefit, but also because it is the simple measures that the trainees will most easily be able to pass on to others by example or instruction.

Training should give the trainee a better understanding of the need to take working conditions and environment into account in his work. In general, the best training for achieving these aims is the training that forms an integral part of general school education, vocational training, higher education or in-service training. Knowledge, attitudes and behaviour that have been acquired at the same time as basic training skills will remain firmly embedded in the young (or adult) trainee. Whilst the principle is fully accepted, such integrated training is far from universal in practice, especially in non-manual occupations; it is nevertheless of the upmost importance that heads of undertakings, technicians, trade unionists, senior civil servants (in Ministries of Planning, Industry, Finance or Agriculture, for example) should be aware of the existence of the human problems involved in work and of the direct or indirect effects that their decisions may have in this regard.

Adaptation to individual roles

When modifying working conditions and environment and, in particular, work organisation and content, one must take care to adapt people to the new responsibilities they have to bear. Wherever possible, the change-over should be preceded by the appropriate training, since many attempts at changing work organisation have failed because workers and management were not fully prepared for their new roles and, especially in the case of management, for introducing methods of supervision that are closer to technical assistance than to authoritarian injunctions.

In-plant training

Training at the workplace obviously has high priority. However, the following comments cannot cover all those who should be trained nor provide a summary of training programmes, but only give some general indications.

Workers' representatives

In many countries, workers' safety delegates or the worker members of occupational safety and health committees are appointed or elected to ensure that safety instructions are followed, that hazards or unsolved safety and health problems are detected and that information on such matters is disseminated. A relatively short training course (which is sometimes a legal requirement) may be sufficient to provide the rudiments for workers to carry out this task. The curriculum may cover the principles of safety and health, the detection and identification of hazards, the assessment of safety and health conditions, legislation and regulations, the implementation of safety measures and the provision of information for workers. In some cases, the safety delegates are also given the task of workers' training.

Other workers' representatives (works committee members, workers' delegates, union delegates, shop stewards) have a wider field of competence and must be able to note problems in working conditions and environment, to understand their inter-relationships, to submit them to the head of the undertaking and to discuss them with him. The training they need is essentially one of stimulating awareness and providing for the acquisition of certain basic principles.

Workers in general

Every worker should be familiar with the main hazards of his job. In some countries the employer is required by law to provide this information. However, a knowledge of the hazards is not sufficient, and workers should be trained in safe working practices so that they do not compromise their own safety or imperil other workers and so that they can contribute to ensuring that the working environment is safe. Special training is required for those doing dangerous work and should, where possible, be integrated into basic vocational training and apprenticeship courses. Similar training should be made available to those (such as young rural workers, migrants, temporary workers, etc.) who for one reason or another have not received vocational training. Short courses should be organised by the employer, by the public authorities or at the sectoral level to update knowledge and stimulate active participation in campaigns to improve accident prevention and working conditions in general; similar activities should also be organised periodically in the less structured sectors such as agriculture. Basic and continuing vocational training, either in-plant or as part of an inter-undertaking arrangement (where the individual undertaking is too small to assume the burden alone, or in trades such as building and transport where workers are scattered), are particularly effective in disseminating basic concepts on safety, health and working conditions, and have produced spectacular results in reducing the number of occupational accidents.[18]

Management, engineers, technicians and foremen

These people have an important role since they have the power to promote or impede change and either participate in or paralyse attempts to improve working conditions and environment; their support is therefore essential. Engineers and technicians should receive a solid grounding in ergonomics during their training. All should be aware of the human and economic significance of working conditions and environment and should be encouraged to integrate these matters into their daily management. Such an approach would affect the way they give orders, supervise or encourage, help them to diagnose work situations, take the necessary measures or inform either their immediate supervisor or, where necessary, the designer of any modifications that are required.

Designers and planners

Designers and planners have a decisive contribution to make in defining and creating the working conditions and environment of manual workers in industry, in particular. Their training must enable them to consider the effects that their designs will have on workers' safety and health; consequently, they need a thorough grounding in ergonomics and should be able to foresee the effects of technical choice on work organisation and job content.

Heads of undertakings

Although heads of undertakings and the chief executives of industrial, agricultural or service establishments do not need to be specialists on working conditions and environment, they must nevertheless give these factors due consideration in managing their undertakings. Their training and information should be given priority, and a flexible approach should be adopted in view of the limited time at their disposal.

First, they should be aware of the importance of working conditions and environment for the lives of the workers they employ and be informed of the main hazards of the work process and their possible effects on human beings. However, a purely humanitarian approach which disregards the undertaking's operating procedures and constraints is not sufficient; managers should be made aware of the high cost of bad working conditions and of how these can reduce productive efficiency. They should realise that the causes of occupational accidents are similar to those of incidents which result in material damage only, that bad working conditions lead to lower reliability and poorer quality, that to reduce human work to the repetition of a few movements that make no allowance for the worker's skills, aptitudes and aspirations is counterproductive, and that changes introduced with the aim of improving working conditions may prove to be profitable.[19]

Heads of undertakings must be made to understand their legal obligations and to realise that not only their moral, and perhaps civil, but also their penal responsibility may be involved; and they must be provided with information about the relevant factors and in particular about their inter-relationship in order to encourage them to make working conditions and environment an integral part of the undertaking's policy.

Finally, they should realise that the importance they give to these problems, to keeping themselves informed and to giving the necessary instructions, will have a decisive effect on the efforts made in this respect at different levels in the hierarchy.

Training of specialists

Labour inspectorate staff

Labour and factory inspectors and labour supervisors require skills which are not normally provided by a university education. They are responsible for enforcing legislation and regulations and for encouraging endeavours to improve working conditions and environment, and should therefore have a knowledge of safety and health hazards, problems and principles, economic and social problems, labour legislation, the protection of workers and, in particular, the position of the wage earner and the State's role and functions in this respect. They should also have a commitment to public service and an awareness of their duties. They need special training which will make due allowance for their basic educational qualifications and which, rather than instilling abstract scientific or legal concepts, will aim mainly at helping them to carry out their role as consultants and representatives of the central authority and at stimulating their interest in occupational safety and health and working conditions. This training should also enable them to interest management and workers' representatives in working conditions and environment and, where possible, to provide them with training.

Occupational health and hygiene staff

Industrial physicians should receive training that equips them to carry out effectively their tasks of prevention and consultancy in order to avoid harmful effects on workers' health by monitoring their health status and working conditions and by involvement in attempts to improve them (see also Chapter 2).[20] Postgraduate doctors specialising in industrial medicine should be given training on the most common occupational hazards and on ergonomics and the working environment. Both training requirements and the work of industrial physicians and occupational hygienists are complicated by rapid advances in biology and in production processes; consequently, continuing and refresher education programmes will be necessary.

Industrial nurses will also need initial basic training and, where possible, specialist training. There is a growing demand for nurses with special occupational health and hygiene skills in such fields as toxicology, and the necessary training should be provided by universities and higher education establishments.

Finally, although its nature will vary from country to country, a policy of training paramedical workers to collaborate closely with the industrial physician may be introduced. In general, not enough of this kind of training is provided at present.

Specialists in occupational safety and ergonomics

These persons should have a sound basic training in the technical problems facing the industries in which they will be involved and should receive instruction in other disciplines, particularly ergonomics. They should learn about other specialist activities and about industrial organisation; consequently, further training and retraining programmes are desirable. Advanced training programmes should update their technical knowledge and should prepare them for consultancy work (identification of safety defects, specialist advice on safety measures and devices), workplace inspection, accident investigations, training programmes and the development of teaching materials, and so forth. They should also know when it is necessary to have recourse to specialists (such as industrial physicians, ergonomists, occupational physiologists, labour inspectors, chemists, etc.) or experts on more specific subjects such as fire and explosion hazards, hoisting apparatus, and the like.

Dissemination of scientific training and the multidisciplinary approach

Although it is true that there are not enough experts specialising in the subject of working conditions and environment to meet current demand, it should not be forgotten that the disciplines in this field of study – usually considered as specialist subjects – should above all and increasingly be incorporated into the training received by other professions. The principle of educational integration should be applied not only to general, primary and secondary technical education: the acquisition of specific technical knowledge should also be a feature of higher educational programmes and training programmes for the various professions (see "Training in teaching establishments", below). What is needed, in fact, is not only a small number of highly qualified people specialising in working conditions and environment but also ever larger numbers of people in different fields of activity with a basic knowledge of working conditions and environment as a whole.

The study of man at work cannot be compartmentalised to observe the divisions of the scientific disciplines currently taught in universities. As yet there is still no "faculty of working conditions" (covering, for instance, hours of work, work organisation and job content); yet a knowledge of a wide range of disciplines is basic to the global approach to working conditions and environment. Some of these disciplines are taught as sciences (occupational psychology and sociology, anthropology, agronomics, nutrition, labour legislation, etc.), others are techniques borne of necessity within the undertaking (as used by the organisation and methods engineer, the personnel manager, etc.), while still others result from the need to bring together various areas of knowledge (ergonomics, socio-technology). Since it is difficult for a single "generalist" to have adequate knowledge of all these different disciplines, there is a growing need for multidisciplinary teams.

Finally, practitioners have a leading role to play in enriching and renewing knowledge on working conditions and environment, and they may even be classed as specialists in their own fields – labour relations, labour inspection, trade union activities, management and administration. Their training cannot be bounded by a corpus of theoretical knowledge, since it derives from day-to-day contact with the reality of working life and the acquisition of various disciplines. Above all, an exchange between practitioners, specialists and educators will benefit not only those concerned but many others.

Training in teaching establishments

Vocational and technical training

Reference has already been made to the positive action that adequate vocational training can have on working conditions and environment. Proper vocational training teaches good working practices, including those related to safety and working conditions. These practices are so intertwined that it is difficult to separate them: for example, the correct installation of roof supports in a mine is necessary both for the miners' safety and for normal production work; again, suitable, well-maintained tools are a part of good working practices and also ensure that the work is done safely, rapidly and correctly. Bad working conditions are often the result of a lack of vocational or technical training or apprenticeships (rural workers, casual workers, foreign migrant workers), or of inadequate training which has not been adapted to changes in the job or trade, or of incomplete training due to a misplaced concern for efficiency.

The Joint ILO/WHO Committee on Occupational Health has stated that "the programmes for students or apprentices should contain health and safety aspects: dangerous work habits; the hazards of tools, machinery and chemicals; safety rules; principles of occupational hygiene, e.g. requirements relating to climate, lighting, control techniques for health hazards, principles of ergonomics (working time, workload, work posture, lifting of weights, fatigue, etc.)".[21]

This implies the adequate training of teachers in technical training colleges and the updating of their knowledge by short refresher courses.

Training of other professionals

Everybody who may directly or indirectly influence working conditions and environment should be aware of their role and responsibilities. "It is essential that teaching establishments provide their students with the basic elements of occupational safety and health. It was concluded that there is an urgent need ... to institute teaching programmes on occupational safety and health, adapted to each particular profession, to ensure that future professionals are fully aware of their responsibility in the protection of workers." [22] This observation summarises well some relatively recent and particularly desirable developments in a number of countries: namely, the introduction of occupational safety and health training in all higher educational establishments, courses on environmental health and the working environment in an engineering college programme, courses on working conditions and ergonomics in various higher education establishments, in particular for architects, physicists, chemists, administrators, economists, lawyers, psychologists, machine designers, etc.[23] On this particularly important point it should be emphasised that "the task of building in safety and health at the design stage of new equipment and technology depends on the engineers', designers' and technicians' knowledge of ergonomics and occupational safety and health. Existing education and training systems in this field should be improved." [24] There is a need to involve people in the safety process, thus making them

PANEL 51

If education in occupational safety and health is to form part of a general health and safety policy in all kinds of human activity, it should start as early as possible to create appropriate habits from childhood. The way in which parents behave in their jobs sets an example to their children. In many countries, schools have introduced into their curricula the teaching of traffic safety; it can be easily extended to basic education in the safe handling of chemicals, machinery, or apparatus now largely used in households or agriculture or industry. In subjects like physics and chemistry, emphasis should be placed on health and safety considerations whenever appropriate (risks of wounds, electrical injuries, toxicity of chemicals, etc.).

WHO: *Education and training in occupational health, safety and ergonomics*, Eighth report of the Joint ILO/WHO Committee on Occupational Health, Technical Report Series, No. 663 (Geneva, 1981), pp. 35-36.

active participants and not just powerless, or even worse, indifferent spectators.[25]

Training in schools

Safety consciousness should be aroused at the earliest age: the ILO/WHO Joint Committee has emphasised the need to teach a number of basic concepts in primary schools (see Panel 51).

Methods should be worked out with care. It has rightly been emphasised that, rather than laying down "prohibitions", it is preferable to create an awareness on the part of the pupils, so that safety "becomes a permanent, intelligent and considered creation".[26] Particular emphasis should be laid on the use of leisure activities for the age groups in which play is a major factor; organised or free leisure may be considered the first step in risk apprenticeship: guiding the taste for risk rather than countering it should be the object of an education which, from the earliest age, should act on the causes of accidents, on the principle that prevention is better than cure.

Training policies and training the trainers

How does one widely disseminate concepts – at least the most basic ones – on working conditions and environment? In the race against the proliferation of work situations which make no allowance for the worker, today's decisions are crucial for tomorrow. The major approaches adopted by the decision-makers and planners will have significant effects on situations many years from now; training programmes call for trainers, who must therefore themselves be trained first.

Decision-makers (senior civil servants and planners)

Providing information for people in this group is a particularly delicate operation. The aim should be to increase their awareness of the implications of the decisions they take. They should be informed, through an exchange of experience at the national and international level, of the effects – on the national budget, on productivity, on the efficiency of equipment, on workers' participation in the endeavour to increase growth, on the social climate and on the country's socio-cultural equilibrium – of any decisions concerning working conditions and work organisation that do not take into account the workers, their characteristics and their aspirations. The information given to them should stress positive action that they may initiate, especially as regards ensuring that jobs created in new plants are safe and compatible with good conditions of life (for instance, new undertakings should be required to build housing for possible occupation by their workers). Emphasis should also be laid on the importance of their decisions in the import of new technology, in working conditions and workers' participation, and in providing support for training about working conditions and environment in educational establishments of all kinds.

Employers' and workers' organisations

Since these organisations have both a consultancy and a training function, they can play an important intermediary role. As advisers to heads of undertakings, representatives of employers' organisations should be informed and trained about the dissemination of basic information to their members (labour legislation, principles of occupational safety and health and working conditions, sources of documentation, cost of bad working conditions), about the role and procedures of safety and health committees, for example, and also about the types of training available from the management training institutes that have been set up in various countries.

Workers' organisations have a similar information and training role, and their representatives should be trained to pass on this information in the light of the fact that "workers are experts in their own right. They know the daily occurrences of their work better than anyone else – they know the shortfalls, they know the intimate details of their work – and, given the appropriate education, they can participate effectively in finding practical solutions to health and safety problems in their workplaces".[27] The workers' education programmes to which the ILO gives its support

PANEL 52

The role of workers' education

One important aspect which workers' education should not overlook is the impact of innovation on health and safety. The enormous and tragic consequences of change, in terms of accidents alone, call for vigorous and continuing education. How many cases can we not recall in which workers have suffered serious injury or death, sometimes because they were not sufficiently aware of new dangers? I think of new chemicals which cause cancer, of workers killed by explosives, others drowned on oil-drilling rigs or trapped by machines. Technological change brings benefits to many, but in too many cases the workers pay too high a price for progress. . . . In ILO terms [workers' education] is an integral part of any satisfactory programme for economic development, a means for the improvement of the lives of working people and not an end in itself.

C. Poloni, in *Labour Education* (Geneva, ILO), 1981/1, No. 45, p. 4.

have an important role in improving working conditions and environment (see Panel 52).

Public information

The public information media – radio, press and television – could, in general, be used more widely to transmit the idea of improved working conditions and environment. In some countries, governments or employers' and workers' organisations have taken action of this type, and have got across to a broad cross-section of the public a simple message usually aimed at increasing general awareness. This type of activity may have a major effect: "the very act of talking about occupational safety helps to change the status of accidents from that of a virtually fatal problem to one on which action can and must be taken".[28] Radio and television campaigns may also be organised at peak listening and viewing times. It may be possible to do more than just increase awareness; information may be given, for example, on matters such as joint safety and health committees. Finally, televised courses on working conditions and environment may be organised.

Training the trainers

To a large extent, a major training effort is the only way of achieving a lasting improvement in working conditions and environment, and of reducing the gap between reality and national and international objectives.

This statement gives some indication of the importance of setting up training programmes for trainers.[29] This approach has three advantages.

First, training trainers has a multiplier effect, by rapidly disseminating the ideas, awareness and basic knowledge that the extent and urgent nature of the need with which we are confronted call for.

Second, it allows the content and methodology of training to be adapted to the final audience, by using the local language and by giving due consideration to local customs, local living and working conditions, the cultural background and attitudes of trainees, the teaching procedures and the programme content. Adapting the message to the audience is a decisive factor in audience comprehension, receptivity and reaction.

Third, training trainers is the most direct way of ensuring that developing countries have control over the transmission of knowledge: many of these countries want international aid to "help them to do without aid". It is the most rapid and effective way possible of prolonging the effects of international technical co-operation. Countries can take over responsibility for their own training and adapt the content, form, programmes and audience to meet their own needs. Training trainers gives them the basic elements of the content and methodology of training for them to use in continuing to disseminate the message.

Who are these current or potential trainers? First, there are all those who teach, in one way or another: the advantages of integrated training have already been mentioned several times. Educational programmes should be used as a tool in any policy for improving working conditions and environment, whether they are traditional educational programmes for young people (technical education and apprenticeship, universities, engineering schools, national administration colleges, etc.) or short-term adult training courses (vocational or management training, workers' education, etc.) (see Panel 53).

The training that teachers should receive will depend on their knowledge and that of their audience: for professionals such as those in technical education and in apprenticeship schools, it should be sufficient to update knowledge and provide some guide-lines on the need to emphasise safety, health and working conditions. For teachers in general education, it may prove necessary to stimulate their awareness and also to establish special programmes.

Second, there are the part-time trainers, including those who carry out training activities either as part of their normal work or on a part-time or regular basis if they have previously received educational training: labour inspectors required to help to train younger colleagues, engineers, safety officers or trade union workers or safety and health supervisors in social security funds who give courses to management or workers, and so on.

PANEL 53

The Occupational Safety and Health Convention, 1981 (No. 155), stipulates in Article 14 that measures should be taken by Members "with a view to promoting . . . the inclusion of questions of occupational safety and health and the working environment at all levels of education and training".

PANEL 54

Training trainers in a rural environment

Having reviewed activities in development programmes (educators for farmers, women, youth movements, national civic services, teachers and trainers, etc.), the group considers it essential to use existing resources to implement safety and health measures in the informal rural sector. If these educators are to play their role in the control of occupational risks, they need adequate training. [. . .]

The group considers that each country should have a body or a centre responsible for training trainers in occupational safety and health and for the co-ordination of all safety programmes.

Joint African and Mauritian Organisation, Working Group on Occupational Accident Hazards: *Séminaire sur l'amélioration des conditions et du milieu de travail des travailleurs ruraux en Afrique, Cotonou, 3 au 7 décembre 1979*, organised by the Joint African and Mauritian Organisation, WHO and the ILO.

When looking for suitable trainers, we should avoid becoming hidebound by a narrow concept of training. Training is not necessarily a formal structure composed of teachers and pupils. This is so in agriculture in particular, where farming methods have been taught by co-operative movement pioneers; these people may also be excellent choices for disseminating the basic principles of improving working conditions and environment (see Panel 54).

Similar considerations apply to health service personnel and primary health care workers, provided that they themselves receive the basic information. This was recommended in particular by the Joint ILO/WHO Committee on Occupational Health (see Panel 55). In the case of small and medium-sized undertakings – which, because of their large

PANEL 55

The ILO and WHO are urged to help in developing manuals and other publications for the education and training of primary health care workers in occupational health and safety activities, as it is felt that such workers would greatly strengthen the usually scanty manpower resources in occupational health and safety and would extend the reach of such services. [. . .]

The ILO and WHO are urged to develop dynamic programmes for training of trainers in occupational health, safety and ergonomics.

WHO: *Education and training in occupational health, safety and ergonomics*, op. cit., p. 46.

number and their dispersion, are particularly difficult to reach – management and production consultants and advisers, whose work is an integral part of policies for the support and development of these undertakings, should be trained in basic concepts of working conditions and environment (see "Coercive and incentive measures", p. 279).

These part-time trainers should receive training that supplements their knowledge and teaches them educational methodology, so that they can effectively pass on the knowledge they have received.

The basic purposes of training the trainers are to give priority to the dissemination of fundamental concepts and measures that are readily understandable by the majority of the population, and to instil in as many people as possible an awareness that each person can act in a responsible manner in relation to his own situation and his role, as part of a general concept that there should be genuine and lasting concern for improving working conditions and environment at all stages of training and in every workplace.

Notes

[1] J. I. Husband: *Labour administration: A general introduction* (Geneva, ILO, 1980), p. 7.

[2] E. Marín Quijada: "Limitations of legislation in improving working conditions: The Venezuelan experience", in *International Labour Review*, Jan.-Feb. 1979, pp. 113-122.

[3] ibid.

[4] Many sections of this publication deal explicitly or implicitly with labour inspection (see in particular Chapter 2, "Enforcement of regulations and advisory activities" (pp. 91-95). However, many of the problems of labour inspection cannot be dealt with in this book, and we shall restrict ourselves to the following quotation:

"If they are to remain true to their vocation, labour inspection services must retain or recover, as the case may be, a dynamic outlook that will enable them to find new solutions for the new problems arising in the changing world of labour and to provide the driving force for such reforms.

"The obstacle to be overcome, particularly in developing countries, is often a lack of resources – lack of funds (in certain African countries the share of the national budget allotted to the ministry of labour is falling steadily and in some cases is less than 1 per cent) and a shortage of qualified staff; the one, of course, is a consequence of the other. It is essential, then, that governments should give fresh thought to labour inspection and acknowledge the value of its contribution to the general welfare; they must recognise the weight of the new responsibilities the labour inspection service has to bear, and must provide it with the financial means and the administrative structure it needs and without which it will become entirely ineffective."

(ILO: *Labour inspection: Purposes and practice* (Geneva, 1973), p. 202).

[5] See the Collective Bargaining Convention, 1981 (No. 154), which specifies, in particular, that collective bargaining should be made possible for all employers and workers and progressively extended to cover conditions of work and terms of employment.

[6] Vast areas remain inadequately researched, especially in the developing countries (e.g. the effects of climate, nutrition, local customs and cultures on working conditions, or the consequences of technology transfer); productivity, workers' welfare and social equilibrium all have much to gain from an extension of our knowledge.

[7] "Work unit" is used to cover industrial, commercial and agricultural undertakings, mines, plantations, artisanal workshops; the term "undertaking" is also used here in its widest sense.

[8] ILO: *Introduction to work study* (Geneva, 3rd (revised) ed., 1979). The chapter from which the quotation is taken comes in the first part of the book, entitled "Productivity and work study". The other parts are entitled "Method study"; "Work measurement"; and "From analysis to synthesis: New forms of work organisation".

[9] For action on workers' housing, see idem: *Housing, medical and welfare facilities and occupational safety and health on plantations*, Report III, Committee on Work on Plantations, Seventh Session, Geneva, 1976, p. 101.

[10] A. E. Louzine: "Improving working conditions in small enterprises in developing countries", in *International Labour Review*, July-Aug. 1982, pp. 443-454.

[11] See, for example, ILO: *Social problems in the construction industry rising out of the industrialisation of developing countries*, Report III, Building, Civil Engineering and Public Works Committee, Eighth Session, Geneva, 1971 (Geneva, 1968), p. 47.

[12] For more detailed information on these subjects, see Chapter 2, "Machinery and plant" (p. 30), "Harmful agents in the work environment" (pp. 30-32), "Plant and equipment hazards" (pp. 47-49), "Dangerous substances" (pp. 49-50), and "Hazards related to working procedures and work organisation" (pp. 50-51); Chapter 5; Chapter 3, "The importance of working time" (p. 101), "Excessive hours in poorly regulated sectors" (p. 115) and "Reasons for shift work" (pp. 122-124); and Chapter 4, "An equitable wage" (pp. 172-174) and "Payment by results" (pp. 175-176).

[13] ILO: *Declaration of Principles and Programme of Action adopted by the Tripartite World Conference on Employment, Income Distribution and Social Progress, and the International Division of Labour*, Geneva, 4-17 June 1976, para. 51.

[14] S. Forssmann and G. Coppée: *Occupational health problems of young workers*, Occupational Safety and Health Series, No. 26 (Geneva, ILO, 1973). See also WHO: *Health needs of adolescents*, Report of a WHO Expert Committee, 28 September-4 October 1976, Technical Report Series, No. 609 (Geneva, 1977).

[15] R. De Grazia: *Clandestine employment* (Geneva, ILO, 1984).

[16] ILO: *Education and Training Policies in Occupational Safety and Health and Ergonomics – International Symposium*, Occupational Safety and Health Series, No. 47 (Geneva, 1982). In August 1981, this International Symposium was organised by the Directorate of Labour Inspection of Norway, in collaboration with the ILO and WHO, in Sandefjord (Norway). The proceedings of this Symposium contain the 79 reports and papers that were presented. Some of the information on the following pages has been drawn from the work of this Symposium.

[17] J.-M. Clerc: "Training as an instrument of a strategy for improvement of working conditions and environment", in *International Labour Review*, Sep.-Oct. 1982, pp. 565-575.

[18] See, for example, A. Eika: "Training and crane-driver safety", in ILO: *Education and Training Policies in Occupational Safety and Health and Ergonomics* . . ., op. cit., pp. 140-142 (in French).

[19] Louzine, op. cit.

[20] WHO: *Education and training in occupational health, safety and ergonomics*, Eighth report of the Joint ILO/WHO Committee on Occupational Health, Technical Report Series, No. 663 (Geneva, 1981).

[21] idem, p. 36.

[22] D. Pupo Nogueira, in ILO: *Education and Training Policies in Occupational Safety and Health and Ergonomics* . . ., op. cit., pp. 316-317.

[23] Various papers, ibid.

[24] Y. I. Kundiev, ibid., p. 29.

[25] F. Jérôme, ibid., p. 305 (in French).

[26] G. Bresson, ibid., p. 321 (in French).

[27] A. Le Serve: "Health and safety in the workplace – A need for workers' education programmes", in *Labour Education* (Geneva, ILO), 1981/1, No. 45, p. 10.

[28] Bresson, in ILO: *Education and Training Policies in Occupational Safety and Health and Ergonomics* . . ., op. cit., p. 360 (in French).

[29] Clerc, op. cit.

CONCLUSION

All that now remains, at the end of this book, is to express the hope that this publication, which is the outcome of wide-ranging collaboration and contains a wealth of varied experience, will, in turn, give rise to effective action, spread the conviction that changes to improve working conditions are possible and, perhaps, be followed by other publications on specific problems, situations and functions.

Improving working conditions and environment is essential for the maintenance of human welfare and dignity, and justifies mobilising efforts for a three-pronged campaign against ignorance:

- to disseminate knowledge about well-identified risks amongst the largest possible number of people;
- to find and implement methods of passing on to the smallest of undertakings and to the most remote rural enterprises the fundamentals of occupational safety and health; and finally,
- to meet the challenge of technological change which is, day by day, preparing and moulding the working conditions of tomorrow.

If, as a result of this book and of the endeavours of those who use it, there is a growth of interest in working conditions and an expansion of knowledge on this subject, a major step will have been taken towards improving those conditions; for knowledge gives the power to control.

APPENDICES

A. THE INTERNATIONAL LABOUR ORGANISATION

The International Labour Organisation works to eradicate poverty and unemployment in the world, to satisfy the basic needs of the very poor, and to create a new world of work. Since its founding in 1919 it has sought to generate action that will promote social justice and reform.

Economic growth has not done enough for the world's poor. Hundreds of millions of men and women are caught in a poverty trap marked by wretched incomes and appalling living conditions. Despite international development strategies and national plans, their plight is growing ever more serious.

The ILO has put into practice a new concept of development. It encourages every country to enable the poorest to reach minimum standards of living as soon as the economic development of the country will allow. These minimum standards include basic material needs – food, clothing and housing – as well as education, health care and protection from disease.

They include opportunities for gainful and productive employment, and the right to act freely and without constraint.

A number of United Nations agencies have joined with the ILO in support of objectives which link growth to the basic needs of the poor.

For the ILO, employment creation is not enough. The jobs created must be good jobs. Parallel with employment, a programme has been created by the ILO to improve working conditions and the working environment. In this area, the ILO provides the international framework for action at the national level and indeed at the workplace, for safer, better and more satisfying jobs.

Success of measures being taken to deal with the world's problems depends to a great extent on the ability of all productive forces to participate fully in the framing of social and economic policies. It is for this reason that a third important ILO activity lies in the strengthening of tripartism and of industrial relations systems.

Protection and promotion of human rights has always been an important ILO function, especially where the social and economic well-being of working people is concerned. This activity takes the form of Conventions and Recommendations relating to basic human rights, employment and training policy, conditions of work, social security, industrial relations and a variety of other social matters. So far, 159 Conventions and 169 Recommendations have been adopted and the ILO has special arrangements to promote their implementation.

A major part of the ILO's work consists in the provision of expert advice and technical assistance to individual countries. Much of this operational activity lies in such fields as vocational training, management development and manpower planning; also in the development of co-operatives and small-scale industries, social security and workers' education.

The International Labour Conference meets annually in Geneva. Its principal functions include the adoption of international labour standards, the final approval of the ILO budget and – every third year – the election of the Governing Body. National delegations are composed of two Government delegates, one Employers' delegate and one Workers' delegate. Delegates speak and vote independently.

The Governing Body of the International Labour Office functions as the executive board of the Organisation. It meets several times a year and is composed of 28 Government members, 14 Employer members and 14 Worker members. It appoints the Director-General, supervises the Office, proposes the ILO budget to the Conference, and performs other functions delegated to it by the ILO Constitution.

The International Labour Office is the Organisation's secretariat, operational headquarters, research and publishing house.

The ILO has at present 150 member countries.

Director-General: Francis Blanchard.

B. THE INTERNATIONAL PROGRAMME FOR THE IMPROVEMENT OF WORKING CONDITIONS AND ENVIRONMENT (PIACT)[1]

The International Labour Organisation, from its very inception in 1919, has been actively concerned with problems of conditions of work and life and occupational safety and health. Under its Constitution, it has a solemn obligation to promote social justice as a precondition of universal and lasting peace, in particular by seeking to improve "conditions of labour" that involve injustice, hardship and privation. Not surprisingly, the first international labour instrument which the ILO adopted after its foundation deals with conditions of work – the Hours of Work (Industry) Convention (No. 1), 1919. Since 1919, a great number of international labour Conventions and Recommendations have been adopted by the ILO which deal, inter alia, with various aspects of conditions of employment and occupational safety and hygiene.

The ILO has given a new orientation and a new impetus to its action in this field by embarking on an International Programme for the Improvement of Working Conditions and Environment (PIACT). The Programme had its origin in the Report of the Director-General to the 60th Session (1975) of the International Labour Conference, entitled *Making work more human.* The Report aimed at reinvigorating both ILO action and action within member States on an issue which, as the Director-General observed, there had been a temptation "to put off to a better tomorrow". The Conference, welcoming this initiative, unanimously adopted a resolution[2] reaffirming "that the improvement of working conditions and environment and the well-being of workers remains the first and permanent mission of the ILO". The resolution supported the world-wide action suggested by the Director-General in the form of a new international programme for the improvement of working conditions and environment designed to promote or support activities of member States in this field.

After thorough technical preparation and discussion with members of the ILO's tripartite constituency, consultation of intergovernmental organisations and specialists from various circles, detailed proposals on this Programme were submitted to the Governing Body of the International Labour Office, which at its November 1976 Session approved the broad lines of the Programme.

[1] Called "PIACT", from the French acronym of the Programme ("*P*rogramme *I*nternational pour l'*A*mélioration des *C*onditions et du milieu de *T*ravail").

[2] Resolution concerning future action of the International Labour Organisation in the field of working conditions and environment: adopted unanimously on 24 June 1975.

The general objective of PIACT – making work more human – can be divided under three main heads. It seeks to ensure that:

– work respects the worker's life and health;
– work leaves him free time for rest and leisure;
– work enables him to serve society and achieve self-fulfilment by developing his personal capacities.

The main thrust of PIACT is at present directed at six technical areas of emphasis; safety and health and the working environment; working time; work organisation and content; working conditions and choice of technology; ergonomics; and the living environment in relation to work.

C. INTERNATIONAL LABOUR CONVENTIONS AND RECOMMENDATIONS CONCERNING WORKING CONDITIONS AND ENVIRONMENT

Chapter 2: Occupational safety and health

General

Medical Examination of Young Persons (Industry) Convention, 1946 (No. 77), and Recommendation, 1946 (No. 79).

Medical Examination of Young Persons (Non-Industrial Occupations) Convention, 1946 (No. 78).

Protection of Workers' Health Recommendation, 1953 (No. 97).

Prevention of Accidents (Seafarers) Convention, 1970 (No. 134), and Recommendation, 1970 (No. 142).

Working Environment (Air Pollution, Noise and Vibration) Convention, 1977 (No. 148), and Recommendation, 1977 (No. 156).

Occupational Safety and Health Convention, 1981 (No. 155), and Recommendation, 1981 (No. 164).

Occupational Health Services Convention, 1985 (No. 161), and Recommendation, 1985 (No. 171).

Protection against specific hazards

(a) Toxic substances and agents

White Lead (Painting) Convention, 1921 (No. 13).

Radiation Protection Convention, 1960 (No. 115), and Recommendation, 1960 (No. 114).

Benzene Convention, 1971 (No. 136), and Recommendation, 1971 (No. 144).

Occupational Cancer Convention, 1974 (No. 139), and Recommendation, 1974 (No. 147).

(b) Machinery

Guarding of Machinery Convention, 1963 (No. 119), and Recommendation, 1963 (No. 118).

(c) Maximum weight to be carried

Maximum Weight Convention, 1967 (No. 127), and Recommendation, 1967 (No. 128).

Safety in specific branches of industry

Safety Provisions (Building) Convention, 1937 (No. 62), and Recommendation, 1937 (No. 53).

Medical Examination (Seafarers) Convention, 1946 (No. 73).

Labour Inspection (Mines and Transport) Recommendation, 1947 (No. 82).

Accommodation of Crews Convention (Revised), 1949 (No. 92).

Ships' Medicine Chests Recommendation, 1958 (No. 105).

Medical Advice at Sea Recommendation, 1958 (No. 106).

Medical Examination (Fishermen) Convention, 1959 (No. 113).

Hygiene (Commerce and Offices) Convention, 1964 (No. 120), and Recommendation, 1964 (No. 120).

Medical Examination of Young Persons (Underground Work) Convention, 1965 (No. 124)

Labour Inspection (Agriculture) Convention, 1969 (No. 129), and Recommendation, 1969 (No. 133).

Occupational Safety and Health (Dock Work) Convention, 1979 (No. 152), and Recommendation, 1979 (No. 160).

Chapter 3: Working hours

Hours of Work (Commerce and Offices) Convention, 1930 (No. 30).

Hours of Work and Rest Periods (Road Transport) Convention, 1979 (No. 153).

Reduction of Hours of Work Recommendation, 1962 (No. 116).

Weekly Rest (Industry) Convention, 1921 (No. 14).

Weekly Rest (Commerce and Offices) Convention, 1957 (No. 106), and Recommendation, 1957 (No. 103).

Holidays with Pay Convention (Revised), 1970 (No. 132).

Paid Educational Leave Convention, 1974 (No. 140), and Recommendation, 1974 (No. 148).

Night Work (Women) Convention (Revised), 1948 (No. 89).

Night Work of Young Persons (Industry) Convention (Revised), 1948 (No. 90).

Night Work of Young Persons (Non-Industrial Occupations) Convention, 1946 (No. 79), and Recommendation, 1946 (No. 80).

Chapter 4: Remuneration

Minimum Wage-Fixing Machinery Convention, 1928 (No. 26).

Protection of Wages Convention, 1949 (No. 95).

Equal Remuneration Convention, 1951 (No. 100), and Recommendation, 1951 (No. 90).

Minimum Wage Fixing Convention, 1970 (No. 131).

Chapter 6: Workers' welfare facilities

Welfare Facilities Recommendation, 1956 (No. 102).

Workers' Housing Recommendation, 1961 (No. 115).

Chapter 7: Workers in the rural and urban informal sectors in developing countries

Rural Workers' Organisations Convention, 1975 (No. 141).

Tenants and Share-croppers Recommendation, 1968 (No. 132).

Labour Inspection (Agriculture) Convention, 1969 (No. 129).

Chapter 8: How to improve working conditions and environment

Freedom of Association and Protection of the Right to Organise Convention, 1948 (No. 87).

Labour Inspection Convention, 1947 (No. 81), and Recommendation, 1947 (No. 81).

Maternity Protection Convention, 1919 (No. 3).

Night Work (Women) Convention, 1919 (No. 4).

Night Work (Women) Convention (Revised), 1934 (No. 41).

Underground Work (Women) Convention, 1935 (No. 45).
Night Work (Women) Convention (Revised), 1948 (No. 89).
Maternity Protection Convention (Revised), 1952 (No. 103), and Recommendation, 1952 (No. 95).
Discrimination (Employment and Occupation) Convention, 1958 (No. 111), and Recommendation, 1958 (No. 111).
Equality of Treatment (Social Security) Convention, 1962 (No. 118).
Workers with Family Responsibilities Convention, 1981 (No. 156), and Recommendation, 1981 (No. 165).
Night Work of Young Persons (Non-Industrial Occupations) Convention, 1946 (No. 79), and Recommendation, 1946 (No. 80).
Night Work of Young Persons (Industry) Convention (Revised), 1948 (No. 90).
Conditions of Employment of Young Persons (Underground Work) Recommendation, 1965 (No. 125).
Minimum Age Convention, 1973 (No. 138), and Recommendation, 1973 (No. 146).
Migration for Employment Convention (Revised), 1949 (No. 97), and Recommendation (Revised), 1949 (No. 86).
Migrant Workers (Supplementary Provisions) Convention, 1975 (No. 143).
Migrant Workers Recommendation, 1975 (No. 151).
Older Workers Recommendation, 1980 (No. 162).
Right to Organise and Collective Bargaining Convention, 1949 (No. 98).
Social Policy (Basic Aims and Standards) Convention, 1962 (No. 117).
Workers' Representatives Convention, 1971 (No. 135).
Labour Administration Convention, 1978 (No. 150).
Collective Bargaining Convention, 1981 (No. 154).

D. GUIDE TO FURTHER READING

The ILO has published many books on working conditions and environment. Rather than merely listing them, it was thought preferable to select those essential for a better understanding of the subject.

Except for some general works, this bibliography does not contain publications dealing with specific points which are mentioned as notes in the main body of the text.

Finally, the reader may obtain further bibliographical guidance by requesting the following brochures: [1]

- *Publications and documents on conditions of work and welfare facilities;*
- *Essential publications on occupational safety and health;*
- *ILO Catalogue of publications in print* (periodically updated).

Abbreviations

The following abbreviations have been used in the bibliography:

ILO/BIT/OIT:
International Labour Office
Bureau international du Travail
Oficina Internacional del Trabajo

Doc.: document
FAO: Food and Agriculture Organisation of the United Nations
GB: Governing Body
IILS: International Institute for Labour Studies
ILC: International Labour Conference
ILR *International Labour Review* (Geneva, ILO)
OB *Official Bulletin* (Geneva, ILO)
OSH: Occupational Safety and Health Series
Vol.: volume
WHO: World Health Organisation

[1] ILO publications can be obtained through major booksellers or ILO local offices in many countries, or direct from ILO Publications, International Labour Office, CH-1211 Geneva 22, Switzerland. The brochures requested and a catalogue or list of new publications will be sent free of charge from the above address.

Asterisks preceding the titles of documents have the following significance:
* = documents published in French only.
** = documents published in Spanish only.

Chapter 1

Bolinder, E.; Gerhardsson, G. "A better environment for the worker", in *ILR*, June 1972, pp. 495-506.

Christensen, E. H. *Man at work.* Studies on the application of physiology to working conditions in a sub-tropical country. (OSH, No. 4, Geneva, ILO, 1964). 55 pp.

De Givry, J. "The ILO and the quality of working life. A new international programme: PIACT", in *ILR*, May-June 1968, pp. 261-271.

Dubos, R. "Progressive degradation", in *World Health* (Geneva, WHO), July 1975, pp. 8-15.

ILO. *Constitution of the International Labour Organisation and Standing Orders of the International Labour Conference.* Geneva, new edition, 1982.

—. *The ILO and the world of work.* Geneva, 1984. 74 pp.

—. *Improving working conditions and environment: An international programme* (PIACT), Geneva, 1984. 133 pp.

—. *Making work more human. Working conditions and environment.* Offprint of the Report of the Director-General/Part I, ILC, 60th Session, Geneva, 1975, 79 pp.; and parallel resolution adopted.

—. *Resolution concerning the working environment*, ILC, 59th Session, Geneva, 1974, in *OB*, 1979, No. 1, pp. 43-44.

—. *Resolution concerning future action of the International Labour Organisation in the field of working conditions and environment*, ILC, 60th Session, Geneva, 1975, in *OB*, 1975, No. 1, pp. 81-84.

Chapter 2

ILO. *The improvement of working conditions and working environment in the iron and steel industry*, Report III, Iron and Steel Committee, Tenth Session, Geneva, 1981. 86 pp.

—. *Working conditions and working environment in the petroleum industry, including offshore activities*, Report III, Petroleum Committee, Ninth Session, Geneva, 1980. 116 pp.

— *Encyclopaedia of occupational health and safety.* Geneva, 3rd ed., 1983. 2 Vols. 2,579 pp.

—. *Occupational exposure to airborne substances harmful to health.* An ILO code of practice. Geneva, 1980. 44 pp.

—. *Guide to health and hygiene in agricultural work.* Geneva, 1979. 309 pp.

—. *Guide to safety in agriculture.* Geneva, 1969. 247 pp.

—. *Guide to safety and health in forestry work.* Geneva, 1968. 223 pp.

—. *The chemical industries and the working environment*, Report II, Chemical Industries Committee, Eighth Session, Geneva, 1976. 78 pp.

—. *Labour inspection. Purposes and practice.* Geneva, 1973. 234 pp.

—. *Housing, medical and welfare facilities and occupational safety and health on plantations*, Report III, Committee on Work on Plantations, Seventh Session, Geneva, 1976. 103 pp.

—. *Accident prevention.* A workers' education manual. Geneva, 2nd ed., 1983. 175 pp.

—. *Occupational safety and health problems in the timber industry*, Report II, Third Tripartite Technical Meeting for the Timber Industry, Geneva, 1981. 78 pp.

—. *Protection of workers against noise and vibration in the working environment*. An ILO code of practice. Geneva, 3rd imp. (with modifications), 1984. 74 pp.

—. *Role of medical inspection of labour*. Geneva, 1968. 111 pp.

—. *Safety and health in coal mines*, Report III, Coal Mines Committee, Tenth Session, Geneva, 1976. 128 pp.

—. *Safety and health in the textile industry*, Report III, Textiles Committee, Ninth Session, Geneva, 1973. 47 pp.

—. *Safety and health in agricultural work*. An ILO code of practice. Geneva, 1965. 132 pp.

—. *Safety and health in building and civil engineering work*. An ILO code of practice. Geneva, 1972. 386 pp.

—. *Safety and health in forestry work*. An ILO code of practice. Geneva, 1969. 165 pp.

—. *Safety, health and working environment in the metal trades: New approaches*, Report II, Metal Trades Committee, Tenth Session, Geneva, 1977. 68 pp.

—. *Safety in the use of asbestos*. An ILO code of practice. Geneva, 1984. 129 pp.

Occupational Safety and Health Series

ILO. *The organisation of occupational health services in developing countries*. OSH, No. 7, Geneva, 1967. 187 pp.

Himbury, S. *Kinetic methods of manual handling in industry*. OSH, No. 10, Geneva, ILO, 1967. 38 pp.

Forssman, S.; Coppée, G. H. *Occupational health problems of young workers*, OSH, No. 26, Geneva, ILO, 1973. 143 pp.

ILO. * *Médecine du travail, protection de la maternité et santé de la famille*. OSH, No. 29, Geneva, 1975. 58 pp.

—. *Asbestos: Health risks and their prevention*. OSH, No. 30, Geneva, 1975. 99 pp.

—. *Dust control in the working environment (silicosis)*. OSH, No. 36, Geneva, 1977. 161 pp.

—. *Occupational exposure limits for airborne toxic substances*. OSH, No. 37, Geneva, 2nd (revised) ed., 1980. 290 pp.

—. *Safe use of pesticides: Guidelines*. OSH, No. 38, Geneva, 1977. 42 pp.

—. *Occupational cancer: Prevention and control*. OSH, No. 39, Geneva, 1977. 36 pp.

—. *Building work: A compendium of occupational safety and health practice*. OSH, No. 42, Geneva, 1979. 256 pp.

—. *Ergonomic principles in the design of hand tools*. OSH, No. 44, Geneva, 1980. 93 pp.

Chapter 3

Allenspach, H. *Flexible working hours*. Geneva, ILO, 1975. 64 pp.

Carpentier, J.; Cazamian, P. *Night work: Its effects on the health and welfare of the worker*. Geneva, ILO, 2nd imp. (with modifications), 1978. 82 pp.

Cuvillier, R. *The reduction of working time: Scope and implications in industrialised market economy countries.* Geneva, ILO, 1984. 150 pp.

Evans, A. A. *Hours of work in industrialised countries.* Geneva, ILO, 1975. 164 pp.

ILO. *Paid educational leave*, Report IV (1), ILC, 59th Session, Geneva, 1974. 42 pp.

—. *Hours of work and rest periods in road transport*, Report VII (1), ILC, 64th Session, Geneva, 1978. 77 pp.

—. *Management of working time in industrialised countries.* Geneva, 1978. 124 pp.

Marić, D. *Adapting working hours to modern needs: The time factor in the new approach to working conditions.* Geneva, ILO, 1977. 50 pp.

—. * *La durée du travail dans les pays en voie de développement.* Geneva, ILO, 1981. 138 pp.

Maurice, M. *Shift work: Economic advantages and social costs.* Geneva, ILO, 1975. 146 pp.

Chapter 4

Beckerman, W.; van Ginneken, W.; Szal, R.; Garzuel, M. *Poverty and the impact of income maintenance programmes in four developed countries.* Geneva, ILO, 1979. 90 pp.

ILO. *General survey of the Reports relating to the Equal Remuneration Convention (No. 100) and Recommendation (No. 90), 1951*, Report III (4 B), ILC, 60th Session, Geneva, 1975. 86 pp.

—. * *Indexation des salaires dans les pays industrialisés à économie de marché.* Geneva, 1978. 90 pp.

—. *Wages.* A workers' education manual. Geneva, 3rd ed., 1982. 255 pp.

—. * *Les salaires dans les pays d'Afrique francophone.* Compte rendu des travaux d'un séminaire régional OIT-DANIDA (Yaoundé, mai 1977). Labour-Management Relations Series, No. 55, Geneva, 1978. 130 pp.

—. *Minimum wage fixing and economic development.* Studies and Reports, New Series, No. 72, Geneva, 1968. 217 pp.

—. *Remuneration and conditions of work in relation to economic development including plant-level welfare facilities and the workers' standard of living*, Report IV, Ninth Conference of American States Members of the ILO, Caracas, 1970. 123 pp.

—. *Wage determination in English-speaking Caribbean countries*, Record of proceedings of, and documents submitted to, an ILO/DANIDA Regional Seminar (Kingston, Jamaica, 1-7 March 1978). Labour-Management Relations Series, No. 57, Geneva, 1979. 121 pp.

Starr, G. *Minimum wage fixing: An international review of practices and problems.* Geneva, ILO, 1981. 203 pp.

Suzuki, H. "Wage indexation in industrialised market economies", in *ILR*, July-Aug. 1980, pp. 455-466.

Chapter 5

Bluestone, I. "Creating a new world of work", in *ILR*, Jan.-Feb. 1977, pp. 1-10.

Burbidge, J. L. *Group production methods and humanisation of work: The evidence in industrialised countries.* Research Series No. 9, Geneva, IILS, 1976. 18 pp.

Carpentier, J. "Organisational techniques and the humanisation of work", in *ILR*, Aug. 1974, pp. 93-116.

ILO. *Bibliography on major aspects of the humanisation of work and the quality of working life*. Geneva, 2nd ed., 1978. 300 pp.

–. *Introduction to work study*. Geneva, 3rd (revised) ed., 1979. 442 pp.

–. *New forms of work organisation*. Vol. 1. Denmark, Norway and Sweden, France, Federal Republic of Germany, United Kingdom, United States. Geneva, 1979. 174 pp.

–. *New forms of work organisation*. Vol. 2. German Democratic Republic, India, Italy, USSR. Economic costs and benefits. Geneva, 1979. 145 pp.

–. *Technology to improve working conditions in Asia*. Geneva, 1979. 154 pp.

Kanawaty, G. (ed.) *Managing and developing new forms of work organisation*. Management Development Series, No. 16, Geneva, ILO, 2nd (revised) ed., 1981. 206 pp.

Tchobanian, R. "Trade unions and the humanisation of work", in *ILR*, Mar. 1975, pp. 199-217.

Chapter 6

Dumont, C. ** *Servicios sociales de las empresas de América Latina*. Geneva, ILO, 1974. 120 pp.

ILO. ** *Las empresas de América Latina ante el problema de la vivienda obrera*. Geneva, 1972. 67 pp.

–. *The welfare of workers in mines other than coal mines*, Report III, Third Tripartite Technical Meeting for Mines Other Than Coal Mines, Geneva, 1975. 70 pp.

–. *Housing, medical and welfare facilities and occupational safety and health on plantations*, Report III, Committee on Work on Plantations, Seventh Session, Geneva, 1976. 103 pp.

Chapter 7

Bequele, A.; Freedman, D. "Employment and basic needs: An overview", in *ILR*, May-June 1979, pp. 315-329.

ILO. *Employment, incomes and equality: A strategy for increasing productive employment in Kenya*. Report of an inter-agency team financed by the United Nations Development Programme and organised by the ILO, Geneva, 1972. 600 pp.

–. *Rural employers' and workers' organisations and participation*, Report III, Advisory Committee on Rural Development, Ninth Session, Geneva, 1979. 56 pp.

–. *Special services of rural workers' organisations*. A workers' education manual. Geneva, 1978. 89 pp.

–. *Poverty and employment in rural areas of the developing countries*, Report II, Advisory Committee on Rural Development, Ninth Session, Geneva, 1979. 55 pp.

Sethuraman, S. V. "The urban informal sector: Concept, measurement and policy", in *ILR*, July-Aug. 1976, pp. 69-81.

–. *The urban informal sector in developing countries: Employment, poverty and environment*. Geneva, 1981. 225 pp.

Weeks, J. "Policies for expanding employment in the informal urban sector of developing countries", in *ILR*, Jan. 1975, pp. 1-13.

Chapter 8

Dy, F. J. "Technology to make work more human", in *ILR*, Sep.-Oct. 1978, pp. 543-555.

Gustavsen, B. "Improving the work environment: A choice of strategy", in *ILR*, May-June 1980, pp. 271-286.

ILO. *Declaration of principles. Summary of proceedings of the World Employment Conference*. Geneva, 1976. 26 pp.

—. *Tripartite declaration of principles concerning multinational enterprises and social policy*. Geneva, 1977. 16 pp.

—. *Asian development in the 1980s: Growth, employment and working conditions*, Report of the Director-General (Part 1), Ninth Asian Regional Conference, Manila, 1980. 100 pp.

—. *Human dignity, economic growth and social justice in a changing Africa: An ILO agenda for Africa*. Report of the Director-General (Part 1), Fourth African Regional Conference, Nairobi, 1973. 66 pp.

—. *Workers' education*. Geneva, 1981 (Doc. No. 45, 1981/1). 24 pp.

—. *The impact of international labour Conventions and Recommendations*. Geneva, 1976. 104 pp.

—. *Labour inspection: Purposes and practice*. Geneva, 1973. 234 pp.

—. *Introduction to work study*. Geneva, 3rd (revised) ed., 1979. 442 pp.

—. *Freedom of association, labour relations and development in Asia*, Report II, Ninth Asian Regional Conference, Manila, 1980. 69 pp.

—. *International labour standards*. A workers' education manual. Geneva, 2nd ed., 2nd imp. (with modifications), 1983. 98 pp.

—. *Policies and practices for the improvement of working conditions and working environment in Europe*, Report III, Third European Regional Conference, Geneva, 1979. 78 pp.

—. *Social problems in the construction industry arising out of the industrialisation of developing countries*, Report III, Building, Civil Engineering and Public Works Committee, Eighth Session, Geneva, 1971. 67 pp.

—. *Report of the Director-General: Human values in social policy. An agenda for Europe*, Report I, Second European Regional Conference, Geneva, 1974. 89 pp.

—. *The role of labour law in developing countries*, Record of proceedings of, and documents submitted to, a Round Table (Geneva, 10-14 September, and Selva di Fasano, Italy, 17-19 September 1974), 1975. Labour-Management Relations Series, No. 49, Geneva, 1975. 266 pp.

—. *Man in his working environment*. A workers' education manual. Geneva, 1979. 142 pp.

—. *Development for social progress: A challenge for the Americas*, Report I/Part 1, Tenth Conference of American States Members of the ILO, Mexico City, 1974. 43 pp.

—. *Growth, employment and basic needs in Latin America and the Caribbean*, Report I/Part 1, Eleventh Conference of American States Members of the ILO, Medellín, 1979. 88 pp.

—. *Public labour administration and its role in economic and social development*, Report II, Eleventh Conference of American States Members of the ILO, Medellín, 1979. 63 pp.

—. *Technology to improve working conditions in Asia*. Geneva, 1979. 154 pp.

Stoddart, L. *Conditions of work and quality of working life: A directory of institutions*. Geneva, ILO, 1981. 255 pp.

WHO. *Education and training in occupational health, safety and ergonomics*. Eighth Report of the Joint ILO/WHO Committee on Occupational Health. Technical Report Series No. 663, Geneva, 1981. 48 pp.

Women workers

ILO. *Equality of opportunity and treatment for women workers*, Report VIII, ILC, 60th Session, Geneva, 1975. 124 pp.

—. *Equal opportunities and equal treatment for men and women workers: Workers with family responsibilities*, Report VI (1), ILC, 66th Session, Geneva, 1980, 69 pp., and Report VI (2), ILC, 66th Session, Geneva, 1980, 126 pp.

—. *Equal opportunities and equal treatment for men and women workers: Workers with family responsibilities*, Report V (1), ILC, 67th Session, Geneva, 1981, 84 pp., and Report V (2), ILC, 67th Session, Geneva, 1981, 88 pp.

—. *Employment of women with family responsibilities.* Summary of reports on Recommendation No. 123 (article 19 of the Constitution), Report III (2), ILC, 64th Session, 1978. 89 pp.

—. *Problems of women non-manual workers: Work organisation, vocational training, equality of treatment at the workplace, job opportunities*, Report III, Advisory Committee on Salaried Employees and Professional Workers, Eighth Session, Geneva, 1981. 92 pp.

—. * *Médecine du travail, protection de la maternité et santé de la famille.* OSH, No. 29, Geneva, 1975. 58 pp.

—. *Selected standards and policy statements of special interest to women workers adopted under the auspices of the International Labour Organisation.* Geneva, 1980. 132 pp.

—. *Women workers and society: International perspectives.* Geneva, 1976. 211 pp.

—. *Conditions of work, vocational training and employment of women*, Report III, Eleventh Conference of American States Members of ILO, Medellín, 1979. 92 pp.

Young workers

Forssman, S.; Coppée, G. H. *Occupational health problems of young workers.* OSH, No. 26, Geneva, ILO, 1973. 143 pp.

ILO. *Minimum age for admission to employment*, Report IV (1), ILC, 57th Session, Geneva, 1972. 44 pp.

—. *Young people and work*, Report II, Third European Regional Conference, Geneva, 1979. 66 pp.

—. *Young people in their working environment.* Geneva, 1977. 37 pp.

Working children

Mendelievich, E. (ed.) *Children at work.* Geneva, 2nd imp. (with modifications), 1980. 176 pp.

Rodgers, G.; Standing, G. *Child work, poverty and underdevelopment: Issues for research in low-income countries.* Geneva, ILO, 1981. 310 pp.

Migrant workers

ILO.* *Logement des travailleurs migrants et de leur famille dans les pays d'Europe occidentale.* Geneva, 1978. 91 pp.

Disabled workers

ILO. *Adaptation of jobs and the employment of the disabled.* Geneva, 1984. 112 pp.

Older workers

ILO. *Older workers: Work and retirement*, Report VI (1), ILC, 65th Session, Geneva, 1979. 95 pp.

—. *Older workers: Work and retirement*, Report IV (1), ILC, 66th Session, Geneva, 1980. 60 pp.

Home, temporary and clandestine workers, and workers without job security

De Grazia, R. "Clandestine employment: A problem of our times", in *ILR*, Sep.-Oct. 1980, pp. 549-563.

—. *Clandestine employment: The situation in the industrialised market economy countries.* Geneva, ILO, 1984. 118 pp.

Valticos, N. "Temporary work agencies and international labour standards", in *ILR*, Jan. 1973, pp. 43-56.

Veldkamp, G. M.; Raetsen, M. J. E. H. "Temporary work agencies and Western European social legislation", in *ILR*, Feb. 1973, pp. 117-131.

The following documents may also be consulted:

1. *Social and Labour Bulletin* (quarterly)
2. Reports prepared for Industrial Committees and Tripartite Technical Meetings, especially in the following areas:
   - Building, civil engineering and public works
   - Chemical industries
   - Civil aviation
   - Coal mines
   - Food and drink industries
   - Forestry and wood industries
   - Hotels, restaurants and similar establishments
   - Inland transport
   - Iron and steel
   - Leather and footwear industry
   - Metal trades
   - Mines other than coal mines
   - Petroleum
   - Plantations
   - Salaried employees and professional workers
   - Textiles